Lecture Notes in Computer Science 2920
Edited by G. Goos, J. Hartmanis, and J. van Leeuwen

Springer
Berlin
Heidelberg
New York
Hong Kong
London
Milan
Paris
Tokyo

Holger Karl Andreas Willig
Adam Wolisz (Eds.)

Wireless Sensor Networks

First European Workshop, EWSN 2004
Berlin, Germany, January 19-21, 2004
Proceedings

Volume Editors

Holger Karl
Adam Wolisz
TU Berlin, Fachgebiet Telekommunikationsnetze
Einsteinufer 25, 10587 Berlin, Germany
E-mail: {hkarl/awo}@ieee.org

Andreas Willig
Universität Potsdam, Hasso-Plattner-Institut
Prof.-Dr.-Helmert-Str. 2-3, 14482 Potsdam, Germany
E-mail: awillig@ieee.org

Cataloging-in-Publication Data applied for

A catalog record for this book is available from the Library of Congress.

Bibliographic information published by Die Deutsche Bibliothek
Die Deutsche Bibliothek lists this publication in the Deutsche Nationalbibliografie; detailed bibliographic data is available in the Internet at <http://dnb.ddb.de>.

CR Subject Classification (1998): C.2.4, C.2, F.2, D.1.3, D.2, E.1, H.4, C.3

ISSN 0302-9743
ISBN 3-540-20825-9 Springer-Verlag Berlin Heidelberg New York

Springer-Verlag is a part of Springer Science+Business Media

springeronline.com

Printed in Germany

Typesetting: Camera-ready by author, data conversion by PTP-Berlin, Protago-TeX-Production GmbH
Printed on acid-free paper SPIN: 10976470 06/3142 5 4 3 2 1 0

Preface

With great pleasure we welcomed the attendees to EWSN 2004, the 1st European Workshop on Wireless Sensor Networks, held in the exciting and lively city of Berlin.

Wireless sensor networks are a key technology for new ways of interaction between computers and the physical environment which surrounds us. Compared to traditional networking technologies, wireless sensor networks are faced with a rather unique mix of challenges: scalability, energy efficiency, self-configuration, constrained computation and memory resources in individual nodes, data-centricity, and interaction with the physical environment, to name but a few. The goal of this workshop is to create a forum for presenting new results in the flourishing field of wireless sensor networks. By bringing together academia and industry we hope to stimulate new opportunities for collaborations.

In compiling the scientific program we have been quite selective. Thanks to the efforts of 90 reviewers who delivered 252 reviews for the 76 papers originally submitted from all over the world, a strong selection of the 24 best contributions was made possible. The Technical Program Committee created an outstanding program covering the broad scope of this highly interdisciplinary field: from distributed signal processing through networking and middleware issues to application experience.

Running such a workshop requires dedication and much work from many people. We want to thank in particular Petra Hutt, Irene Ostertag and Heike Klemz for their valuable and esteemed help in the local organization of this workshop.

We hope that you enjoy this volume, and if you were lucky enough to attend we hope that you enjoyed the discussions with colleagues working in this fascinating area.

Adam Wolisz
Holger Karl
Andreas Willig

Committees

General Chair

Adam Wolisz, Technische Universität Berlin

Program Committee

Holger Boche,	Heinrich-Hertz-Institute, Berlin
Erdal Cayirci,	Yeditepe University, Istanbul
Kurt Geihs,	Technische Universität Berlin
Paul Havinga,	University of Twente
Holger Karl,	Technische Universität Berlin
Thomas Lentsch,	Infineon
Ian Marshall,	BT exact
Friedemann Mattern,	ETH Zürich
Chiara Petrioli,	University "La Sapienza," Rome
Radu Popescu-Zeletin,	FhG Fokus
Jan Rabaey,	University of California at Berkeley
Herbert Reichl,	FhG IZM
Lionel Sacks,	University College London
Jochen Schiller,	Freie Universität Berlin
Thomas Sikora,	Technische Universität Berlin
Martin Vetterli,	Swiss Federal Institute of Technology, Lausanne (EPFL)
Andreas Willig,	Hasso-Plattner-Institut, Potsdam
Adam Wolisz,	Technische Universität Berlin
Michele Zorzi,	Università degli Studi di Ferrara
Richard Zurawski,	ISA Corp

Organizing Committee

Holger Karl,	Technische Universität Berlin
Andreas Willig,	Hasso-Plattner-Institut, Potsdam

Reviewers

Bengt Ahlgren, SICS, Sweden
Matthias Anlauff, Kestrel Institute, USA
Stefano Basagni, Northeastern University, USA
Sebnem Baydere, Yeditepe University, Turkey
Baltasar Beferull-Lozano, Swiss Federal Institute of Technology,
Lausanne (EPFL), Switzerland
Ljubica Blazevic, ST Microelectronics, USA
Holger Boche, Heinrich-Hertz-Institut für Nachrichtentechnik Berlin GmbH,
Germany
Holger Bock, Infineon Technologies, Austria
Jürgen Bohn, ETH Zürich, Switzerland
Matthew Britton, University College London, UK
Sonja Buchegger, IC-LCA, Swiss Federal Institute of Technology,
Lausanne (EPFL), Switzerland
Nirupama Bulusu, National Information Communication Technology
Australia Ltd., Australia
Srdan Capkun, Swiss Federal Institute of Technology,
Lausanne (EPFL), Switzerland
Erdal Cayirci, Istanbul Technical University, Turkey
Alberto Cerpa, University of California at Los Angeles, USA
Shyam Chakraborty, Helsinki University of Technology, Finland
Carla-Fabiana Chiasserini, Politecnico di Torino, Italy
Jim Chou, University of California at Berkeley, USA
Edward Coyle, Purdue University, USA
Razvan Cristescu, Swiss Federal Institute of Technology,
Lausanne (EPFL), Switzerland
Svetlana Domnitcheva, ETH Zürich, Switzerland
Stefan Dulman, University of Twente, The Netherlands
Jean-Pierre Ebert, Technische Universität Berlin, Germany
Claudia Eckert, TU Darmstadt, Germany
Andras Farago, University of Texas at Dallas, USA
Laura Feeney, Swedish Institute of Computer Science, Sweden
Michael Gastpar, University of California at Berkeley, USA
Kurt Geihs, Technische Universität Berlin, Germany
Mario Gerla, University of California at Los Angeles, USA
Silvia Giordano, SUPSI, Lugano, Switzerland
Peter Gober, FhG Fokus, Germany
James Gross, Technische Universität Berlin, Germany
Joerg Haehner, University of Stuttgart, Germany
Robert Hahn, Fraunhofer IZM, Germany
Vlado Handziski, Technische Universität Berlin, Germany
Paul Havinga, University of Twente, The Netherlands
Hans-Ulrich Heiss, Technische Universität Berlin, Germany

Christian Hoene, Technische Universität Berlin, Germany
Daniel Hollos, Technische Universität Berlin, Germany
Phillip Hünerberg, FhG Fokus, Germany
Friedrich Jondral, Karlsruhe University, Germany
Holger Karl, Technische Universität Berlin, Germany
Wolfgang Kastner, TU Vienna, Austria
Wolfgang Kellerer, DoCoMo EuroLaboratories, Germany
Ralph Kling, Intel, USA
Manish Kochhal, Wayne State University, USA
Rolf Kraemer, IHP Microelectronics, Frankfurt/Oder, Germany
Bhaskar Krishnamachari, University of Southern California, USA
Srikanth Krishnamurthy, University of California at Riverside, USA
Koen Langendoen, Technical University of Delft, The Netherlands
Jangwon Lee, University of Texas at Austin, USA
Thomas Lentsch, Infineon, Austria
Jason Liu, Dartmouth College, USA
Mingyan Liu, University of Michigan, USA
Ian Marshall, BT, UK
Friedemann Mattern, ETH Zürich, Switzerland
Gianluca Mazzini, University of Ferrara, Italy
Suman Nath, Carnegie Mellon University, USA
Sven Ostring, University of Cambridge, UK
Sergio Palazzo, University of Catania, Italy
Seung-Jong Park, Georgia Institute of Technology, USA
Chiara Petrioli, University of Rome "La Sapienza", Italy
Jan Rabaey, University of California at Berkeley, USA
Ilja Radusch, Technische Universität Berlin, Germany
Kay Römer, ETH Zürich, Switzerland
Hartmut Ritter, Freie Universität Berlin, Germany
Michael Rohs, ETH Zürich, Switzerland
Lionel Sacks, University College London, UK
Asuman Sünbül, Kestrel Institute, USA
Guenter Schaefer, Technische Universität Berlin, Germany
Jochen Schiller, Freie Universität Berlin, Germany
Morten Schläger, Technische Universität Berlin, Germany
Thomas Schoch, ETH Zürich, Switzerland
Katja Schwieger, TU Dresden, Germany
Rahul Shah, University of California at Berkeley, USA
Thomas Sikora, Technische Universität Berlin, Germany
Rainer Steinwandt, Universität Karlsruhe, Germany
John Sucec, Rutgers University, USA
Zhi Tian, Michigan Technological University, USA
Petia Todorova, FhG Fokus, Germany
Robert Tolksdorf, Free University Berlin, Germany
Jana van Greunen, University of California at Berkeley, USA
Harald Vogt, ETH Zürich, Switzerland
Dirk Westhoff, NEC Europe Ltd., Germany

Andreas Willig, Hasso-Plattner-Institut, Potsdam, Germany
Ibiso Wokoma, University College London, UK
M. Juergen Wolf, Fraunhofer IZM, Germany
Paul Wright, University of California at Berkeley, USA
Jin Xi, Technical University of Munich, Germany
Zixiang Xiong, Texas A&M University, USA
Feng Zhao, Parc, USA
Holger Ziemek, Fraunhofer Institute for Open Communication
Systems (FOKUS), Germany
Michele Zorzi, Università degli Studi di Ferrara, Italy

Table of Contents

Power Sources for Wireless Sensor Networks

Shad Roundy[1], Dan Steingart[2], Luc Frechette[3], Paul Wright[2], and Jan Rabaey[2]

[1]Australian National University, Department of Engineering, Canberra ACT 0200, Australia
shad.roundy@anu.edu.au
[2]University of California at Berkeley, Department of {Mechanical Engineering, Electrical Engineering and Computer Science}, Berkeley CA 94720, USA
steinda@uclink.berkeley.edu,
pwright@kingkong.me.berkeley.edu,
jan@eecs.berkeley.edu
[3]Columbia University, Department of Mechanical Engineering , 228 S.W. Mudd Bldg, Mail Code 4703, 500 West 120th Street, New York, NY 10027-6699, USA
lf307@columbia.edu

Abstract. Wireless sensor networks are poised to become a very significant enabling technology in many sectors. Already a few very low power wireless sensor platforms have entered the marketplace. Almost all of these platforms are designed to run on batteries that have a very limited lifetime. In order for wireless sensor networks to become a ubiquitous part of our environment, alternative power sources must be employed. This paper reviews many potential power sources for wireless sensor nodes. Well established power sources, such as batteries, are reviewed along with emerging technologies and currently untapped sources. Power sources are classified as energy reservoirs, power distribution methods, or power scavenging methods, which enable wireless nodes to be completely self-sustaining. Several sources capable of providing power on the order of 100 $\mu W/cm^3$ for very long lifetimes are feasible. It is the authors' opinion that no single power source will suffice for all applications, and that the choice of a power source needs to be considered on an application-by-application basis.

1 Introduction

The vast reduction in size and power consumption of CMOS circuitry has led to a large research effort based around the vision of ubiquitous networks of wireless sensor and communication nodes [1-3]. As the size and cost of such wireless sensor nodes continues to decrease, the likelihood of their use becoming widespread in buildings, industrial environments, automobiles, aircraft, etc. increases. However, as their size and cost decrease, and as their prevalence increases, effective power supplies become a larger problem.

The issue is that the scaling down in size and cost of CMOS electronics has far outpaced the scaling of energy density in batteries, which are by far the most prevalent power sources currently used. Therefore, the power supply is usually the largest and most expensive component of the emerging wireless sensor nodes being proposed and designed. Furthermore, the power supply (usually a battery) is also the limiting

H. Karl, A. Willig, A. Wolisz (Eds.): EWSN 2004, LNCS 2920, pp. 1–17, 2004.

factor on the lifetime of a sensor node. If wireless sensor networks are to truly become ubiquitous, replacing batteries in every device every year or two is simply cost prohibitive.

The purpose of this paper, then, is to review existing and potential power sources for wireless sensor networks. Current state of the art, ongoing research, and theoretical limits for many potential power sources will be discussed. One may classify possible methods of providing power for wireless nodes into three groups: store energy on the node (i.e. a battery), distribute power to the node (i.e. a wire), scavenge available ambient power at the node (i.e. a solar cell). Power sources that fall into each of these three categories will be reviewed.

A direct comparison of vastly different types of power source technologies is difficult. For example, comparing the efficiency of a solar cell to that of a battery is not very useful. However, in an effort to provide general understanding of a wide variety of power sources, the following metrics will be used for comparison: power density, energy density (where applicable), and power density per year of use. Additional considerations are the complexity of the power electronics needed and whether secondary energy storage is needed.

2 Energy Reservoirs

Energy storage, in the form of electrochemical energy stored in a battery, is the predominant means of providing power to wireless devices today. However, several other forms of energy storage may be useful for wireless sensor nodes. Regardless of the form of the energy storage, the lifetime of the node will be determined by the fixed amount of energy stored on the device. The primary metric of interest for all forms of energy storage will be usable energy per unit volume (J/cm^3). An additional issue is that the instantaneous power that an energy reservoir can supply is usually dependent on its size. Therefore, in some cases, such as micro-batteries, the maximum power density (μW/cm^3) is also an issue for energy reservoirs.

2.1 Macro-Scale Batteries

Primary batteries are perhaps the most versatile of all small power sources. Table 1 shows the energy density for a few common primary battery chemistries. Note that while zinc-air batteries have the highest energy density, their lifetime is very short, and so are most useful for applications that have constant, relatively high, power demands.

Table 1. Energy density of three primary battery chemistries.

Chemistry	Zinc-air	Lithium	Alkaline
Energy (J/cm^3)	3780	2880	1200

Because batteries have a fairly stable voltage, electronic devices can often be run directly from the battery without any intervening power electronics. While this may not be the most robust method of powering the electronics, it is often used and is advantageous in that it avoids the extra power consumed by power electronics.

Macro-scale secondary (rechargeable) batteries are commonly used in consumer electronic products such as cell phones, PDA's, and laptop computers. Table 2 gives the energy density of a few common rechargeable battery chemistries. It should be remembered that rechargeable batteries are a *secondary* power source. Therefore, in the context of wireless sensor networks, another primary power source must be used to charge them.

Table 2. Energy density of three secondary battery chemistries.

Chemistry	Lithium	NiMHd	NiCd
Energy (J/cm^3)	1080	860	650

2.2 Micro-Scale Batteries

The size of batteries has only decreased mildly when compared to electronic circuits that have decreased in size by orders of magnitude. One of the main stumbling blocks to reducing the size of micro-batteries is power output due to surface area limitations of micro-scale devices. The maximum current output of a battery depends on the surface area of the electrodes. Because micro-batteries are so small, the electrodes have a small surface area, and their maximum current output is also very small.

The challenge of maintaining (or increasing) performance while decreasing size is being addressed on multiple fronts. Bates *et al* at Oak Ridge National Laboratory have created a process by which a primary thin film lithium battery can be deposited onto a chip [4]. The thickness of the entire battery is on the order of 10's of μm, but the areas studied are in the cm^2 range. This battery is in the form of a traditional Volta pile, with alternating layers of Lithium Manganese Oxide (or Lithium Cobalt Oxide), Lithium Phosphate Oxynitride and Lithium metal. Maximum potential is rated at 4.2 V with Continuous/Max current output on the order of 1 mA/cm^2 and 5 mA/cm^2 for the $LiCoO_2$ – Li based cell.

Work is being done on thick film batteries with a smaller surface area by Harb *et al* [5], who have developed micro-batteries of Ni/Zn with an aqueous NaOH electrolyte. Thick films are on the order of 0.1 mm, but overall thicknesses are minimized by use of three-dimensional structures. While each cell is only rated at 1.5 V, geometries have been duty-cycle optimized to give acceptable power outputs at small overall theoretical volumes (4 mm by 1.5 mm by 0.2 mm) with good durability demonstrated by the electrochemical components of the battery. The main challenges lie in maintaining a microfabricated structure that can contain an aqueous electrolyte.

Radical three dimensional structures are also being investigated to maximize power output. Hart *et al* [6] have theorized a three dimensional battery made of series alternating cathode and anode rods suspended in a solid electrolyte matrix. Theoretical power outputs for a three dimensional microbattery are shown to be many times larger than a two dimensional battery of equal size because of higher electrode surface area to volume ratios and lower ohmic losses due to lower ionic transport distances. However, it should be noted that the increased power density comes at a lower energy density because of the lower volume percentage of electrolyte.

2.3 Micro-Fuel Cells

Hydrocarbon based fuels have very high energy densities compared to batteries. For example, methanol has an energy density of 17.6 kJ/cm^3, which is about 6 times that of a lithium battery. Like batteries, fuel cells produce electrical power from a chemical reaction. A standard fuel cell uses hydrogen atoms as fuel. A catalyst promotes the separation of the electron in the hydrogen atom from the proton. The proton diffuses through an electrolyte (often a solid membrane) while the electron is available for use by an external circuit. The protons and electrons recombine with oxygen atoms on the other side (the oxidant side) of the electrolyte to produce water molecules. This process is illustrated in Figure 1. While pure hydrogen can be used as a fuel, other hydrocarbon fuels are often used. For example, in Direct Methanol Fuel Cells (DFMC) the anode catalyst draws the hydrogen atoms out from the methanol.

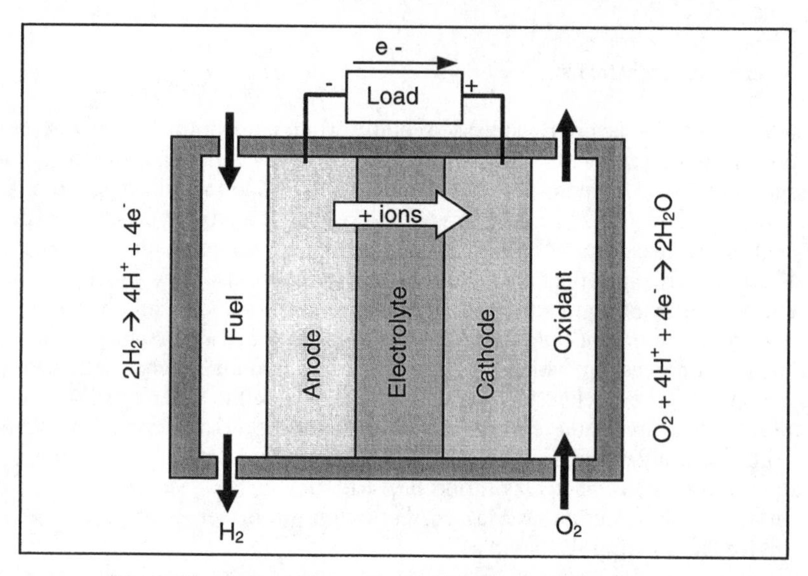

Fig. 1. Illustration of how a standard hydrogen fuel cell works.

Most single fuel cells tend to output open circuit voltages around 1.0 – 1.5 volts. Of course, like batteries, the cells can be placed in series for higher voltages. The voltage is quite stable over the operating lifetime of the cell, but it does fall off with increasing current draw. Because the voltage drops with current, it is likely that some additional power electronics will be necessary if replacing a battery with a fuel cell.

Large scale fuel cells have been used as power supplies for decades. Recently fuel cells have gained favor as a replacement for consumer batteries [7]. Small, but still macro-scale, fuel cells are likely to soon appear in the market as battery rechargers and battery replacements [8].

The research trend is toward micro-fuel cells that could possibly be closely integrated with wireless sensor nodes. Like micro-batteries, a primary metric of comparison in micro-fuel cells is power density in addition to energy density. As with micro-batteries, the maximum continuous current output is dependent on the electrode surface area. Efficiencies of large scale fuel cells have reached approximately 45%

electrical conversion efficiency and nearly 90% if cogeneration is employed [9]. Efficiencies for micro-scale fuel cells will certainly be lower. The maximum obtainable efficiency for a micro-fuel cell is still uncertain. Demonstrated efficiencies are generally below 1% [10].

Many research groups are working on microfabricated partial systems that typically include an electrolyte membrane, electrodes, and channels for fuel and oxidant flow. Recent examples include the hydrogen based fuel cells developed by Hahn *et al* [11] and Lee *et al* [12]. Both systems implement microfabricated electrodes and channels for fuel and oxidant flow. The system by Hahn *et al* produces power on the order of 100 mW/cm^2 from a device 0.54 cm^2 in size. The system by Lee *et al* produces 40 mW/cm^2. It should be noted that the fundamental characteristic here is power per unit area rather than power per unit volume because the devices are fundamentally planar. Complete fuel storage systems are not part of their studies, and therefore an energy or power per unit volume metric is not appropriate. Fuel conversion efficiencies are not reported.

Hydrogen storage at small scales is a difficult problem that has not yet been solved. Primarily for this reason, methanol based micro-fuel cells are also being investigated by numerous groups. For example, Holloday *et al* [10] have demonstrated a methanol fuel processor with a total size on the order of several mm^3. This fuel processor has been combined with a thin fuel cell, 2 cm^2 in area, to produce roughly 25 mA at 1 volt with 0.5% overall efficiency. They are targeting a 5% efficient cell.

Given the energy density of fuels such as methanol, fuel cells need to reach efficiencies of at least 20% in order to be more attractive than primary batteries. Nevertheless, at the micro scale, where battery energy densities are also lower, a lower efficiency fuel cell may still be attractive. Finally, providing for sufficient fuel and oxidant flows is a very difficult task in micro-fuel cell development. The ability to microfabricate electrodes and electrolytes does not guarantee the ability to realize a micro-fuel cell. To the authors' knowledge, a self-contained, on-chip fuel cell has yet to be demonstrated.

2.4 Micro Heat Engines

At large scales, fossil fuels are the dominant source of energy used for electric power generation, mostly due to the low cost per joule, high energy density (gasoline has an energy density of 12.7 kJ/cm^3), abundant availability, storability and ease of transport. To date, the complexity and multitude of components involved have hindered the miniaturization of heat engines and power generation approaches based on combustion of hydrocarbon fuels. As the scale of a mechanical system is reduced, the tolerances must reduce accordingly and the assembly process becomes increasingly challenging. This results in increasing costs per unit power and/or deteriorated performance.

The extension of silicon microfabrication technology from microelectronics to micro-electromechanical systems (or MEMS) is changing this paradigm. In the mid-1990's, Epstein *et al* proposed that microengines, i.e. dime-size heat engines, for portable power generation and propulsion could be fabricated using MEMS technology [13]. The initial concept consisted of using silicon deep reactive ion etching, fusion wafer bonding, and thin film processes to microfabricate and integrate high speed turbomachinery, with bearings, a generator, and a combustor within a cubic centimeter volume. An application-ready power supply would also require auxiliary components, such as a fuel tank, engine and fuel controller, electrical power conditioning with short term storage, thermal management and packaging. Expected performance is 10-20 Watt of electrical power output at thermal efficiencies on the order of 5-20%. Figure 2 shows a microturbine test device used for turbomachinery and air bearing development.

Multiple research groups across the globe have also undertaken the development of various micro heat engine-based power generators. Approaches ranging from micro gas turbine engines to thermal-expansion-actuated piezoelectric generators and micro-thermophotovoltaic systems are being investigated [13–20].

Most of these and similar efforts are at initial stages of development and performance has not been demonstrated. However, predictions range from 0.1-10W of electrical power output, with typical masses ~1-5 g and volumes ~1 cm^3. Given the relatively large power level, a single microengine would only need to operate at low duty-cycles (less than 1% of the time) to periodically recharge a battery. Microengines are not expected to reduce further in size due to manufacturing and efficiency constraints. At small scales, viscous drag on moving parts and heat transfer to the ambient and between components increase, which adversely impacts efficiency.

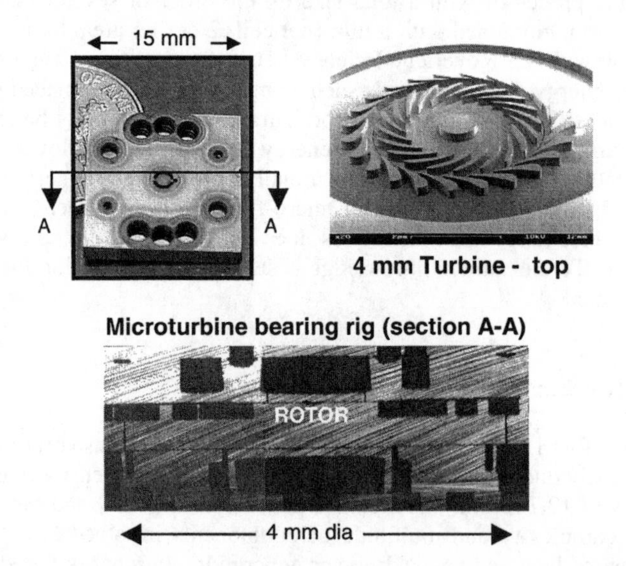

Fig. 2. Micro-turbine development device, which consists of a 4 mm diameter single crystal silicon rotor enclosed in a stack of five bonded wafers used for micro air bearing development.

Overall, the greatest benefits of micro heat engines are their high power density (0.1-2 W/g, without fuel) and their use of fuels allowing high density energy storage for compact, long duration power supplies. For low power applications, the power density is not as important as efficiency. Microengines will therefore require many years of development before reaching the expected efficiencies and being applicable for wireless sensor network applications.

2.5 Radioactive Power Sources

Radioactive materials contain extremely high energy densities. As with hydrocarbon fuels, this energy has been used on a much larger scale for decades. However, it has not been exploited on a small scale as would be necessary to power wireless sensor networks. The use of radioactive materials can pose a serious health hazard, and is a highly political and controversial topic. It should, therefore, be noted that the goal here is neither to promote nor discourage investigation into radioactive power sources, but to present their potential, and the research being done in the area.

The total energy emitted by radioactive decay of a material can be expressed as in equation 1.

$$E_t = A_c E_e T \tag{1}$$

where E_t is the total emitted energy, A_c is the activity, E_e is the average energy of emitted particles, and T is the time period over which power is collected. Table 3 lists several potential radioisotopes, their half-lives, specific activities, energy densities, and power densities based on radioactive decay. The half-life of the material has been used as the time over which power would be collected.

Table 3. Comparison of radio-isotopes.

Material	Half-life (years)	Activity volume density (Ci/cm^3)	Energy density (J/cm^3)	Power density (mW/cm^3)
^{238}U	4.5 X 10^9	6.34 X 10^{-6}	2.23 X 10^{10}	1.6 X 10^{-4}
^{63}Ni	100.2	506	1.6 X 10^8	50.6
^{32}Si	172.1	151	3.3 X 10^8	60.8
^{90}Sr	28.8	350	3.7 X 10^8	407
^{32}P	0.04	5.2 X 10^5	2.7 X 10^9	2.14 X 10^6

While the energy density numbers reported for radioactive materials are extremely attractive, it must be remembered that efficient methods of converting this power to electricity at small scales do not exist. Therefore, efficiencies would likely be extremely low.

Recently, Li and Lal [21] have used the ^{63}Ni isotope to actuate a conductive cantilever. As the beta particles (electrons) emitted from the ^{63}Ni isotope collect on the conductive cantilever, there is an electrostatic attraction. At some point, the cantilever contacts the radioisotope and discharges, causing the cantilever to oscillate. Up to this point the research has only demonstrated the actuation of a cantilever, and not electric power generation. However, electric power could be generated from an oscillating cantilever. The reported power output, defined as the change over time in the combined mechanical and electrostatic energy stored in the cantilever, is 0.4 pW from a 4mm X 4mm thinfilm of ^{63}Ni. This power level is equivalent to 0.52 µW/cm^3. However, it should be noted that using 1 cm^3 of ^{63}Ni is impractical. The reported efficiency of the device is 4 X 10^{-6}.

3 Power Distribution

In addition to storing power on a wireless node, in certain circumstances power can be distributed to the node from a nearby energy rich source. It is difficult to characterize the effectiveness of power distribution methods by the same metrics (power or energy density) because in most cases the power received at the node is more a function of how much power is transmitted rather than the size of the power receiver at the node. Nevertheless an effort is made to characterize the effectiveness of power distribution methods as they apply to wireless sensor networks.

3.1 Electromagnetic (RF) Power Distribution

The most common method (other than wires) of distributing power to embedded electronics is through the use of RF (Radio Frequency) radiation. Many passive electronic devices, such as electronic ID tags and smart cards, are powered by a nearby energy rich source that transmits RF energy to the passive device. The device then uses that energy to run its electronics [22-23]. This solution works well, as evidenced by the wide variety of applications where it is used, if there is a high power scanner or other source in very near proximity to the wireless device. It is, however, less effective in dense ad-hoc networks where a large area must be flooded with RF radiation to power many wireless sensor nodes.

Using a very simple model neglecting any reflections or interference, the power received by a wireless node can be expressed by equation 2 [24].

$$P_r = \frac{P_0 \lambda^2}{4\pi R^2} \qquad (2)$$

where P_o is the transmitted power, λ is the wavelength of the signal, and R is the distance between transmitter and receiver. Assume that the maximum distance between the power transmitter and any sensor node is 5 meters, and that the power is being transmitted to the nodes in the 2.4 – 2.485 GHz frequency band, which is the unlicensed industrial, scientific, and medical band in the United States. Federal regulations limit ceiling mounted transmitters in this band to 1 watt or lower. Given a 1 watt transmitter, and a 5 meter maximum distance the power received at the node would be 50 μW, which is probably on the borderline of being really useful for wireless sensor nodes. However, in reality the power transmitted will fall off at a rate faster than $1/R^2$ in an indoor environment. A more likely figure is $1/R^4$. While the 1 watt limit on a transmitter is by no means general for indoor use, it is usually the case that some sort of safety limitation would need to be exceeded in order to flood a room or other area with enough RF radiation to power a dense network of wireless devices.

3.2 Wires, Acoustic, Light, Etc.

Other means of transmitting power to wireless sensor nodes might include wires, acoustic emitters, and light or lasers. However, none of these methods are appropriate for wireless sensor networks. Running wires to a wireless communications device defeats the purpose of wireless communications. Energy in the form of acoustic waves has a far lower power density than is sometimes assumed. A sound wave of 100 dB in sound level only has a power level of approximately 1 $\mu W/cm^2$. One could also imagine using a laser or other focused light source to direct power to each of the nodes in the sensor network. However, to do this is a controlled way, distributing light energy directly to each node, rather than just flooding the space with light, would likely be too complex and not cost effective. If a whole space is flooded with light, then this source of power becomes attractive. However, this situation has been classified as "power scavenging" and will be discussed in the following section.

4 Power Scavenging

Unlike power sources that are fundamentally energy reservoirs, power scavenging sources are usually characterized by their power density rather than energy density. Energy reservoirs have a characteristic energy density, and how much average *power* they can provide is then dependent on the lifetime over which they are operating. On the contrary, the *energy* provided by a power scavenging source depends on how long the source is in operation. Therefore, the primary metric for comparison of scavenged sources is power density, not energy density.

4.1 Photovoltaics (Solar Cells)

At midday on a sunny day, the incident light on the earth's surface has a power density of roughly 100 mW/cm^2. Single crystal silicon solar cells exhibit efficiencies of 15% - 20% [25] under high light conditions, as one would find outdoors. Common indoor lighting conditions have far lower power density than outdoor conditions. Common office lighting provides about 100 $\mu W/cm^2$ at the surface of a desk. Single crystal silicon solar cells are better suited to high light conditions and the spectrum of light available outdoors [25]. Thin film amorphous silicon or cadmium telluride cells offer better efficiency indoors because their spectral response more closely matches that of artificial indoor light. Still, these thin film cells only offer about 10% efficiency. Therefore, the power available from photovoltaics ranges from about 15 mW/cm^2 at midday outdoors to 10 $\mu W/cm^2$ indoors.

A single solar cell has an open circuit voltage of about 0.6 volts. Individual cells are easily placed in series, especially in the case of thin film cells, to get almost any desired voltage needed. A current vs. voltage (I-V) curve for a typical five cell array (wired in series) is shown below in Figure 3. Unlike the voltage, current densities are directly dependent on the light intensity.

Solar cells provide a fairly stable DC voltage through much of their operating space. Therefore, they can be used to directly power electronics in cases where the current load is such that it allows the cell to operate on high voltage side of the "knee" in the I-V curve and where the electronics can tolerate some deviation in source voltage. More commonly solar cells are used to charge a secondary battery. Solar cells can be connected directly to rechargeable batteries through a simple series diode to prevent the battery from discharging through the solar cell. This extremely simple circuit does not ensure that the solar cell will be operating at its optimal point, and so power production will be lower than the maximum possible. Secondly, rechargeable batteries will have a longer lifetime if a more controlled charging profile is employed. However, controlling the charging profile and the operating point of the solar cell both require more electronics, which use power themselves. An analysis needs to be done for each individual application to determine what level of power electronics would provide the highest net level of power to the load electronics. Longevity of the battery is another consideration to be considered in this analysis.

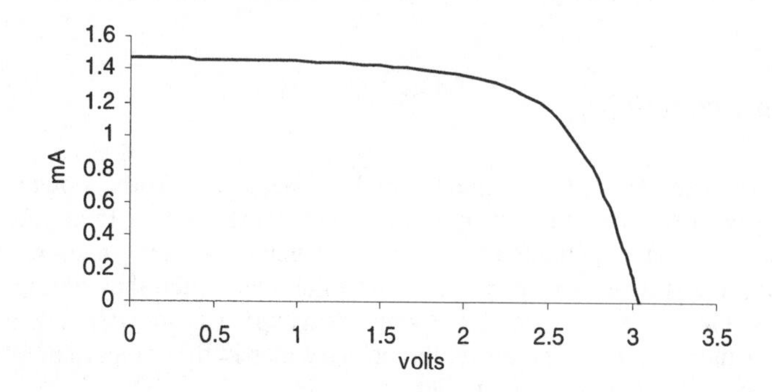

Fig. 3. Typical I-V curve from a cadmium telluride solar array (Panasonic BP-243318).

4.2 Temperature Gradients

Naturally occurring temperature variations can also provide a means by which energy can be scavenged from the environment. The maximum efficiency of power conversion from a temperature difference is equal to the Carnot efficiency, which is given as equation 3.

$$\eta = \frac{T_{high} - T_{low}}{T_{high}} \tag{3}$$

Assuming a room temperature of 20 °C, the efficiency is 1.6% from a source 5 °C above room temperature and 3.3% for a source 10 °C above room temperature.

A reasonable estimate of the maximum amount of power available can be made assuming heat conduction through silicon material. Convection and radiation would be quite small compared to conduction at small scales and low temperature differentials. The amount of heat flow (power) is given by equation 4.

$$q' = k\frac{\Delta T}{L} \tag{4}$$

where k is the thermal conductivity of the material and L is the length of the material through which the heat is flowing. The conductivity of silicon is approximately 140 W/mK. Assuming a 5 °C temperature differential and a length of 1 cm, the heat flow is 7 W/cm^2. If Carnot efficiency could be obtained, the resulting power output would be 117 mW/cm^2. While this is an excellent result compared with other power sources, one must realize demonstrated efficiencies are well below the Carnot efficiency.

A number of researchers have developed systems to convert power from temperature differentials to electricity. The most common method is through thermoelectric generators that exploit the Seebeck effect to generate power. For example Stordeur and Stark [26] have demonstrated a micro-thermoelectric generator capable of generating 15 μW/cm^2 from a 10 °C temperature differential. Recently, Applied Digital Solutions have developed a thermoelectric generator soon to be marketed as a commercial product. The generator is reported as being able to produce 40 μW of power from a 5 °C temperature differential using a device 0.5 cm^2 in area and a few millimeters thick [27]. The output voltage of the device is approximately 1 volt. The thermal-expansion actuated piezoelectric generator referred to earlier [17] has also been proposed as a method to convert power from ambient temperature gradients to electricity.

4.3 Human Power

An average human body burns about 10.5 MJ of energy per day. (This corresponds to an average power dissipation of 121 W.) Starner has proposed tapping into some of this energy to power wearable electronics [28]. The conclusion of studies undertaken at MIT suggests that the most energy rich and most easily exploitable source occurs at the foot during heel strike and in the bending of the ball of the foot [29]. This research has led to the development of piezoelectric shoe inserts capable of producing an average of 330 μW/cm^2 while a person is walking. The shoe inserts have been used to power a low power wireless transceiver mounted to the shoes. While this power source is of great use for wireless nodes worn on a person's foot, the problem of how to get the power from the shoe to the point of interest still remains.

The sources of power mentioned above are passive power sources in that the human doesn't need to do anything to generate power other than what they would normally do. There is also a class of power generators that could be classified as active human power in that they require the human to perform an action that they would not normally perform. For example Freeplay [30] markets a line of products that are powered by a constant force spring that the user must wind up. While these types of products are extremely useful, they are not very applicable to wireless sensor networks because it would be impractical and not cost efficient to individually wind up every node.

4.4 Wind / Air Flow

Wind power has been used on a large scale as a power source for centuries. Large windmills are still common today. However, the authors' are unaware of any efforts to try to generate power at a very small scale (on the order of a cubic centimeter) from air flow. The potential power from moving air is quite easily calculated as shown in equation 5.

$$P = \frac{1}{2}\rho A v^3 \tag{5}$$

where P is the power, ρ is the density of air, A is the cross sectional area, and v is the air velocity. At standard atmospheric conditions, the density of air is approximately 1.22 kg/m^3. Figure 4 shows the power per square centimeter versus air velocity.

Large scale windmills operate at maximum efficiencies of about 40%. Efficiency is dependent on wind velocity, and average operating efficiencies are usually about 20%. Windmills are generally designed such that maximum efficiency occurs at wind velocities around 8 m/s (or about 18 mph). At low air velocity, efficiency can be significantly lower than 20%. Figure 4 also shows power output assuming 20% and 5% efficiency in conversion. As can be seen from the graph, power densities from air velocity are quite promising. As there are many possible applications in which a fairly constant air flow of a few meters per second exists, it seems that research leading to the development of devices to convert air flow to electrical power at small scales is warranted.

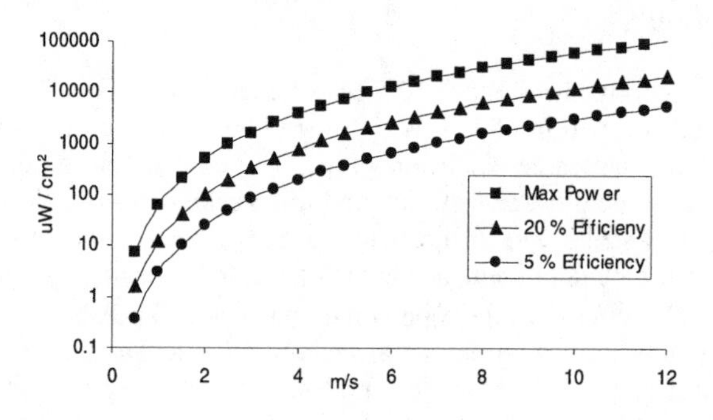

Fig. 4. Maximum power density from air flow. Power density assuming 20% and 5% conversion efficiencies are also shown.

4.5 Vibrations

Low level mechanical vibrations are present in many environments. Examples include HVAC ducts, exterior windows, manufacturing and assembly equipment, aircraft, automobiles, trains, and small household appliances. The results of measurements performed by the authors on many commonly occurring vibration sources suggest that the dominant frequency is generally between 60 to 200 Hz at amplitudes ranging from 1 to 10 m/s^2.

A simple general model for power conversion from vibrations has been presented by Williams *et al* [31]. The final equation for power output from this model is shown here as equation 6.

$$P = \frac{m\zeta_e A^2}{4\omega(\zeta_e + \zeta_m)^2} \tag{6}$$

where P is the power output, m is the oscillating proof mass, A is the acceleration magnitude of the input vibrations, ω is the frequency of the driving vibrations, ζ_m is the mechanical damping ratio, and ζ_e is an electrically induced damping ratio. In the derivation of this equation, it was assumed that the resonant frequency of the oscillating system matches the frequency of the driving vibrations. While this model is oversimplified for many implementations, it is useful to get a quick estimate on potential power output from a given source. Three interesting relationships are evident from this model.

1. Power output is proportional to the oscillating mass of the system.
2. Power output is proportional to the square of the acceleration amplitude.
3. Power is inversely proportional to frequency.

Point three indicates that the generator should be designed to resonate at the lowest frequency peak in the vibrations spectrum provided that higher frequency peaks do not have a higher acceleration magnitude. Many spectra measured by Roundy *et al* [32] verify that generally the lowest frequency peak has the highest acceleration magnitude.

Figures 5 and 6 provide a range of power densities that can be expected from vibrations similar to those listed above. The data shown in the figures are based on calculations from the model of Williams *et al* and do not consider the technology that is used to convert the mechanical kinetic energy to electrical energy.

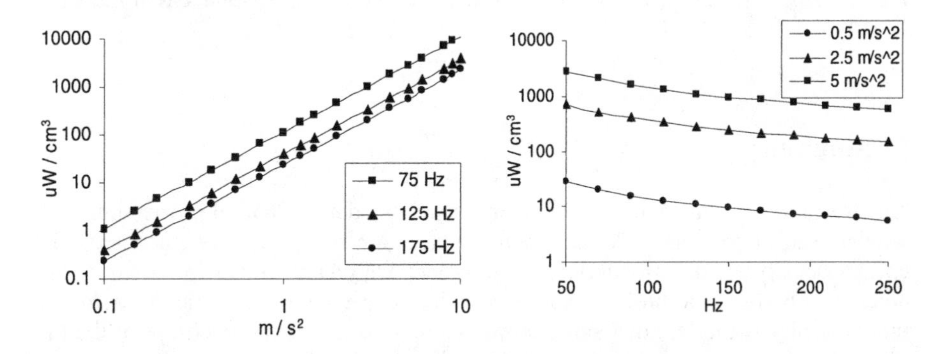

Fig. 5. Power density vs. vibration amplitude for three frequencies.

Fig. 6. Power density vs. frequency of vibration input for three amplitudes.

Several researchers have developed devices to scavenge power from vibrations [31-34]. Devices include electromagnetic, electrostatic, and piezoelectric methods to convert mechanical motion into electricity. Theory, simulations, and experiments performed by the authors suggest that for devices on the order of 1 cm^3 in size, piezoelectric generators will offer the most attractive method of power conversion. Piezoelectric conversion offers higher potential power density from a given input, and produces voltage levels on the right order of magnitude. Roundy *et al* [35] have demonstrated a piezoelectric power converter of 1cm^3 in size that produces 200 μW from input vibrations of 2.25 m/s^2 at 120 Hz. Both Roundy *et al* and Ottman *et al* [34-35] have demonstrated wireless transceivers powered from vibrations. Figure 7 shows the generator, power circuit, and transceiver developed by Roundy *et al.*

The power signal generated from vibration generators needs a significant amount of conditioning to be useful to wireless electronics. The converter produces an AC voltage that needs to be rectified. Additionally the magnitude of the AC voltage depends on the magnitude of the input vibrations, and so is not very stable. Although more power electronics are needed compared with some other sources, commonly occurring vibrations can provide power on the order of hundreds of microwatts per cubic centimeter, which is quite competitive compared to other power scavenging sources.

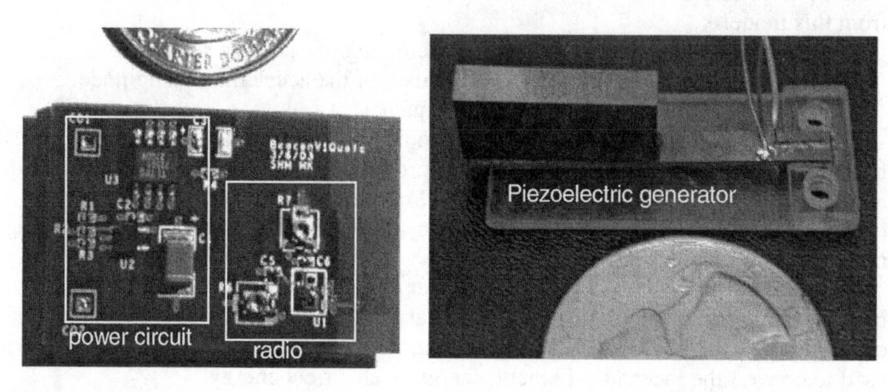

Fig. 7. Piezoelectric generator, power circuit, and radio powered from vibrations of 2.25 m/s^2 at 120 Hz.

5 Summary

An effort has been made to give an overview of the many potential power sources for wireless sensor networks. Because some sources are fundamentally characterized by energy density (such as batteries) while others are characterized by power density (such as vibrations) a direct comparison with a single metric is difficult. Adding to this difficulty is the fact that some power sources do not make much use of the third dimension (such as solar cells), so their fundamental metric is power per square centimeter rather than power per cubic centimeter. Nevertheless, in an effort to compare all possible sources, a summary table is shown below as Table 4. Note that power density is listed as μW/cm^3, however, it is understood that in certain instances the number reported really represents μW/cm^2. Such values are marked with a "*". Note also that, with only one exception, values listed are numbers that have been demonstrated or are based on experiments rather than theoretical optimal values. The one exception is power from air flow, which has been italicized to indicate that it is a theoretical value. In many cases the theoretical best values are explained in the text above.

Almost all wireless sensor nodes are presently powered by batteries. This situation presents a substantial roadblock to the widespread deployment of wireless sensor networks because the replacement of batteries is cost prohibitive. Furthermore, a battery that is large enough to last the lifetime of the device would dominate the overall system size and cost, and thus is not very attractive. It is therefore essential that alternative power sources be considered and developed.

This paper has attempted to characterize a wide variety of such sources. It is the authors' opinion that no single alternative power source will solve the problem for all, or even a large majority of cases. However, many attractive and creative solutions do exist that can be considered on an application-by-application basis.

Table 4. Comparison of various potential power sources for wireless sensor networks. Values shown are actual demonstrated numbers except in one case which has been italicized.

Power Source	P/cm^3 (μW/cm^3)	E/cm^3 (J/cm^3)	P/cm^3/yr (μW/cm^3/Yr)	Secondary Storage Needed	Voltage Regulation	Comm. Available
Primary Battery	-	2880	90	No	No	Yes
Secondary Battery	-	1080	34	-	No	Yes
Micro-Fuel Cell	-	3500	110	Maybe	Maybe	No
Heat engine	-	3346	106	Yes	Yes	No
Radioactive(^{63}Ni)	0.52	1640	0.52	Yes	Yes	No
Solar (outside)	15000 *	-	-	Usually	Maybe	Yes
Solar (inside)	10 *	-	-	Usually	Maybe	Yes
Temperature	40 * †	-	-	Usually	Maybe	Soon
Human Power	330	-	-	Yes	Yes	No
Air flow	*380 ††*	-	-	Yes	Yes	No
Vibrations	200	-	-	Yes	Yes	No

* Denotes sources whose fundamental metric is power per **square** centimeter rather than power per **cubic** centimeter.
† Demonstrated from a 5 °C temperature differential.
†† Assumes air velocity of 5 m/s and 5 % conversion efficiency.

References

[1] Rabaey J, Ammer J, Karalar T, Li S, Otis B, Sheets M, Tuan T, (2002) PicoRadios for Wireless Sensor Networks: The Next Challenge in Ultra-Low-Power Design. *Proceedings of the International Solid-State Circuits Conference*, 2002.

[2] Warneke B, Atwood B, Pister KSJ (2001) Smart Dust Mote Forerunners, *Fourteenth Annual International Conference on Micro-electromechanical Systems (MEMS 2001)*.

[3] Hill J, Culler D (2002) Mica: A Wireless Platform for Deeply Embedded Networks, *IEEE Micro*. 22(6) : 12–24.

[4] Bates J, Dudney N, Neudecker B, Ueda A, Evans CD (2000) Thin-film lithium and lithium-ion batteries. *Solid State Ionics* 135: 33–45.

[5] Harb JN, LaFollete RM, Selfridge RH, Howell LL (2002) Mircobatteries for self-suststained hybrid micropower supplies. *Journal of Power Sources*, 104: 46–51
[6] Hart RW, White HS, Dunn B, Rolison DR (2003) 3-D Microbatteries. *Electrochemistry Communications*, 5:120–123.
[7] Heinzel A, Hebling C, Muller M, Zedda M, Muller C (2002) Fuel cells for low power applications. *Journal of Power Sources* 105: 250 – 255.
[8] www.toshiba.co.jp/about/press/2003_03/pr0501.htm, 2003.
[9] Kordesh K, Simader G (2001) *Fuel cells and their applications.* VCH Publishers, New York
[10] Holloday JD, Jones EO, Phelps M, Hu J, (2002) Microfuel processor for use in a miniature power supply, *Journal of Power Sources*, 108:21–27.
[11] www.pb.izm.fhg.de/hdi/040_groups/group4/fuelcell_micro.html
[12] Lee SJ, Chang-Chien A, Cha SW, O'Hayre R, Park YI, Saito Y, Prinz FB (2002) Design and fabrication of a micro fuel cell array with "flip-flop" interconnection. *Journal of Power Sources* 112: 410–418.
[13] Epstein, AH, *et al* (1997) Micro-Heat Engine, Gas Turbine, and Rocket Engines – The MIT Microengine Project, *AIAA 97-1773, 28th AIAA Fluid Dynamics Conf.*, 1997.
[14] Lee C, Arslan S, Liu Y.-C, Fréchette LG, (2003) Design of a Microfabricated Rankine Cycle Steam Turbine for Power Generation. *ASME IMECE*, 2003.
[15] Fu K, Knobloch AJ, Martinez FC, Walther DC, Fernandez-Pello C, Pisano AP, Liepmann D (2001) Design and Fabrication of a Silicon-Based MEMS Rotary Engine, *ASME IMECE*, 2001.
[16] Toriyama T, Hashimoto K, Sugiyama S, (2003) Design of a Resonant Micro Reciprocating Engine for Power Generation. *Transducers'03*, 2003.
[17] Whalen S, Thompson M, Bahr D, Richards C, Richards R (2003) Design, Fabrication and Testing of the P3 Micro Heat Engine, *Sensors and Actuators*, 104(3), 200–208.
[18] Schaevitz SB, Franz AJ, Jensen KF, Schmidt MA (2001) A Combustion-based MEMS Thermoelectric Power Generator, *Transducers'01*, Munich, 2001, pp. 30–33.
[19] Zhang C, Najafi K, Bernal LP, Washabaugh PD (2003) Micro Combustion-Thermionic Power Generation: Feasibility, Design and Initial Results. *Transducers'03*, 2003.
[20] Nielsen OM, Arana LR, Baertsch CD, Jensen KF, Schmidt MA, (2003) A Thermophotovoltaic Micro-Generator for Portable Power Applications. *Transducers'03*, 2003.
[21] Li H, Lal M, (2002) Self-reciprocating radio-isotope powered cantilever, *Journal of Applied Physics*, 92(2) : 1122 – 1127.
[22] Friedman D, Heinrich H, Duan D-W, (1997) A Low-Power CMOS Integrated Circuit for Field-Powered Radio Frequency Identification. *Proceedings of the 1997 IEEE Solid-State Circuits Conference*, p. 294 – 295, 474.
[23] www.hitachi.co.jp/Prod/mu-chip/, 2003.
[24] Smith AA, (1998) *Radio frequency principles and applications : the generation, propagation, and reception of signals and noise*, IEEE Press, New York, 1998.
[25] Randall JF (2003) *On ambient energy sources for powering indoor electronic devices*, Ph.D Thesis, Ecole Polytechnique Federale de Lausanne, Switzerland, May 2003.
[26] Stordeur M, Stark I (1997) Low Power Thermoelectric Generator – self-sufficient energy supply for micro systems. *16th International Conference on Thermoelectrics*, 1997, p. 575 – 577.
[27] Pescovitz D (2002) The Power of Small Tech. *Smalltimes*, 2(1).
[28] Starner T (1996) Human-powered wearable computing. *IBM Systems Journal*, 35 (3) : 618–629.
[29] Shenck N S, Paradiso JA, (2001) Energy Scavenging with Shoe-Mounted Piezoelectrics, *IEEE Micro*, 21 (2001) 30–41.
[30] http://www.freeplay.net, 2003
[31] Williams CB, Yates RB, (1995) Analysis of a micro-electric generator for Microsystems. *Transducers 95/Eurosensors IX*, (1995) 369 – 372.

[32] Roundy S, Wright PK, Rabaey J (2003) A Study of Low Level Vibrations as a Power Source for Wireless Sensor Nodes, *Computer Communications*. 26(11) : 1131–1144.
[33] Meninger S, Mur-Miranda JO, Amirtharajah R, Chandrakasan AP, Lang JH, (2001) Vibration-to-Electric Energy Conversion. *IEEE Trans. VLSI Syst.*, 9 : 64–76.
[34] Ottman GK, Hofmann HF, Lesieutre GA (2003) Optimized piezoelectric energy harvesting circuit using step-down converter in discontinuous conduction mode. *IEEE Transactions on Power Electronics*. 18(2) : 696–703.
[35] Roundy S, Otis B, Chee, Y-H, Rabaey J, Wright PK (2003) A 1.9 GHz Transmit Beacon using Environmentally Scavenged Energy. *ISPLED 2003*.

Matrix Pencil for Positioning in Wireless Ad Hoc Sensor Network

Liang Song, Raviraj Adve, and Dimitrios Hatzinakos

Edward S. Rogers Sr. Department of Electrical and Computer Engineering
University of Toronto, 10 King's College Road
Toronto, ON M5S 3G4, Canada
{songl,rsadve,dimitris}@comm.utoronto.ca

Abstract. Wireless ad-hoc sensor networks (WASN) are attracting research interest recently because of their various applications in distributed monitoring, communication, and computing. In this paper, we concentrate on the range error problem of WASN positioning algorithms. A new matrix pencil based time of arrival (TOA) positioning method (MPP) is proposed for WASN. The new scheme employs matrix pencil for multipath time-delay estimation, and triangulation for absolute positioning. Simulations in a square-room scenario show that the positioning performance is generally robust to multipath effect and the positioning error is limited to around one meter.

1 Introduction

Wireless ad-hoc sensor networks (WASN) are being developed for use in various applications, ranging from monitoring of the environmental characteristics, to home networking, medical applications and smart battlefields. Many such services provided by WASN rely heavily on the acquisition of sensor nodes' position information. In addition, position assisted routing [8] and minimum energy [7] schemes have also been proposed for general ad-hoc networks and can greatly enhance the throughput performance and lifetime of WASN. Thus, developing a distributed practical positioning algorithm is probably the most important and challenging task in WASN. The term, "practical", suggests that such algorithms should be versatile for diverse environments, rapidly deployable, of low energy consumption, and low cost.

A fundamental problem in WASN positioning is range error, which is defined according to different measurement methods. In RSSI [5] (Received Signal Strength Indicator) and TOA or TDOA (Time of Arrival / Time Difference of Arrival) [3], range measurement is the associated distance estimate, and positioning can be carried through triangulation if more than three anchors are available [6]. In AOA (angle of arrival), it is the angle estimation. Among these measurement methods, AOA requires costly antenna arrays at each node, and is hard to implement. Although RSSI suffers significant range error (as high as over 50%) due to multi-path fading, it is deemed as the primary candidate for WASN range measurement and is well researched [2,4,9]. Various propagation methods

H. Karl, A. Willig, A. Wolisz (Eds.): EWSN 2004, LNCS 2920, pp. 18–27, 2004.

were proposed in [4] to achieve a crude localization. A two-phase positioning algorithm was proposed in [2] to compensate high range error. However, these methods either suffer from low accuracy or high communication and computation load. Based on RSSI, RADAR [9] was proposed for indoor localization and tracking. This method, however, is centralized and depends on a considerable preplanning effort, which is not appropriate for rapidly deployable WASN.

In TOA approaches, traditional code acquisition and tracking radiolocation, [19], meets significant problems, since the short distances between sensor nodes can demand an unacceptably wide system bandwidth for measurement. TDOA also suffers from this problem. An alternative is the combination of RF and acoustic signals [3,20], where time-of-flight of acoustic signal is used to calculate the distance. Experiments of such systems as AHLos [3] demonstrated fairly accurate positioning. However, acoustic signals are usually temperature dependent and require an unobstructed line of sight. Furthermore, the maximum range of acoustic signals is usually small, around 10 meters.

Instead of code tracking and acoustic signals, a proposed TOA approach [1] is based on applying high-resolution techniques, such as MUSIC [18] and matrix pencil algorithms [15,16], on channel response function. MUSIC suffers some severe limitations when compared with matrix pencil, which includes covariance matrix estimation and the assumption of white noise. Both these two assumptions are usually not available in the estimated channel response. In [1], the authors compared MUSIC and direct matrix pencil for multipath delay estimation, and matrix pencil was found much better in terms of both accuracy and computation cost.

In this paper, a matrix pencil based positioning (MPP) algorithm is proposed for WASN, when the transmission time of RF signal (TTR) is assumed to be available at receiver. Based on simulations with a square-room multipath model, we demonstrate that the range measurement error of MPP is much lower than RSSI. By least squares (LS) triangulation [6], the localization error can be generally limited to as low as one meter. When transmission ranges in simulations are considered, the positioning error percent is around 5%. The TTR assumption of our approach may be realized by MAC (Medium Access Control) layer designs, where RF signals can be transmitted on fixed time slots. In the following, Section 2 describes the channel model and the estimation method. The MPP algorithm is then proposed and analyzed in Section 3. We describe our indoor simulation model and present the simulation results of MPP in Section 4. Finally, conclusions are drawn in Section 5.

2 Channel Model and Estimation

Propagation of RF signals is through a quasi-fading channel with complex additive white Gaussian noise, and multipath effect of a maximum propagation delay D. Then, the channel response function can then be written as,

$$h(t) = \sum_{m=1}^{M} \alpha_m \delta\left(t - \tau_m\right), \tag{1}$$

and in the frequency domain,

$$H(\omega) = \sum_{m=1}^{M} \alpha_m e^{-j\omega\tau_m}, \tag{2}$$

where M is the number of multipaths, τ_m and α_m are the associated time delay and gain respectively.

Assuming that $X(\omega)$ and $Y(\omega)$ denote the frequency representations of transmitted and received signal respectively, we have,

$$Y(\omega) = X(\omega)H(\omega) + N(\omega), \tag{3}$$

where $N(\omega)$ denotes the additive Gaussian white noise.

Most advanced channel estimation algorithms, as [10], are based on the symbol rate or chip rate in CDMA system. However, these methods fail in WASN settings where nodes are densely deployed. In WASN, most multipath delays τ_m are less than one symbol/chip time T_S, and the maximum propagation delay D is usually of the order of T_S. As an example, the typical bandwidth of IEEE 802.11 wireless LAN is $F_B = 11$MHz, thus $T_S = 1/F_B = 0.09\mu$s. A typical distance between two nodes in WASN is around 10 meters, with the associated LOS (Line of Sight) delay $\tau_{LOS} = 0.03\mu$s. The maximum multipath delay will then be $D \approx T_S$. This observation suggests a simple FFT-based channel estimation scheme.

Consider transmitting a training sequence $T_R(n)$ of length $K \cdot L$, designed as,

$$\begin{cases} T_R(lK+1) = 1, & l = 0 \ldots L-1 \\ T_R(lK+i) = 0, & l = 0 \ldots L-1, i = 2 \ldots K \end{cases}. \tag{4}$$

At the receiving node, the RF training signal is divided into L segments, where each contain one '1' and $K-1$ '0's. K should be chosen such that $(K-1) \cdot T_S > D$. In the described settings above, K is practically small and can be set to 2. Then let $Y_l(\omega_n)$ denote the FFT of the lth such segment, and let $G(\omega)$ denote the transmit filter response. According to (3), we get,

$$\begin{cases} Y_l(\omega_n) = H(\omega_n) \cdot G(\omega_n) + N_l(\omega_n) \\ \omega_n = 2\pi n \cdot F_B/N - C, n = 0 \ldots N-1 \end{cases}, \tag{5}$$

where N is the number of FFT points, and C is a constant shift. The channel estimation can be obtained by,

$$\begin{aligned} \widehat{H}(n) &= \tfrac{1}{L} \textstyle\sum_{l=1}^{L} \frac{Y_l(\omega_n)}{G(\omega_n)} \\ &= H(\omega_n) + \tfrac{1}{L} \textstyle\sum_{l=1}^{L} \frac{N_l(\omega_n)}{G(\omega_n)}, \\ &= H(\omega_n) + v_n \end{aligned} \tag{6}$$

where v_n denotes noise component in channel estimation and can be suppressed by increasing the training length L.

By equations (2), (5), and (6), we have,

$$\begin{cases} \widehat{H}(n) = \sum_{m=1}^{M} \beta_m e^{-jn \cdot (2\pi F_B \cdot \tau_m / N)} + v_n \\ \beta_m = \alpha_m \cdot e^{jC\tau_m}, n = 0 \ldots N-1 \end{cases}. \tag{7}$$

3 MPP Algorithm

Once channel estimates $\left\{\widehat{H}(n)|n=0\ldots N-1\right\}$ are obtained via (6), MPP can be performed for positioning. We define a group of $L_2 \times L_1$ matrices $\mathbf{X}_n$as,

$$\mathbf{X}_n = \begin{bmatrix} \widehat{H}(n) & \widehat{H}(n+1) & \ldots & \widehat{H}(n+L_1-1) \\ \widehat{H}(n+1) & \widehat{H}(n+2) & \ldots & \widehat{H}(n+L_1) \\ \vdots & \vdots & \ddots & \vdots \\ \widehat{H}(n+L_2-1) & \widehat{H}(n+L_2) & \ldots & \widehat{H}(n+L_1+L_2-2) \end{bmatrix}, \qquad (8)$$
$$n=0\ldots N-L_1-L_2+1$$

where L_1 and L_2 satisfy,

$$\begin{matrix} L_1, L_2 \geq M \\ L_1+L_2 \leq N \end{matrix}. \qquad (9)$$

In a noise free environment, the rank of $\mathbf{X}_n$ is M, i.e. $v_n = 0$, however, it is greater than M when noise components are considered. To reduce the effect of noise, the M-truncated SVD (Singular Value Decomposition) of $\mathbf{X}_n$ is obtained by,

$$\widetilde{\mathbf{X}}_n = \mathbf{U}_n \cdot \mathbf{\Sigma}_n \cdot \mathbf{V}_n^H, \qquad (10)$$

where $\mathbf{\Sigma}_n$ is the M-by-M diagonal matrix of the M principal singular values of $\mathbf{X}_n$. The $\mathbf{U}_n$ consists of the M principal left singular vectors of $\mathbf{X}_n$, and the $\mathbf{V}_n$ consists of the M principal right singular vectors of $\mathbf{X}_n$. Based on relation (7), it was shown in [16] that the M multipath delays can be estimated by the eigenvalues of any M-by-M $\mathbf{Q}$ matrices,

$$\begin{cases} \mathbf{Q}_n = \mathbf{\Sigma}_{n+1}^{-1}(\mathbf{U}_{n+1}^H \mathbf{U}_n \mathbf{\Sigma}_n \mathbf{V}_n^H \mathbf{V}_{n+1}) \\ n = 0\ldots N-L_1-L_2 \end{cases}. \qquad (11)$$

Thus, $N-L_1-L_2+1$ $\mathbf{Q}$ matrices can be averaged to reduce noise interference. The delay estimation is obtained by,

$$\widehat{\tau_m} = \frac{\ln(z_m)\cdot N}{-j2\pi F_{\mathrm{B}}}, \qquad (12)$$

where $\{z_m|m=1\ldots M\}$ are the M eigenvalues of,

$$\mathbf{Q}_{\mathrm{avg}} = \frac{1}{N-L_1-L_2+1}\sum_{n=0}^{N-L_1-L_2}\mathbf{Q}_n. \qquad (13)$$

Since the LOS path corresponds to the shortest delay, distance between two nodes is decided by,

$$\widehat{d} = c\cdot\min\{\widehat{\tau_m}, m=1\ldots M\}, \qquad (14)$$

where c is the speed of EM waves, 3×10^8m/sec. Once distances to more than three anchor nodes are available, the desired absolute position can be estimated by LS triangulation localization [6].

An important related question is how to decide coefficients M, and L_1, L_2 at sensor nodes. Decision on M involves a well-researched area of estimating the sinusoids number [17]. However, most these approaches assume white noise components, which are generally not true in our scenario. Alternatively, a simple way is adopted to estimate M. Assume that the maximum singular value of $\mathbf{X}_n$ is ϑ_n , and M_n denotes the number of singular values greater than $\delta \cdot \vartheta_n$, where δ is a threshold coefficient related to channel condition. M is decided by averaging M_n over $n = 0 \ldots N - L_1 - L_2 + 1$.

L_1, L_2 are then chosen under condition (9). However, analyzing the optimal choice of L_1, L_2 is a difficult problem. In [15], one similar such coefficient was treated as a free parameter. In our scenario, it is more complex, since two parameters are introduced. Basically, choosing smaller values for both L_1 and L_2 will increase the number of $\mathbf{Q}$ matrices and suppress noise degradations, however, it will also make $\mathbf{X}_n$ more ill-conditioned. We simply point out the tradeoff here and leave it for future research.

4 Simulations and Results

Simulations are performed to test the positioning accuracy of MPP in WASN. In obtaining the channel model coefficients in (2), the simulation environmental model is set as a square room, with $(M-1)$ scatterers uniform-randomly distributed in it. Thus, including LOS, there are altogether M multipaths in one channel. Assume that $\{\mathbf{p}_m | m = 1 \ldots M-1\}$ denote the $(M-1)$ scatterers' coordinates, and $\mathbf{rx}, \mathbf{tx}$ denote the coordinates of receive node and transmit node respectively. Then, the associated time-delays with $M-1$ scattered paths are,

$$\tau_m = \frac{\{\|\mathbf{p}_m - \mathbf{rx}\| + \|\mathbf{p}_m - \mathbf{tx}\|\}}{c}, m = 1 \ldots M-1, \tag{15}$$

while the LOS delay is,

$$\tau_M = \tau_{\text{LOS}} = \frac{\|\mathbf{tx} - \mathbf{rx}\|}{c}. \tag{16}$$

Being consistent with some indoor propagation experimental results [12,13], the coefficients α_m are modeled as following,

$$\alpha_m = u_m \cdot \tau_{\text{LOS}} / \tau_m, m = 1 \ldots M, \tag{17}$$

where $\{u_m | m = 1 \ldots M\}$ are independent zero-mean unit-variance complex Gaussian random variables.

In our simulations, all nodes are assumed to be within the transmission range of each other. System bandwidth is set the same as IEEE 802.11, $F_{\text{B}} = 11\text{MHz}$. The transmitting filter uses a raised cosine pulse. The training length L in (5) is set as 1000, and K varies according to different room sizes. For all cases, $K \leq 4$, thus the total length of training sequence is less than 4000. White Gaussian noise with variance σ_n^2 is added to RF signals on receiver. Peak signal to noise ratio is thus defined as,

$$\mathrm{SNR} = \frac{|\alpha_{\mathrm{LOS}}|^2}{\sigma_n^2}. \tag{18}$$

The number of FFT points is set to $N = 25$. To simplify our problem, L_1 is set to $M + 2$, and L_2 is chosen as $N - L_1$. Thus, only one $\mathbf{Q}$ matrix is obtained, which also suggests our simulation results can be further improved. The M-decision threshold δ is set as 0.01.

Range measurement error percents are simulated in different rooms when either M or SNR varies. It is defined as,

$$\mathrm{RangeErrorPercent} = \left| \frac{\|\mathbf{rx} - \mathbf{tx}\| - \widehat{d}}{\|\mathbf{rx} - \mathbf{tx}\|} \right| \times 100. \tag{19}$$

Transmit and receive nodes are randomly deployed in the room. The average of 1000 Monte-Carlo runs is plotted. In Fig.1, M is fixed as 3 and SNR varies from -4dB to 20dB. In Fig.2, SNR is fixed as 10dB, while M varies from 1 to 7. In both figures, simulations in larger rooms show better performance. Larger rooms are more likely to have widespread multipath delays, which prevent ill-conditioning of the $\mathbf{Q}$ matrix. Another observation is that generally smaller M suggests better performance, especially, $M = 1$ results in far lower error percent than others. The reasons can be found by Cramer-Rao bound analysis [14]. However, the degradation with increasing M is not steep. At 10dB SNR, the range error is still within 30%, even with $M = 8$. The fluctuations of these curves can be due to our non-optimal choice of L_1 and L_2. Compared with RSSI, which may exhibit a range error as large as 50% [5], our results show a significant improvement.

Furthermore, positioning errors are simulated with different size rooms and different multipath number M, when the number of anchor nodes varies. SNR is fixed at 10dB. All nodes are randomly deployed over the room. A simple strategy similar to [11] is employed to discover and discard NLOS measurements, and 10 MPP iterations are averaged for each positioning. Results of 200 Monte-Carlo runs are then averaged and plotted in Fig.3-Fig.5 for different size rooms. The results show robustness to multipath number M, when $M > 1$. However, when $M = 1$, much better performance is achieved due to much smaller range error. When enough anchors are available, the positioning error is around one meter. Considering different transmission ranges in different rooms, the positioning error percent is within 5%. In an extreme case with only three anchors, it is around 10%. When positioning error percents are considered, larger rooms still have better performance, which is due to the same reason as in range error percent.

5 Conclusions and Future Work

In this paper, we have proposed a new positioning algorithm for wireless adhoc sensor network, MPP. The new method depends on matrix pencil approach

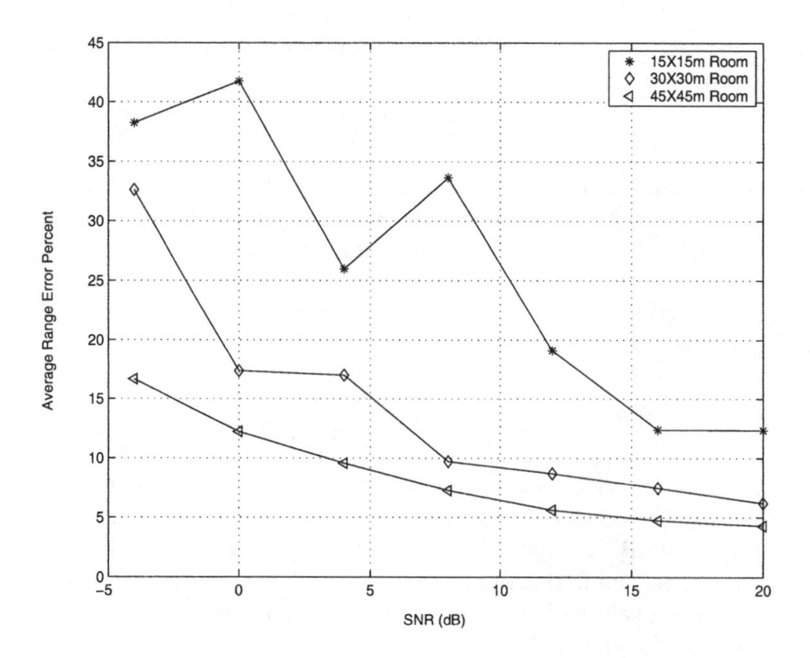

Fig. 1. $M = 3$, range error percent simulations in square rooms

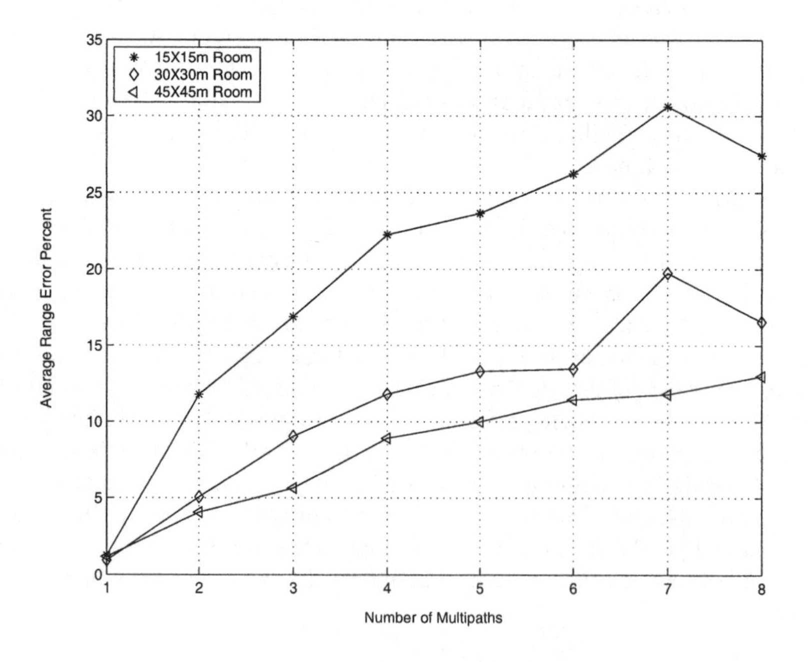

Fig. 2. $SNR = 10dB$, range error percent simulations in square rooms

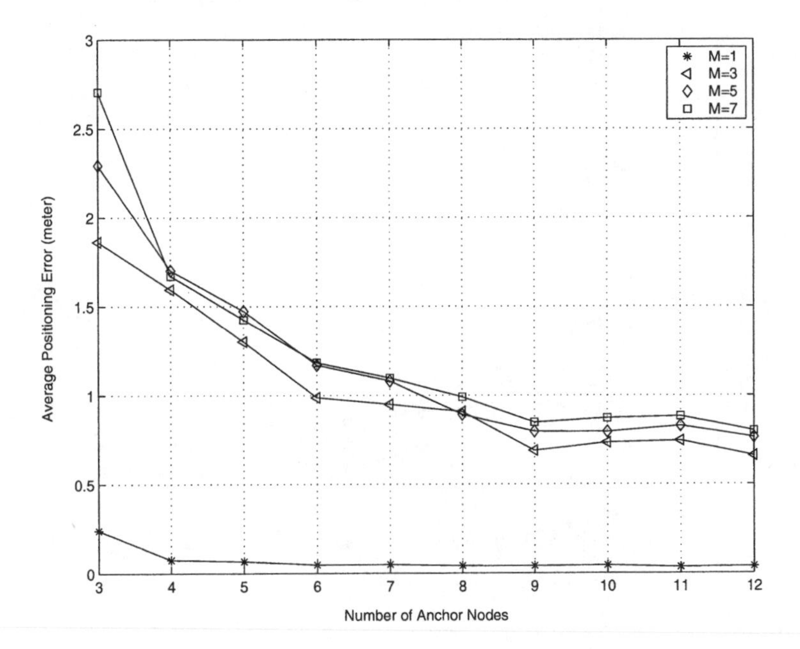

Fig. 3. Mean positioning error simulations in $15 \times 15m$ Room

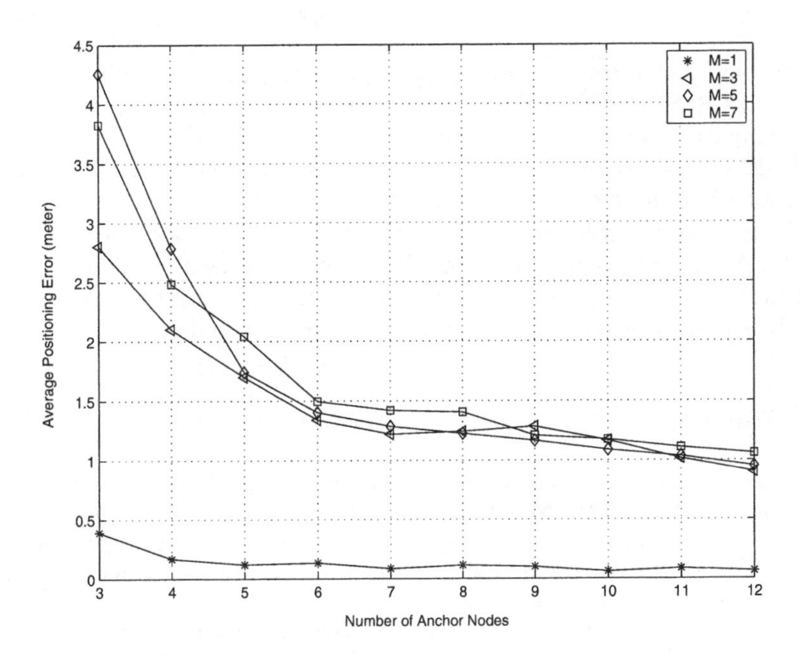

Fig. 4. Mean positioning error simulations in $30 \times 30m$ Room

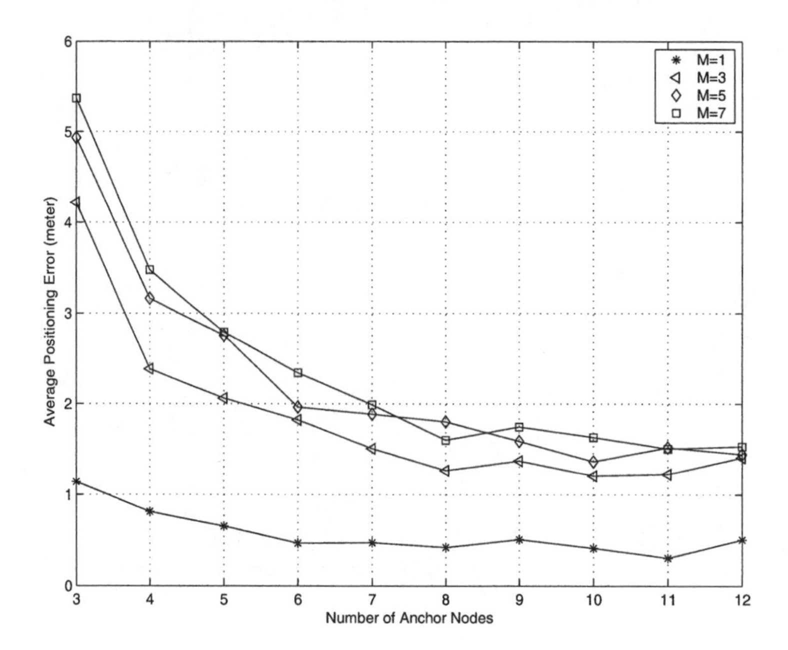

Fig. 5. Mean positioning error simulations in $45 \times 45m$ Room

to perform TOA estimation and on triangulation to obtain positioning. Compared with RSSI, the TOA range measurement by MPP is much more accurate, thus complex compensation algorithms are avoided. Compared with AHLos like methods, our approach avoids the disadvantages of acoustic signal and can be more robust to diverse environments.

Future works should consider an optimal decision theory of coefficients L_1 and L_2. Also, the implementation of MPP depends on the important assumption that the transmission time is known. More research on this TTR assumption in WASN is needed before it can be practically implemented.

References

1. N.Dharamdial, R.Adve, and R.Farha, Multipath Delay Estimations using Matrix pencil, Proc. IEEE WCNC'03, New Orleans, 2003.
2. C.Savarese, J.Rabaey, and K.Langendoen, Robust Positioning Algorithm for Distributed Ad-hoc Wireless Sensor Networks, USENIX Technical Annual Conference, Monterey, CA, June 2002.
3. A.Savvides, C.C.Han, and M.Srivastava, Dynamic fine-grained localization in ad-hoc networks of sensors, Proc. ACM MobiCom'01, Aug. 2001.
4. D.Niculescu and B.Nath, Ad-hoc positioning system, Proc. IEEE GlobeCom'01, Nov. 2001.
5. C.Savarese, J.Rabaey, and J.Beutel, Locationing in distributed ad-hoc wireless sensor networks, Proc. IEEE ICASSP'01, May, 2001.

6. J.Beutel, Geolocation in a picoradio environment, M.S. thesis, ETH Zurich, Electronics Lab, 1999.
7. V.Rodoplu and T.H.Meng, Minimum energy mobile wireless networks, IEEE Journal on Selected Areas in Communications, Vol. 17, No.8, Aug. 1999.
8. K.Amouris, S.Papavassiliou, and M.Li, A position-based multi-zone routing protocol for wide area mobile ad-hoc networks, Proceedings of VTC, 1999.
9. P.Bahl and V.Padmanabhan, RADAR: an in-building RF-based user location and tracking system, Proc. IEEE InfoCom'00, Mar. 2000.
10. N.Nefedov and M.Pukkila, Iterative channel estimation for GPRS, Proc. 11th international symposium on PIMRC, Vol. 2, 2000.
11. A.Ward, A.Jones, A.Hopper, A new location technique for the active office, IEEE Personal Communication, Oct. 1997.
12. K.Pahlavan, P.Krishnamurthy, and J.Beneat, Wideband radio propagation modeling for indoor geolocation applications, IEEE Commun. Magazine, Apr. 1998.
13. S.S.Ghassemzadeh, L.J.Greenstein, and V.Tarokh, The Untra-wideband indoor multipath model, Project: IEEE P802.15 Working Group for Wireless Personal Area Netsworks (WPANs), July 2002.
14. P.Stoica, and A.Nehorai, Music, maximum likelihood, and Cramer-Rao bound, IEEE Trans. on Acoust., Speech, Signal Processing, Vol. 37, No. 5, May 1989.
15. Y.Hua, and T.K.Sarkar, Matrix pencil method for estimating parameters of exponentially damped/undamped sinusoids in noise, IEEE Trans. on Acoust., Speech, Signal Processing, Vol. 38, No. 5, May 1990.
16. Y.Hua, and T.K.Sarkar, On SVD for estimating generalized eigenvalues of singular matrix pencil in noise, IEEE Trans. on Signal Processing, Vol. 39, No. 4, Apr. 1991.
17. M.Wax, T.Kailath, Detection of signals by information theoretic criteria, IEEE Trans. on Acoust., Speech, Signal Processing, Vol. ASSP-33, No. 2, Apr. 1985.
18. R.O.Schmidt, Multiple emitter location and signal parameter estimation, Proc. RADC Spectral Estimation Workshop, 1979.
19. J.Spilker, Delay-lock tracking of binary signals, IEEE Trans. on Space Electronics and Telemetry, Vol. SET-9, Mar. 1963.
20. N.Priyantha, A.Chakraborthy, and H.Balakrishnan, The Cricket location-support system, Proc. MobiCom'00, Aug. 2000.

Tracking Real-World Phenomena with Smart Dust

Kay Römer

Institute for Pervasive Computing
Dept. of Computer Science
ETH Zurich, Switzerland
roemer@inf.ethz.ch

Abstract. So-called "Smart Dust" is envisioned to combine sensing, computing, and wireless communication capabilities in an autonomous, dust-grain-sized device. Dense networks of Smart Dust should then be able to unobtrusively monitor real-world processes with unprecedented quality and scale. In this paper, we present and evaluate a prototype implementation of a system for tracking the location of real-world phenomena (using a toy car as an example) with Smart Dust. The system includes novel techniques for node localization, time synchronization, and for message ordering specifically tailored for large networks of tiny Smart Dust devices. We also point out why more traditional approaches developed for early macro prototypes of Smart Dust (such as the Berkeley Motes) are not well suited for systems based on true Smart Dust.

1 Introduction

Smart Dust is commonly used as a synonym for tiny devices that combine sensing, computing, wireless communication capabilities, and autonomous power supply within a volume of only few cubic millimeters at low cost. The small size and low per-device cost allows an unobtrusive deployment of large and dense Smart Dust populations in the physical environment, thus enabling detailed in-situ monitoring of real-world phenomena, while only marginally disturbing the observed physical processes. Smart Dust is envisioned to be used in a wide variety of application domains, including environmental protection (identification and monitoring of pollutions), habitat monitoring (observing the behavior of animals in their natural habitats), and military systems (monitoring activities in inaccessible areas). Due to its tiny size, Smart Dust is expected to enable a number of novel applications. For example, it is anticipated that Smart Dust nodes can be moved by winds or can even remain suspended in air, thus supporting better monitoring of weather conditions, air quality, and many other phenomena. Also, it is hard to detect the bare presence of Smart Dust and it is even harder to get rid of it once deployed, which might be helpful for many sensitive application areas.

Current research (cf. [1] for an overview) is mainly focusing on so-called COTS (Commercial Off The Shelf) Dust, early macro prototypes of Smart Dust. COTS Dust nodes such as the Motes [24] developed at UC Berkeley are built from commercially available hardware components and still have a volume of several cubic centimeters. Unfortunately, these devices cannot be simply scaled down to the cubic millimeter size of true Smart Dust. First Smart Dust prototypes [21] demonstrate that the tremendous

H. Karl, A. Willig, A. Wolisz (Eds.): EWSN 2004, LNCS 2920, pp. 28–43, 2004.

volume reduction (factor 1000 and more) may require radical changes in the employed technologies (e.g., use of optical instead of radio communication) compared to COTS Dust. These technological changes have important implications for algorithms, protocols, systems, and infrastructure. Our goal is to examine these implications and develop solutions for the resulting problems. To identify and illustrate these issues, we have developed an object tracking system that makes use of Smart Dust. This system allows tracking the location of "targets" with Smart Dust, using a remote-controlled toy car as a sample target.

Since Smart Dust hardware is currently in a very early prototyping stadium, our implementation still uses COTS Dust. However, our algorithms and protocols are designed to be directly portable to true Smart Dust, once the hardware becomes available. In the following sections we first outline the characteristics of Smart Dust, before presenting the details of our object tracking system. Particularly, this includes novel techniques for synchronizing time among the nodes of the network, for localizing Smart Dust nodes in physical space, and for ordering event notifications according to their time of occurrence. This will be followed by the presentation of some measurements and a discussion.

2 Smart Dust

The envisioned dust-grain size of Smart Dust nodes has a number of important implications with respect to their hardware design. Most importantly, it is hardly possible to fit current radio communication technology into Smart Dust – both size-wise (antennas) and energy-wise (power consumption of current transceivers) [21]. Hence, Smart Dust prototypes developed at UC Berkeley utilize a more power and size efficient passive laser-based communication scheme to establish a bidirectional communication link between dust nodes and a so-called base station transceiver (BST). For downlink communication (BST to dust), the base station points a modulated laser beam at a node. The latter uses a simple optical receiver to decode the incoming message. For uplink communication (dust to BST), the base station points an unmodulated laser beam at a node, which in turn modulates and reflects back the beam to the BST. For this, the dust nodes are equipped with a so-called Corner Cube Retro Reflector (CCR), which is a special Micro Electro-Mechanical System (MEMS) structure consisting of three mutually perpendicular mirrors. The CCR has the property that any incident ray of light is reflected back to the source under certain conditions. If one of the mirrors is misaligned, this retroreflection property is spoiled. The Smart Dust CCR includes an electrostatic actuator that can deflect one of the mirrors at kilohertz rates. Using this actuator, the incident laser beam is "on-off" modulated and reflected back to the BST.

This type of design implies a single-hop network topology, where dust nodes cannot directly communicate with each other, but only with a base station. The base station can be placed quite far away from the nodes, since the employed laser communication works over a range of hundreds of meters, provided a free line-of-sight between the BST and the nodes. Communication may suffer from significant and highly variable delays if the laser beam is not already pointing at a node which is subject to communication with the BST. Smart Dust nodes can be highly mobile, since nodes are small enough to be moved by winds or even to remain suspended in air, buoyed by air currents.

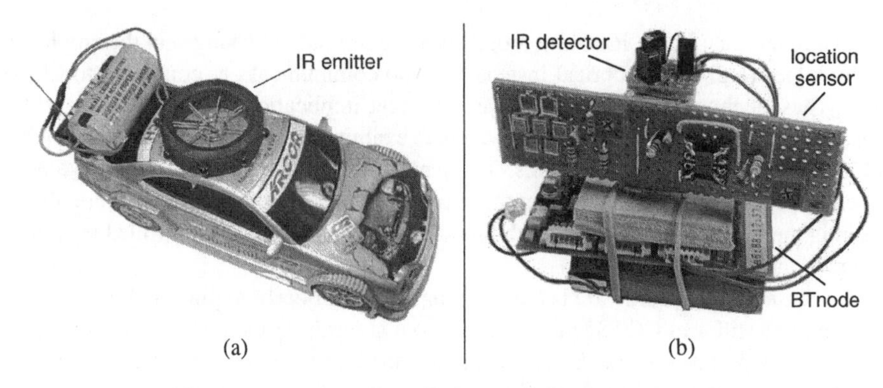

Fig. 1. (a) Remote-controlled car (b) "Smart Dust" node.

Early prototypes of Smart Dust [22] implement the optical receiver, CCR, a light sensor, a solar cell, and a simple control unit within 16 cubic millimeters. Future devices are expected to include a complete processing unit instead of the simple control unit, provide a wider variety of sensors, and will feature further reductions in size and energy consumption.

Due to the limited size, on-board devices for energy storage and harvesting can only provide a tiny amount of energy, thus every hardware and software component has to be optimized for energy efficiency. Despite the passive optical communication scheme, data communication still consumes most of the energy. Hence, communication has to be kept to a minimum. In addition, many thousands of sensors may have to be deployed for a given task – an individual sensor's small effective range relative to a large area of interest makes this a requirement, and its small form factor and low cost makes this possible. Therefore, scalability is another critical factor in the design of the system. Note that Smart Dust is subject to frequent communication failures (e.g., line-of-sight obstructions) and node failures (e.g., destruction due to environmental influences, depleted batteries). Hence, applications must be robust against these types of failures.

While the tiny size of Smart Dust nodes leads to a number of challenging problems as pointed out above, the anticipated high density of Smart Dust deployments may allow to monitor environmental phenomena with unprecedented quality and detail. The tracking system described in the following section has been specifically designed with the above requirements and characteristics in mind.

3 The Tracking System

The purpose of the tracking system is to track the location of real-world phenomena with a network of Smart Dust nodes. We use a remote-controlled toy car (Figure 1 (a)) as a sample target. The current tracking system assumes that there is only one car. Wirelesse sensor nodes are randomly deployed in the area of interest and can change their location after deployment. When they detect the presence of the car (Section 3.1), they send notifications to a base station. The base station fuses these notifications (Section 3.2) in order to estimate the current location of the car. A graphical user interface displays the track and allows to control various aspects of the system. The data fusion process requires that all nodes share a common reference system both in time and space, which necessitates mechanisms for node localization (Section 3.3) and time synchronization (Section 3.4).

Unfortunately, true Smart Dust hardware is not yet available. Some recent prototypes are presented in [22], but they still do not include important components like a full-fledged processing unit. Therefore, we still use COTS Dust for the implementation of our prototype system. Note however, that this is only an intermediate step to demonstrate the feasibility of our approaches. Our ultimate goal is to implement the tracking system using true Smart Dust.

The current sensor node hardware is based on BTnodes (Figure 1 (b)) [25], which provide Bluetooth communication, an ATMEL ATMega 128L microcontroller, and various external interfaces for attaching sensors on a matchbox-sized printed circuit board. The base station consists of a Linux laptop computer equipped with a Bluetooth radio. In analogy to the single-hop network topology of Smart Dust described in Section 2, BTnodes do not directly communicate with each other, but only with the base station. Before communication can take place, the base station has to set up a so-called Bluetooth Piconet containing no more than 7 BTnodes. To support more than 7 nodes, the base station has to periodically switch the Piconet in a round robin fashion, such that eventually every BTnode gets a chance to talk to the base station. Again, this is very similar to Smart Dust, where the base station has to point the (typically slightly defocused) laser beam at a group of nodes in order to enable communication with them.

3.1 Target Detection

Tracking targets with networks of sensors has been an active research topic for many years, [3,8,18] give a good overview of many tracking algorithms. Most of the approaches are optimized for sparse networks, where a high tracking accuracy should be achieved despite a relatively low node density. To achieve this, many approaches make a number of assumptions about the tracked target. Methods which estimate target distance based on signal strength estimates, for example, require knowledge of the intensity of the signal emitted by the target in order to achieve good accuracy. Approaches based on measuring the difference of time of arrival of a signal emitted by the target at different sensor nodes are typically limited to sound or other signal modalities with low propagation speed. Signal modalities with high propagation speeds such as radio waves would require distributed clock synchronization with an accuracy of few nanoseconds, which is typically not available. Other approaches need to know lower and upper bounds of the velocity or the acceleration of the tracked target.

While these assumptions help to achieve good tracking accuracy, they also limit the applicability of the tracking system. In order to make our system applicable to a wide variety of targets, we tried to avoid making assumptions about the target as much as possible. In order to achieve a satisfactory tracking accuracy nevertheless, we exploit the anticipated high density of Smart Dust deployments – which is expected because of the intended small size and low cost of Smart Dust devices.

Our approach assumes that the presence of the target can be detected with an omnidirectional sensor featuring an arbitrary but fixed sensing range r, that is, the sensor

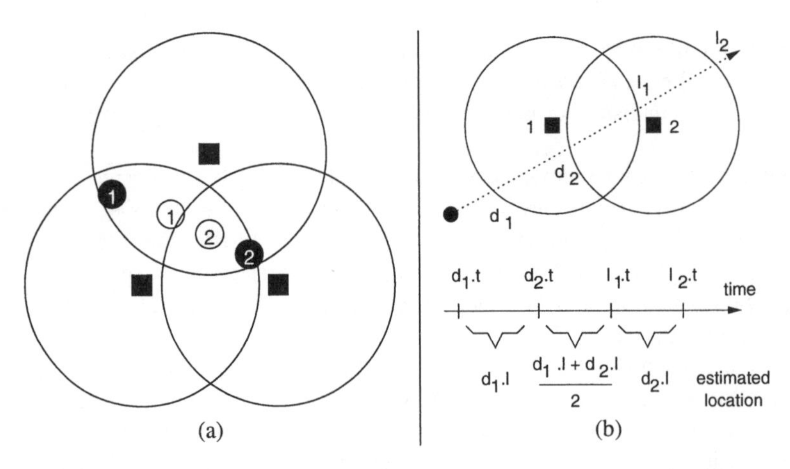

Fig. 2. (a) Estimating car location by centroids (b) Data fusion algorithm.

can "see" the target if and only if the distance to the target is lower than r. The data fusion algorithm presented in the following section needs to know an upper bound R of this sensing range. In many applications, the target cannot be instrumented for tracking purposes (e.g., a cloud of toxic gas, an oil slick, fire). The remote-controlled car (used as a sample target in our tracking system) emits a characteristic acoustic signature which could be used for detection. However, this signature depends on the velocity of the car. To avoid the intricacies with detecting this variable signature, we chose in our experiment a different solution based on infrared (IR) light, leaving detection based on the car's acoustic signature as future work.

In the current version of the prototype, we equipped the car with an omnidirectional IR light emitter consisting of eight IR LEDs mounted on top of the car (Figure 1 (a)). Accordingly, the sensor nodes are equipped with an omnidirectional IR light detector consisting of three IR photo diodes (Figure 1 (b)). The used IR photo diodes include a filter to cancel out visible light. The output of the IR detector is connected to an analog-to-digital (ADC) converter of the BTnode's microcontroller. If the output value of the ADC exceeds a certain threshold, the presence of the car is assumed. Using a low-pass filter, the threshold value is adopted to slowly changing daylight, which also contains IR components. With this setup, the BTnodes can detect the car at a distance of up to approximately half a meter. When a node first detects the car, it sends a "detection notification" to the base station, containing its node ID as well as its time and location at the time of detection. When the node no longer sees the car, it sends a "loss notification" to the base station, which contains its node ID and its current time. If the node changes its location during the presence of the car, a loss notification is emitted, followed by a detection notification with the new node location.

3.2 Data Fusion

Following the argumentation at the beginning of the previous section, we try to avoid making assumptions about the target as much as possible. Therefore, the base station has to derive the current location of the tracked target solely based on detection notifications and loss notifications received from the sensor nodes.

We use an approach that estimates the car's location at time t by the centroid of the locations of the sensor nodes that "see" the car at time t. The centroid of a set of N locations $\{l_i = (x_i, y_i, z_i)\}$ is defined as $\hat{l} := \frac{1}{N}\sum l_i = (\frac{1}{N}\sum x_i, \frac{1}{N}\sum y_i, \frac{1}{N}\sum z_i)$. Consider Figure 2 (a) for an example. Depicted are three sensor nodes (black squares) with their respective sensing ranges, and two car locations (black circles). The hollow circles indicate the respective estimated locations (i.e., centroids).

Figure 2 (b) illustrates an algorithm to calculate the car location estimates given the detection and loss notifications received from the sensor nodes as described in Section 3.1. The figure shows sensor nodes 1 and 2, their respective sensing ranges, and the trajectory of the car (dotted arrow). When the car enters the sensing range of node i, a detection notification d_i is emitted, containing time $d_i.t$ and location $d_i.l$ of node i at the time of detection. Accordingly, node i emits a loss notification l_i when the car leaves the sensing range. In a first step, all notifications are sorted by increasing timestamps $d_i.t$ ($l_i.t$) as depicted on the time axis in the lower half of Figure 2 (b). In a second step, we iterate over these sorted notifications from left to right, recording the active nodes (those that currently see the car) in a set S. If we come across a loss notification l_i, we remove i from S. If we come across a detection message d_i, we add i to S. Additionally, we remove all nodes j from S, whose sensing ranges do not overlap with the detection range of node i, that is, for which $|d_i.l - d_j.l| > 2R$ holds. This is necessary to compensate for missed loss notifications, which would otherwise permanently affect the accuracy of the tracking system by not removing the respective entries from S. A missing enter notification will lead to a temporarily decreased tracking accuracy, but will not otherwise permanently affect the system.

The location of the car during the time interval starting at the currently considered notification and ending at the next notification is estimated by the centroid $\hat{S}$ of the locations of the nodes in S (i.e., $d_1.l$ during $[d_1.t, d_2.t)$, $(d_1.l+d_2.l)/2$ during $[d_2.t, l_2.t)$, and $d_2.l$ during $[l_1.t, l_2.t)$).

The localization accuracy of a similar centroid-based algorithm was examined in [2] in a different context under the assumption that nodes are located on a regular grid. We can interpret their results for our setting as follows. The localization accuracy depends on the sensing range r of the nodes (about 50cm in our case) and the distance d between adjacent nodes. For $r/d = 2$ (i.e., $d \approx 25$cm in our case) the average and maximum localization errors are $0.2d$ (i.e., 5cm) and $0.5d$ (i.e., 12.5cm), respectively. In general, larger r/d values yield better accuracy. Therefore, the accuracy can be improved by increasing the node deployment density, since that reduces d while keeping r constant.

3.3 Node Localization

In order to derive the location of the tracked car from proximity detections as described in Section 3.1, the locations of the sensor nodes have to be estimated. Various systems have been designed for localization purposes. However, most of them are not suited for Smart Dust. Will will first explain why this is the case, before outlining the solution we developed for the localization of Smart Dust nodes.

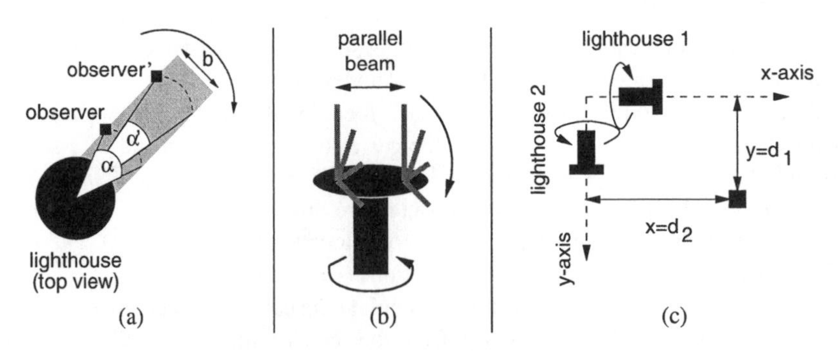

Fig. 3. (a) Lighthouse with parallel beam (b) Lighthouse implementation with two rotating laser beams (c) 2D location system using two lighthouses.

Energy, size, and cost constraints preclude equipping Smart Dust with receivers for localization infrastructures like GPS. With Smart Dust, it might not even be possible to equip sensor nodes with transceivers for radio waves or ultra sound due to the tiny size and energy budget of Smart Dust nodes. Hence, traditional ranging approaches such as ones based on time of flight of ultrasound signals or received radio signal strength might be unusable in the context of Smart Dust.

Many localization systems such as [2,20] depend on an extensive hardware infrastructure. Localization systems based on trilateration, for example, require many spatially distributed and well-placed infrastructure components in order to achieve high accuracy. This is not an adequate solution for Smart Dust, since it contradicts the ad hoc nature of Smart Dust, where nodes may have to be deployed in remote, inaccessible, or unexploited regions. Other localization approaches such as [4,16] require centralized computation, which results in systems that do not scale well to large numbers of nodes. To overcome the limitations of infrastructure-based approaches, various schemes for ad hoc localization have been devised (e.g., [14,15]). However, these schemes depend on inter-node communication, which is not scalable with Smart Dust, since any inter-node communication has to pass through the BST.

An important overhead involved in setting up a localization system is node calibration in order to enforce a correct mapping of sensor readings to location estimates [23]. In systems based on radio signal strength, for example, the received signal strength is mapped to a range estimate. Variations in transmit power and frequency among the nodes can cause significant inaccuracies in the range estimates when used without calibration. Since the cheap low-power hardware used in sensor nodes typically introduces a high variability between nodes, sensor nodes have to be individually calibrated. This, however, may not be feasible in large networks.

To overcome the above limitations, we designed and implemented a novel localization system for Smart Dust. This system consists of a single infrastructure device, which emits certain laser light patterns. By observing these patterns, dust nodes can autonomously estimate their location with high accuracy. Since dust nodes only passively observe light flashes, this system is very energy efficient on the side of the nodes. Moreover, optical receivers consume only little power and can be made small enough to fit in a volume of few cubic millimeters. Since dust nodes do not need to interact with other nodes in order to estimate their location, the system scales to very large networks. Also, node calibration is not necessary due to using differential measurements, where constant offsets cancel out due to using the difference between two measurements that use the same signal path.

To understand our localization approach, consider a lighthouse with a *parallel* beam (i.e., a beam with constant width b) as depicted in Figure 3 (a). Assume that it takes the lighthouse t_{turn} for one rotation. When the parallel beam passes by an observer (black square), the observer will see the lighthouse flash for a certain period of time t_{beam}. Note that t_{beam} and hence the angle $\alpha = 2\pi t_{\text{beam}}/t_{\text{turn}}$ under which the observer sees the beam, depend on the observer's distance d from the lighthouse rotation axis, since the beam is parallel. Using α, the observer's distance d from the lighthouse rotation axis can be expressed as $b/(2\sin(\alpha/2))$.

A parallel beam can be implemented as depicted in Figure 3 (b): two rotating laser beams (at high speeds) define the outline of a wide "virtual" parallel beam, which in turn is rotating around a central axis (at much lower speeds) to create a rotating lighthouse effect. An observer looking at such a lighthouse sees two sequences of short laser flashes as the two "laser light planes" rotate by. t_{beam} can then be obtained by measuring the amount of time elapsed between the two flash sequences.

Using two such lighthouses, a 2D location system can be constructed as depicted in Figure 3 (c). The two lighthouses are assembled such that their rotation axes are mutually perpendicular. The distances d_1 and d_2 to the lighthouse rotation axes then equal the y and x coordinates of the observer in the 2-dimensional coordinate system defined by the lighthouse rotation axes. Accordingly, a 3D location system can be built out of 3 lighthouses with mutually perpendicular rotation axes. Realizing a lighthouse with an exactly parallel beam is very difficult in practice. Therefore we developed a geometrical model of a more realistic lighthouse with an approximately parallel beam in [13]. Using this model, a 2D prototype (see Figure 4) of the system allows to estimate sensor node locations with an accuracy of about 5cm in our tracking system.

Figure 1 (b) shows a BTnode with the hardware for receiving the laser light flashes attached to it. The size of the current prototype of this receiver hardware is about 7cm $\times$ 3cm. Using an ASIC, however, this receiver can be fit within few cubic millimeter. The Smart Dust prototypes presented in [22], for example, contain a similar optical receiver within a volume of 16 cubic millimeters.

3.4 Time Synchronization

The car location estimation described in Section 3.2 assumes that the timestamps contained in notification messages refer to a common physical time scale, requiring synchronization of clocks of the Smart Dust nodes. Various clock synchronization schemes exist for a variety of application domains. We will first outline why these approaches are not suitable for Smart Dust, before presenting our own synchronization approach.

As with node localization, energy, size, and cost constraints preclude equipping Smart Dust with receivers for time infrastructure like GPS [6] or DCF77 [26]. Also, logical time [7] is not sufficient, since it only captures causal relationships between "in system" events, defined by message exchanges between event-generating processes. In contrast, phenomena sensed by dust nodes are triggered by external physical events which are not

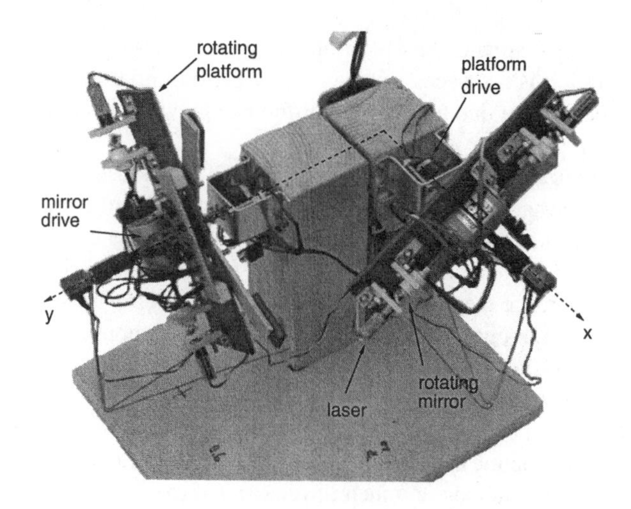

Fig. 4. Lighthouse prototype device for node localization.

defined by in-system message exchanges; physical time must be used to relate events in the physical world.

Many traditional synchronization approaches such as NTP [10] are based on frequently exchanging "time beacons" – messages containing clock readings of the sender at the time of message generation – among nodes to be synchronized. This leads to synchronization algorithms which are typically not energy efficient. For example, the CPU is used continuously to perform frequency disciplining of the oscillator by adding small increments to the system clock. In addition, synchronization beacons are frequently exchanged, which can require significant amounts of energy. Also, the CPU may not be available if the processor is powered down to save energy.

These problems can be solved by rethinking various aspects of a time synchronization service [5]. Energy efficiency, for example, can be significantly improved by exploiting certain characteristics of sensor network applications. As with our tracking system, sensor networks are typically triggered by physical events. Hence, sensor network activity is rather bursty than continuous and rather local than global. This leads to a situation, where synchronized clocks are only required occasionally and only for certain subsets of nodes. One possible way to exploit these characteristics is called post-facto synchronization. There, unsynchronized clocks are used to timestamp sensor events. Only when two timestamps have to be compared by the application, they are reconciled in order to establish a common time scale.

Our time synchronization approach lets the node's clocks run unsynchronized. Timestamps are generated using these unsynchronized clocks. However, if a notification message containing a timestamp is sent to the base station, the contained timestamp is transformed to the receiver's local time. This transformation can often be performed without any additional message exchanges by piggybacking on the notification messages that have to be exchanged anyway.

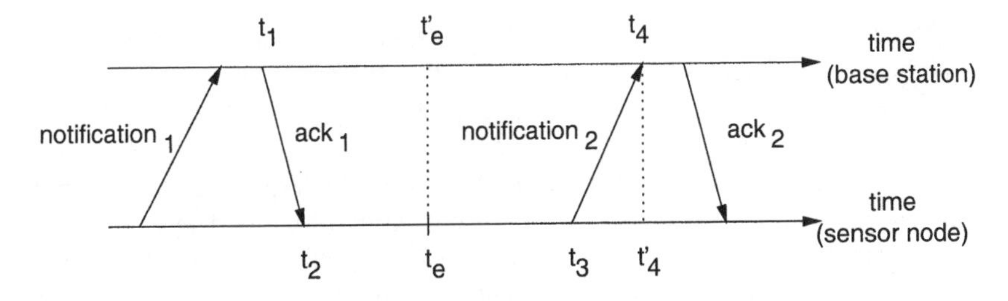

Fig. 5. Time transformation.

Consider Figure 5 to understand how this can be achieved. The figure depicts two consecutive notification message exchanges (and according acknowledgments) between a sensor node and the base station. At sensor-node-time t_e the car enters the sensing range. The respective notification message is sent to the base station at time t_3, containing t_e as the timestamp. We want to determine t'_e, the base-station time that corresponds to sensor-node-time t_e. If we knew t'_4 (i.e., sensor-node time for base-station-time t_4), we could calculate the message delay $D := t'_4 - t_3$. Using D, we could transform t_e to base-station time by calculating $t'_e := t_4 - D - (t_3 - t_e)$.

Unfortunately we don't know t'_4, since the clocks are unsynchronized. However, we can estimate D by $0 \leq D \leq RTT$ with $RTT := (t_4 - t_1) - (t_3 - t_2)$. This gives us the following estimation for t'_e: $t_4 - RTT - (t_3 - t_e) \leq t'_e \leq t_4 - (t_3 - t_e)$. We can thus transform t_e into the time interval $\tilde{t}_e := [t_4 - RTT - (t_3 - t_e), t_4 - (t_3 - t_e)]$ with $t'_e \in \tilde{t}_e$. We use $\tilde{t}_e$ as the notification's transformed timestamp in the base station. Appropriate interval-arithmetic operators can be used to compare such "time-interval stamps" (e.g., $[t^L_A, t^R_A] < [t^L_B, t^R_B] \Leftrightarrow t^R_A < t^L_B$).

Note that $(t_3 - t_2)$ and $(t_3 - t_e)$ can be piggybacked on the notification$_2$ message, such that the base station can perform the time transformation without additional message exchanges. Also note that the base station can be relieved of storing t_1 between two consecutive message exchanges by sending back t_1 to the sensor node as part of the ack$_1$ message. By including t_1 in the following notification$_2$ message, the base station will have t_1 available to calculate the timestamp transformation.

The above transformation rule can be extended to take into account the clock drift, which we demonstrate in [12]. This approach allows us to estimate t'_e with an accuracy of about 10 milliseconds in our tracking system.

3.5 Message Ordering

The data fusion algorithm described in Section 3.2 requires sorting notifications by their timestamps. The time transformation approach described in Section 3.4 enables us to compare and sort timestamps originating from different nodes. However, we still have to ensure that a notification message is not processed by the data fusion algorithm until all earlier notifications arrived at the base station. This is of particular importance for Smart Dust, since messages are subject to long and variable delays as described in Section 2.

One particularly attractive approach to message ordering is based on the assumption that there is a known maximum network latency Δ. Delaying the evaluation of inbound messages for Δ will ensure that out-of-order messages will arrive during this artificial delay and can be ordered correctly using their timestamps. That is, message ordering can be achieved without any additional message exchanges. The literature discusses a number of variants of this basic approach [9,11,17].

However, there is one major drawback of this approach: the assumption of a bounded and known maximum network latency. As discussed earlier, Smart Dust suffers from long and variable network delays. Using a value for Δ which is lower than the actual network latency results in messages being delivered out of order. Using a large value for Δ results in long artificial delays, which unnecessarily decreases the performance of the tracking system.

We therefore introduce a so-called adaptive delaying technique that measures the actual network delay and adapts Δ accordingly. Doing so, it is possible that the estimated Δ is too small and messages would be delivered out of order. Our algorithm detects such late messages and deletes them (i.e., does not deliver them to the application at all). Recall that the data fusion algorithm presented in Section 3.2 was specifically designed to tolerate missing detection and loss notifications. Hence, deleting a message only results in a less accurate track, since then one Smart Dust node less contributes to the estimation of the target location. We argue that this slight decrease of accuracy is acceptable since deleting a message is a rare event, which only occurs at startup or when the maximum network latency increases during operation (i.e., when the value of Δ is lower than the actual maximum network latency). The expected high density of Smart Dust deployments can also compensate this decrease of accuracy. Additionally, our algorithm includes a parameter which can be tuned to trade off tracking latency for tracking accuracy.

Specifically, the adaptive delaying algorithm executed in the BST maintains a variable Δ holding the current delay value, a variable t_{latest} holding the timestamp of the latest message delivered to the application, and a queue which stores messages ordered by increasing timestamps. Initially, Δ is set to some estimate of the maximum network latency, t_{latest} is set to the current time t_{now} in the base station, and the queue is empty.

Upon arrival of a new notification n with timestamp (interval) $n.t$, the actual message delay $d := t_{\text{now}} - n.t^L$ is calculated[1]. Δ is then set to the maximum of Δ and $c \cdot d$. The constant factor c influences the chance of having to delete an out-of-order message and can thus be tuned to trade off tracking latency for tracking accuracy. We use $c = 1.2$ in our prototype. Now we check if $t_{\text{latest}} < n.t^L$ holds, in which case n can still be delivered in order. If so, n is inserted into the queue at the right position according to $n.t$. Otherwise, n is deleted.

The first element n_0 of the queue (i.e., the one with the smallest timestamp) is removed from the queue as soon as the base station's clock (t_{now}) shows a value greater than $n_0.t^R + \Delta$. Now t_{latest} is set to $n_0.t^R$ and n_0 is delivered to the data fusion algorithm.

4 Evaluation

In order to assess the accuracy of the proposed tracking system, we performed a set of initial measurements. Figure 6 shows the setup of our measurements. The lighthouse

[1] t^R (t^L) refers to the right (left) end of the time interval t.

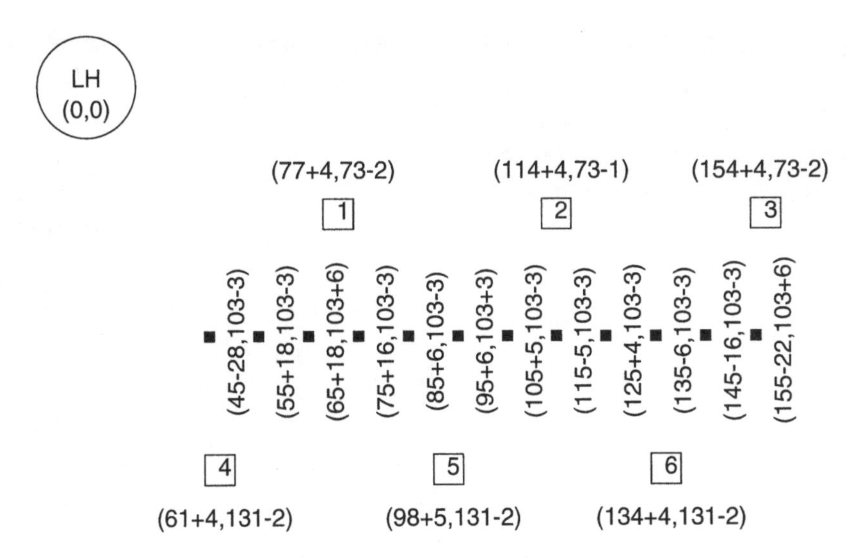

Fig. 6. Measurement setup (not drawn to scale).

device ("LH" in the picture) was placed in the upper left corner and defines the origin (0,0) of a 2D coordinate system. Six sensor nodes (numbered rectangles in the figure) were placed in an area of about one square meter. The car moved then through the sensor field. Location estimates were obtained at 12 positions of the car (indicated by the black squares in the picture). We performed the whole experiment 10 times and calculated averages. The sensor nodes as well as the car locations are annotated with coordinates $(x \pm \Delta_x, y \pm \Delta_y)$, where (x, y) are the ground truth positions in centimeters obtained by a tape measure. $\pm\Delta_x$ and $\pm\Delta_y$ indicate the average errors of the output of the tracking system relative to the ground truth position in the x and y axis, respectively. The *average* error of the sensor node location estimates is $\bar{\Delta}_x = 4.16$cm and $\bar{\Delta}_y = 1.83$cm. We attribute the larger $\bar{\Delta}_x$ value to mechanical problems with one of the lighthouses. The *average* error of the car location estimates is $\bar{\Delta}_x = 12.5$cm and $\bar{\Delta}_y = 3.5$cm. The *maximum* error of the sensor node location estimates is $\hat{\Delta}_x = 5$cm and $\hat{\Delta}_y = 2$cm. The *maximum* error of the car location estimates is $\hat{\Delta}_x = 28$cm and $\hat{\Delta}_y = 6$cm. The difference between the values for the x and y axis is due to the asymmetry of the node arrangement.

The tracking latency is defined as the delay after which the state of the real world is reflected by the output of the tracking system. This delay depends on the following additive components: (1) the sampling interval of the sensors, (2) processing delays in the sensor nodes, (3) the network latency, (4) delays caused by the message ordering algorithm, and (5) delays caused by the algorithm used to compute the target location estimate. The minimum value of (1) heaviliy depends on the sensor and processor hardware. In our implementation, (1) is limited to about 0.1ms by the analog-to-digital converter. Components (2) and (4) are small in our system due to the simplicity of the used algorithms.

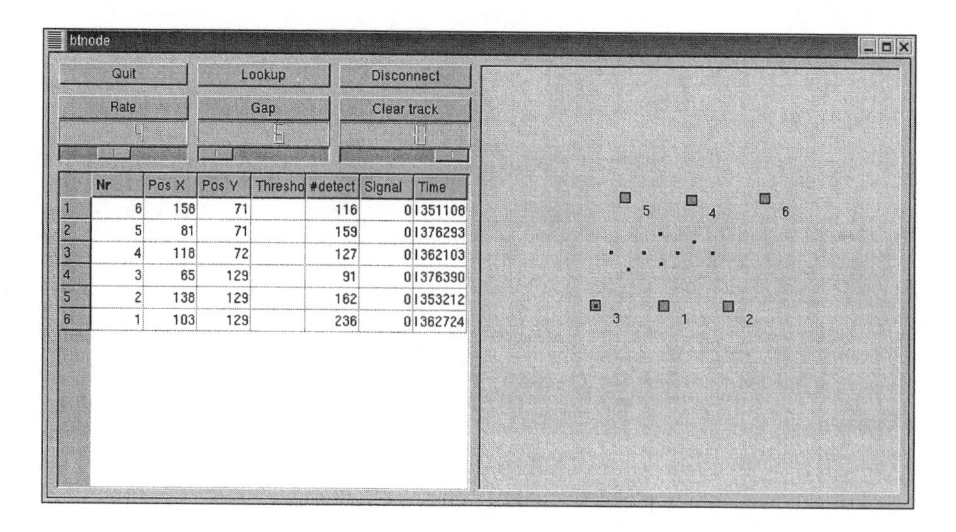

Fig. 7. Measurement setup as shown by the graphical user interface.

To evaluate the tracking latency of our system, we measured the sum of (2), (3), (4), and (5) by calculating the age of each notification after it has been processed by the location estimation algorithm. During the course of our experiment, the average age was 56ms. We also monitored the value of Δ used by the message ordering algorithm. We used an initial guess of $\Delta = 20$ms. At the beginning of the experiment, this value was quickly adapted to 52ms. Recall from Section 3.5 that messages may be dropped by the ordering algorithm if the value used for Δ is lower than the actual maximum network latency. Surprisingly, during our experiments not a single message was dropped. This is due to the fact that the time between arrival of successive notifications at the base station was always greater than the network latency in our experiment. However, this is typically not the case for a real deployment, where the network latency can be significantly larger and where many densely deployed nodes may detect the target almost concurrently and generate according notifications in short intervals.

Note that the above values give only a rough impression of the performance of the tracking system, since we had only 6 sensor nodes available. We plan to build the necessary sensor hardware for a significantly larger number of sensor nodes, which will allow us to perform more thorough measurements.

Figure 7 shows the above measurement setup as depicted by the graphical user interface. In the top left, a number of controls are shown to lookup sensor nodes ("Lookup"), to disconnect from the sensor nodes ("Disconnect"), to adjust the frequency of sensor readout ("Rate"), to control the detection threshold ("Gap"), and to clear the displayed track ("Clear track"). The table below the controls contains one line for each sensor node, showing x and y position, the current detection threshold, number of detections, the currently detected signal strength, and the time of the last detection. On the right, a display of the tracking area is shown, depicting the sensor nodes (larger rectangles) and some of the location estimates of the car (smaller squares) moving from right to left.

5 Discussion

The use of passive optical communication allows the construction of tiny and energy efficient sensor nodes compared to current radio-based COTS devices. However, it must be emphasized that this mode of communication implies a number of drawbacks. Besides requiring a free line of sight for communication, Smart Dust networks have a single-hop topology, where sensor nodes can only communicate with a base station transceiver. That is, Smart Dust nodes cannot talk to each other directly. Additionally, the base station can become a scalability bottleneck, since all communication must pass through the BST.

In the presented tracking system, we tried to achieve scalability despite these limitations. This is achieved in part by strictly avoiding inter-node communication. Additionally, the algorithms employed in the BST are designed to be independent of the actual number of nodes in the network. Instead, the overhead of the base station algorithms depends on the number of *active* nodes – those that currently "see" the tracked target. Also, the base station only has to store state information for active nodes.

Despite some similarities in the communication scheme (both true Smart Dust and our prototype use a single-hop network topology), there is one important difference between our prototype and a system based on true Smart Dust. While in our system nodes can send messages to the BST at any time, communication with a Smart Dust node requires that the BST points its laser beam at that particular node. Even though the developers envision a slightly defocused laser beam to enable the BST to communicate with many nodes at a time, the laser has to sweep over the deployment area to give each node a chance to talk to the BST. We deal with the resulting long and variable network delays by introducing a message ordering technique which adapts to the actual maximum network latency.

The data fusion algorithm used in our system might seem somewhat simplistic compared to many approaches described in the literature. However, it achieves a reasonable accuracy while only making a minimum of assumptions about the tracked target. The loss of accuracy can be compensated by increasing the node density – which is be possible due to the expected small size and low cost of Smart Dust nodes.

The tracking system has been designed to tolerate node failures, since these are likely to happen. Messages lost due to node failures will only affect the accuracy of the estimated track. However, this can be compensated by a higher node deployment density.

6 Related Work

Current sensor network research is mainly focusing on COTS Dust platforms such as [24], thus developing algorithms, protocols, and systems that typically depend on the characteristics of these platforms, and which are therefore often not portable to true Smart Dust. Systems for fine-grained sensor node localization, for example, often rely on direct node-to-node communication or use ultrasound [15]. Approaches for clock synchronization often ignore the fact that time synchronization in networks of sensors is often only needed when and where an "event" occurs, thus wasting energy for continuously synchronizing all clocks [5].

Algorithms for localization and tracking of targets using networks of sensors have been extensively studied in the literature [3,8]. However, there are only few reports on actual tracking systems with complete support for node localization and time synchronization. Many systems like [27], for example, place sensor nodes at well-known positions and cannot deal with node mobility. Systems for target localization like [19] are often too slow to track the location of a moving target in real-time. Moreover, the employed Time-Difference-Of-Arrival (TDOA) method limits this system to targets which emit acoustic or other signals with low propagation speeds. We are not aware of tracking systems that have been specifically designed for true Smart Dust.

7 Conclusion and Outlook

We have presented a complete proof-of-concept system for tracking the location of real-world phenomena with Smart Dust, using a remote-controlled toy car as a sample target. We presented approaches for target location estimation, node localization, time synchronization, and message ordering that match the requirements of Smart Dust. Since target location estimation is solely based on detecting the proximity of the target by individual Smart Dust nodes, the presented tracking system should be applicable to a wide range of possible target types.

As one of the next steps we want to base car detection on the car's acoustic signature, and plan to evaluate the system using a larger number of sensor nodes. The ultimate goal is to reimplement the system using true Smart Dust.

Acknowledgements. The work presented in this paper was supported in part by the National Competence Center in Research on Mobile Information and Communication Systems (NCCR-MICS), a center supported by the Swiss National Science Foundation under grant number 5005-67322.

References

1. I. F. Akyildiz, W. Su, Y. Sankarasubramaniam, and E. Cayirci. Wireless Sensor Networks: A Survey. *Computer Networks*, 38(4):393–422, March 2002.
2. N. Bulusu, J. Heideman, and D. Estrin. GPS-less Low Cost Outdoor Localization for Very Small Devices. *IEEE Personal Communications*, 7(5):28–34, October 2000.
3. J. C. Chen, K. Yao, and R. E. Hudson. Source Localization and Beamforming. *IEEE Signal Processing Magazine*, 19(2):30–39, 2002.
4. L. Doherty, K. S. J. Pister, and L. E. Ghaoui. Convex Position Estimation in Wireless Sensor Networks. In *Infocom 2001*, Anchorage, Alaska, April 2001.
5. J. Elson and K. Römer. Wireless Sensor Networks: A New Regime for Time Synchronization. *ACM SIGCOMM Computer Communication Review (CCR)*, 33(1):149–154, January 2003.
6. B. Hofmann-Wellenhof, H. Lichtenegger, and J. Collins. *Global Positioning System: Theory and Practice, 4th Edition*. Springer-Verlag, 1997.
7. L. Lamport. Time, Clocks, and the Ordering of Events in a Distributed System. *Communications of the ACM*, 21(4):558–565, July 1978.
8. D. Li, K. D. Wong, Y. H. Hu, and A. M. Sayeed. Detection, Classification, and Tracking of Targets. *IEEE Signal Processing Magazine*, 19(2):17–29, 2002.

9. M. Mansouri-Samani and M. Sloman. GEM – A Generalised Event Monitoring Language for Distributed Systems. *IEE/IOP/BCS Distributed Systems Engineering Journal*, 4(25), February 1997.
10. D. L. Mills. Improved algorithms for synchronizing computer network clocks. In *Conference on Communication Architectures (ACM SIGCOMM'94)*, London, UK, August 1994. ACM.
11. G. J. Nelson. *Context-Aware and Location Systems*. PhD thesis, University of Cambridge, 1998.
12. K. Römer. Time Synchronization in Ad Hoc Networks. In *MobiHoc 2001*, Long Beach, USA, October 2001.
13. K. Römer. The Lighthouse Location System for Smart Dust. In *MobiSys 2003*, San Franscisco, USA, May 2003.
14. C. Savarese, J. M. Rabaey, and K. Langendoen. Robust Positioning Algorithms for Distributed Ad-Hoc Wireless Sensor Networks. In *USENIX Annual Technical Conference*, Monterey, USA, June 2002.
15. A. Savvides, C. C. Han, and M. Srivastava. Dynamic Fine-Grained Localization in Ad-Hoc Networks of Sensors. In *Mobicom 2001*, Rome, Italy, July 2001.
16. Y. Shang, W. Ruml, Y. Zhang, and M. Fromherz. Localization from Mere Connectivity. In *ACM MobiHoc 2003*, Annapolis, USA, June 2003.
17. Y. C. Shim and C. V. Ramamoorthy. Monitoring and Control of Distributed Systems. In *First Intl. Conference of Systems Integration*, pages 672–681, Morristown, USA, 1990.
18. R. Viswanathan and P. Varshney. Distributed Detection with Multiple Sensors: I. Fundamentals. *Proceedings of the IEEE*, 85(1):54–63, 1997.
19. H. Wang, J. Elson, L. Girod, D. Estrin, and K. Yao. Target Classification and Localization in Habit Monitoring. In *IEEE ICASSP 2003*, Hong Kong, China, April 2003.
20. R. Want, A. Hopper, V. Falcao, and J. Gibbons. The Active Badge Location System. *ACM Transactions on Information Systems*, 10(1):91–102, 1992.
21. B. Warneke, M. Last, B. Leibowitz, and K. S. J. Pister. Smart Dust: Communicating with a Cubic-Millimeter Computer. *IEEE Computer Magazine*, 34(1):44–51, January 2001.
22. B. A. Warneke, M. D. Scott, B. S. Leibowitz, L. Zhou, C. L. Bellew, J. A. Chediak, J. M. Kahn, B. E. Boser, and K. S. J. Pister. An Autonomous 16 cubic mm Solar-Powered Node for Distributed Wireless Sensor Networks. In *IEEE Sensors*, Orlando, USA, June 2002.
23. K. Whitehouse and D. Culler. Calibration as Parameter Estimation in Sensor Networks. In *Workshop on Wireless Sensor Networks and Applications (WSNA) 02*, Atlanta, USA, September 2002.
24. Berkeley Motes. www.xbow.com/Products/Wireless_Sensor_Networks.htm.
25. BTnodes. www.inf.ethz.ch/vs/res/proj/smart-its/btnode.html.
26. DCF77 Radio Time Signal. www.dcf77.de.
27. The 29 Palms Experiment: Tracking vehicles with an UAV-delivered sensor network. tinyos.millennium.berkeley.edu/29Palms.htm.

Networked Slepian-Wolf: Theory and Algorithms*

Răzvan Cristescu[1], Baltasar Beferull-Lozano[1], and Martin Vetterli[1,2]

[1] Laboratory for Audio-Visual Communications (LCAV),
Swiss Federal Institute of Technology (EPFL),
Lausanne CH-1015, Switzerland
[2] Department of EECS, University of California at Berkeley,
Berkeley CA 94720, USA
{Razvan.Cristescu,Baltasar.Beferull,Martin.Vetterli}@epfl.ch
http://ip7.mics.ch/

Abstract. In this paper, we consider the minimization of a relevant energy consumption related cost function in the context of sensor networks where correlated sources are generated at various sets of source nodes and have to be transmitted to some set of sink nodes. The cost function we consider is given by the product [rate] $\times$ [link weight]. The minimization is achieved by jointly optimizing the transmission structure, which we show consists of a superposition of trees from each of the source nodes to its corresponding sink nodes, and the rate allocation across the source nodes. We show that the overall minimization can be achieved in two concatenated steps. First, the optimal transmission structure has to be found, which in general amounts to finding a Steiner tree and second, the optimal rate allocation has to be obtained by solving a linear programming problem with linear cost weights determined by the given optimal transmission structure. We also prove that, if any arbitrary traffic matrix is allowed, then the problem of finding the optimal transmission structure is NP-complete. For some particular traffic matrix cases, we fully characterize the optimal transmission structures and we also provide a closed-form solution for the optimal rate-allocation. Finally, we analyze the design of decentralized algorithms in order to obtain exactly or approximately the optimal rate allocation, depending on the traffic matrix case. For the particular case of data gathering, we provide experimental results showing a good performance in terms of approximation ratios.

1 Introduction

1.1 Problem Motivation

Consider networks that transport supplies among nodes. This is for instance the case in sensor networks that measure some environmental data. Nodes are

* The work presented in this paper was supported (in part) by the National Competence Center in Research on Mobile Information and Communications Systems (NCCR-MICS), a center supported by the Swiss National Science Foundation under grant number 5005-67322.

H. Karl, A. Willig, A. Wolisz (Eds.): EWSN 2004, LNCS 2920, pp. 44–59, 2004.

supplied amounts of measured data which need to be transmitted to end sites, called sinks, for control or storage purposes. An example is shown in Fig.1, where there are N nodes with sources $X_1, \ldots, X_N$, two of them being the sinks denoted by S_1, S_2, and a graph of connectivity with edges connecting certain nodes. We will use interchangeably the notions of network entity and its graph representation across the paper. Sources corresponding to nodes in the sets V^1, V^2 need to transmit their data, possibly using other nodes as relays, to sinks S_1, S_2 respectively. A very important task in this scenario is to find a rate allocation at nodes and a transmission structure on the network graph that minimizes a cost function of interest (e.g. flow cost [data size] $\times$ [link weight], total distance, etc.). This implies a joint treatment of source coding and optimization of the transmission structure.

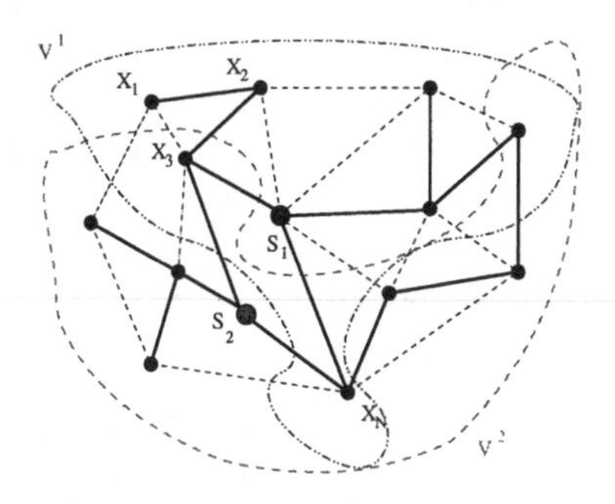

Fig. 1. An example of a network. Sources transmit their data to the sinks. Nodes from the V^1, V^2 set of nodes need to arrive at sink S_1, S_2, respectively. A rate supply R_i is allocated to each node X_i. In solid lines, a chosen transmission structure is shown. In dashed lines, the other possible links are shown

The problem is trivial if the data measured at nodes are statistically independent: each node codes its data independently, and well developed algorithms can be used to solve the minimum cost flow problem [5].

However, in many situations, data at nodes are *not* independent, such as in typical sensor networks. It can be thus expected that approaches that take into account the correlation present in the data can improve over existing algorithms, with regard to optimizing many cost metrics of interest.

1.2 Correlated Data

The source coding approach that takes maximum advantage of the data correlation, at the expense of coding complexity, is based on the important work of [19]. In that work, Slepian and Wolf showed that when nodes measure correlated data, these data can be coded with a total rate not exceeding the joint entropy, even *without* nodes explicitly communicating with each other (under some constraints on the minimal rates given by the Slepian-Wolf region). Their result provides the whole achievable *rate region* for the rate allocation, that is *any* rate in that region is achievable. We describe in more detail the Slepian-Wolf coding in Sect. 2.2.

In addition to encoding the data, these data usually need to be transmitted over the network from the sources to the sinks. In such situations, it is important to study the influence that the transmission structure used to transport the data has on the rate allocation at the nodes. In this work, we consider a joint treatment of both the rate allocation and the chosen transmission structure. We show that when separable joint cost functions are considered (e.g. the [data size] $\times$ [link weight] metric), and Slepian-Wolf coding is used, then the problem of joint optimization separates, in the sense that first an optimal transmission structure needs to be determined, and second the optimal rate allocation is found on this transmission structure. However, the rate allocation is determined by the transmission structure, and thus the respective optimizations are not independent. The optimal rate allocation is in general unique, except in some degenerate cases. Since nodes have limited processing capability and/or battery power, it is necessary that the rate allocation and transmission structure optimization are done locally at each node, in a decentralized manner, by using information available only from nodes in the neighborhood.

In particular, let us consider the case of a network of sensors taking measurements from the environment [2], [12], [15]. Let $\mathbf{X} = (X_1, \ldots, X_N)$ be the vector formed by the random variables representing the sources measured at the nodes $1, \ldots, N$. We assume that the random variables are continuous and that there is a quantizer in each sensor (with the same resolution for all sensors). There are also a number of sinks to where data from different subsets of nodes have to be sent. A rate allocation $(R_1, \ldots, R_N)$ (bits) has to be assigned at the nodes so that the quantized measured information samples are described losslessly. Notice that it is also possible to allocate different rates at each node, depending on which sink it sends its data to, but this involves important additional coding overhead, which might not be always feasible. We consider both cases in this paper. We assume that the spatial correlation between samples taken at the nodes depends in our setting only on the distance distribution across space. In this work, we assume that contention is solved by the upper layers. The transmission topology in our model is assumed to be an undirected graph with point-to-point links. Practical approaches use nearest neighbor connectivity, avoiding thus the complexity of the wireless setting. Since battery power is the scarce resource for autonomous sensors, a meaningful metric to minimize in the case of sensor networks is the total energy consumption. This is essentially given by the sum of products [data size] $\times$ [link weight] for all the links used in the transmission. The weight of the link between two nodes is a function of the distance d of the two nodes (e.g. kd^{α} or $k \exp(\alpha d)$, with k, α constants of the medium).

The novelty of our approach stems from the fact that we consider jointly the optimization of both source coding and transmission structure in the context of sensor networks measuring correlated data. To the best of our knowledge, this is the first research work that addresses jointly Slepian-Wolf lossless source coding and network flow cost optimization.

1.3 Related Work

Progress towards practical implementation of Slepian-Wolf coding has been achieved in [1], [13], [14]. Bounds on the performance of networks measuring correlated data have been derived in [11], [18]. However, none of these works takes into consideration the cost of transmitting the data over the links and the additional constraints that are imposed on the rate allocation by the joint treatment of source coding and transmission.

The problem of optimizing the transmission structure in the context of sensor networks has been considered in [9], [16], where the energy, and the [energy] $\times$ [delay] metric are studied, and practical algorithms are proposed. In these studies, the correlation present in the data is not exploited for the minimization of the metric.

A joint treatment of data aggregation and the transmission structure is considered in [8]. The model in [8] does not take into account possible collaborations among nodes. In our work, we consider the case of collaboration between nodes because we allow nodes to perform (jointly) Slepian-Wolf coding.

1.4 Main Contributions

In this paper, we address the problem of Slepian-Wolf source coding for general networks and patterns of traffic, namely, in terms of our graph representation, for general (undirected) graphs and different sets of source nodes and sinks. We consider the flow cost metric given by [data size]$\times$[link weight], and we assess the complexity of the resulting joint optimization problem, for various network settings. We first prove that the problem can be always be separated into the tasks of transmission structure optimization and rate allocation optimization. For some particular cases, we provide closed-form solutions and efficient approximation algorithms. These particular cases include correlated data gathering where there is only one sink node. In the general case, we prove that the problem is NP-complete.

The rest of this paper is organized as follows. In Sect. 2, we state the optimization problem and describe the Slepian-Wolf source coding approach and the optimal region of rate allocations. In Sect. 3 we study the complexity for the case of a general traffic matrix problem and we prove that finding the optimal transmission structure is NP-complete. We show that, if centralized algorithms were allowed, finding the optimal rate allocation is simple; however, in our sensor network setting, the goal is to find distributed algorithms, and we show that in order to have a decentralized algorithm, we need a substantially large communication overhead in the network. In Sect. 4, we fully solve an important particular case, namely the correlated data gathering problem with Slepian-Wolf source coding. In Sect. 5 we consider other particular cases of interests. Finally, we present some numerical simulations in Sect. 6. We conclude and present directions of further work in Sect. 7.

2 Problem Formulation

2.1 Optimization Problem

Consider a graph $G = (V, E)$, $|V| = N$. Each edge $e \in E$ is assigned a weight w_e. Nodes on the graph are sources of data. Some of the nodes are also sinks. Data has to be transported over the network from sources to sinks. Denote by $S_1, S_2, \ldots, S_M$ the set of sinks and by $V^1, V^2, \ldots, V^M$ the set of subsets $V^j \subseteq V$ of sources; data measured at nodes V^j have to be sent to sink S_j. Denote by S^i the set of sinks to which data from node i have to be sent. Denote by $E^i \subseteq E$ the subset of edges used to transmit data from node i to sinks S^i, which determines the transmission structure corresponding to node i.

Definition 1 (Traffic matrix). *We call the* traffic matrix *of a graph G the $N \times N$ square matrix T that has elements given by:*

$$T_{ij} = 1, \text{ if source } i \text{ is needed at sink } j,$$
$$T_{ij} = 0, \text{ else.}$$

With this notation, $V^j = \{i : T_{ij} = 1\}$ and $S^i = \{j : T_{ij} = 1\}$.

The overall task we consider is to assign an optimal rate allocation $R_i^*, i = 1, \ldots, N$ for the N nodes and to find the optimal transmission structure on the graph G that minimizes the total flow cost [data size] × [link weight]. Thus, the optimization problem is:

$$\{R_i^*, d_i^*\}_{i=1}^N = \arg_{\{R_i, d_i\}} \min \sum_{i=1}^N R_i d_i \qquad (1)$$

where d_i is the total weight of the transmission structure chosen to transmit data from source i to the set of sinks S^i:

$$d_i = \sum_{e \in E^i} w_e .$$

Notice that finding the optimal $\{d_i\}_{i=1}^N$ is equivalent to finding the optimal transmission structure.

In the next Sect. 2.2 we show that, when Slepian-Wolf coding is used, the tasks of finding the optimal $\{d_i\}_{i=1}^N$ and respectively $\{R_i\}_{i=1}^N$ are separated, that is, one can first find the optimal transmission structure, which can be shown to be always a tree, and then find the optimal rate allocation. As a consequence, after finding the optimal transmission structure, (1) can be posed as a linear programming problem in order to find the optimal rate allocation. We study the complexity of solving the overall problem under various scenarios.

2.2 Slepian Wolf Coding

Consider the case of two random sources X_1 and X_2 that are correlated (see Fig. 2(a)). Intuitively, each of the sources can code their data at a rate equal to at least their corresponding entropies, $R_1 = H(X_1)$, $R_2 = H(X_2)$, respectively. If they are able to communicate, then they could coordinate their coding and use together a total rate equal to the joint entropy $R_1 + R_2 = H(X_1, X_2)$. However, Slepian and Wolf [19] showed that two correlated sources can be coded with a total rate $H(X_1, X_2)$ even if they are *not* able to communicate with each other. This can be also easily generalized to the N-dimensional case. Fig. 2(b) shows the Slepian-Wolf rate region for the case of two sources.

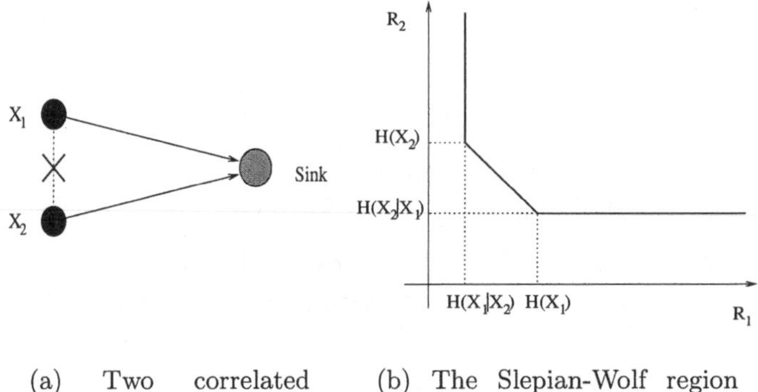

(a) Two correlated sources X_1, X_2 send their data to one sink

(b) The Slepian-Wolf region shows the achievable pairs of rates that can be allocated to sources X_1 and X_2 for lossless data coding

Fig. 2. Two correlated sources, and the Slepian-Wolf region for their rate allocation

Consider again the example shown in Fig. 1. Assume that the set of sources that send their data to sink j, that is the set of nodes denoted $\{X_{j1}, \ldots, X_{j|V^j|}\} \in V^j, j = 1, 2$, know in advance the correlation structure in that set V^j (which depends only on the distance in our model). This is a reasonable assumption to make for localized data requests from the sources (that is, when nodes in V^j are geographically close to each other). Then, nodes in V^j can code their data jointly, without communicating with each other with a total rate of $H(X_{j1}, X_{j2}, \ldots, X_{j|V^j|})$ bits, as long as their individual rates obey the Slepian-Wolf constraints related to the different conditional entropies [6], [19].

Proposition 1 (Separation of source coding optimization and transmission structure optimization). *When Slepian-Wolf coding is used, the transmission structure optimization separates from the rate allocation optimization, in terms of the overall minimization of (1).*

Proof. Once the rate allocation is *fixed*, the best way to transport any amount of data from a given node i to the set of sinks S^i does not depend on the value of the rate. This is true because we consider separable flow cost functions, and the rate supplied at each node does not depend on the incoming flow at that node. Since this holds for any rate allocation, it is true for the minimizing rate allocation and the results follows. □

For each node i, the optimal transmission structure is in fact a tree that spans node i and the sinks S^i to which its data are sent [5]. Thus, the whole optimization problem can be separated into a spanning tree optimization for each source, and the rate allocation optimization. Then, after the optimal tree structure is formed, (1) becomes a problem of rate allocation that can be posed as a linear programming (LP) problem under the usual Slepian-Wolf linear constraints:

$$\begin{aligned} &\min_{\{R_i\}_{i=1}^N} \sum_{i=1}^N R_i d_i^* \\ &\text{under constraints:} \\ &\sum_{l \in \mathbf{Y}_j} R_l \geq H(\mathbf{Y}_j | V^j - \mathbf{Y}_j), (\forall) V^j, \mathbf{Y}_j \subseteq V^j, \end{aligned} \tag{2}$$

that is, first the optimal weights $\{d_i^*\}_{i=1}^N$ are found (which determine in general uniquely the optimal transmission structure), and then the optimal rate allocation is found using the fixed values $\{d_i^*\}_{i=1}^N$ in (2). Note that there is one set of constraints for each set V^j.

Moreover, note that (2) is an LP optimization problem under linear constraints, so if the weights $\{d_i^*\}_{i=1}^N$ can be determined, the optimal allocation $\{R_i\}_{i=1}^N$ can be found easily with a *centralized* simplex algorithm [10]. However, in Sect. 3 we will show that for a general traffic matrix T, finding the optimal coefficients $\{d_i^*\}_{i=1}^N$ is NP-complete. Moreover, in general, even if the optimal structure is found, it is hard to *decentralize* the algorithm that finds the optimal solution $\{R_i^*\}_{i=1}^N$ of (2), as this requires a substantial amount of global knowledge of the network.

In the following sections, for various problem settings, we first show how the transmission structure can be found (i.e. the values of $\{d_i^*\}_{i=1}^N$), and then we discuss the complexity of solving (2) in a decentralized manner.

3 Arbitrary Traffic Matrix

We begin the analysis with the most general case, that is when the traffic matrix T is arbitrary, by showing the following proposition:

Proposition 2 (The optimal transmission structure is a superposition of Steiner trees). *Given an arbitrary traffic matrix T, then, for any i, the optimal value d_i^* in (2) is given by the minimum weight tree rooted in node i and which spans the nodes in S^i; this is exactly the minimum Steiner tree that has node i as root and which spans S^i, which is an NP-complete problem.*

Proof. The proof is straightforward: data from node i has to be send over the minimum weight structure to the nodes in S^i, possibly via nodes in $V - \{i, S^i\}$. This is a minimum Steiner tree problem for the graph G with weights w_e, thus it is NP-complete. □

The approximation ratio of an algorithm that finds a solution for an optimization problem is defined as the guaranteed ratio between the cost of the found solution and the optimal one. If the weights of the graph are the Euclidean distances ($w_e = l_e$ for all $e \in E$), then the problem becomes the Euclidean Steiner tree problem, and it admits a PTAS [3] (that is, for any $\epsilon > 0$, there is a polynomial time approximation algorithm with an approximation ratio of $1 + \epsilon$). However, in general, the link weights are not the Euclidean distances (e.g. if $w_e = l_e^2$ etc.). Then finding the optimal Steiner tree is APX-complete (that is, there is a hard lower bound on the approximation ratio), and is only approximable (with polynomial time in the input instance size) within a constant factor $(1 + \ln 3)/2$ [4], [17].

The approximation ratios of the algorithms for solving the Steiner tree translate into bounds of approximation for our problem. By using the respective approximation algorithms for determining the weights d_i, the cost of the approximated solution for the joint optimization problem will be within the Steiner approximation ratio away from the optimal one.

Once the optimal weights d_i^*'s are found (i.e. approximated by some approximation algorithm for solving the Steiner tree), then, as we mentioned above, (2) becomes a Linear Programming (LP) problem. Consequently, it can be readily solved with a centralized program. The solution of this problem is given by the innermost corner of the Slepian-Wolf region that is tangent to the cost function (see Fig. 3 for an example with two nodes or sources). If global knowledge of the network is allowed, then this problem can be solved computationally in a simple way. However, it is not possible in general to find in closed-form the optimal solution determined by the corner that minimizes the cost function, and consequently the derivation of a decentralized algorithm for the rate allocation, as this involves exchange of network knowledge among the clusters.

Figure 4 shows a simple example (but sufficiently complete) which illustrates the difficulty of this problem. Suppose that the optimal total weights $\{d_i^*\}_{i=1}^3$ in (2) have been approximated by some algorithm. Then the cost function to be minimized is:

$$R_1 w_{11} + R_2(w_{21} + w_{22}) + R_3 w_{32}$$

with $d_1^* = w_{11}, d_2^* = w_{21} + w_{22}, d_3^* = w_{32}$, and the Slepian-Wolf constraints are given by:

$$\begin{aligned} R_1 + R_2 &\geq H(X_1, X_2) \\ R_1 &\geq H(X_1|X_2), \text{ for set } V^1 = \{X_1, X_2\} \\ R_2 &\geq H(X_2|X_1) \end{aligned}$$

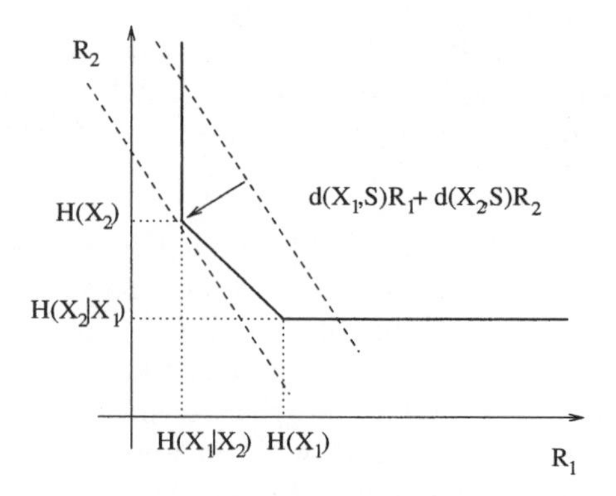

Fig. 3. A simple example with two nodes. The total weights d_1, d_2, from sources X_1, X_2 to the sinks, are respectively $d(X_1, S), d(X_2, S)$, $d(X_1, S) > d(X_2, S)$, in this particular case. In order to achieve the minimization, the cost line $d(X_1, S)R_1 + d(X_2, S)R_2$ has to be tangent to the most interior point of the Slepian-Wolf rate region, given by $(R_1, R_2) = (H(X_1|X_2), H(X_2))$

and respectively,

$$\begin{aligned} R_2 + R_3 &\geq H(X_2, X_3) \\ R_2 &\geq H(X_2|X_3), \text{ for set } V^2 = \{X_2, X_3\} \\ R_3 &\geq H(X_3|X_2). \end{aligned}$$

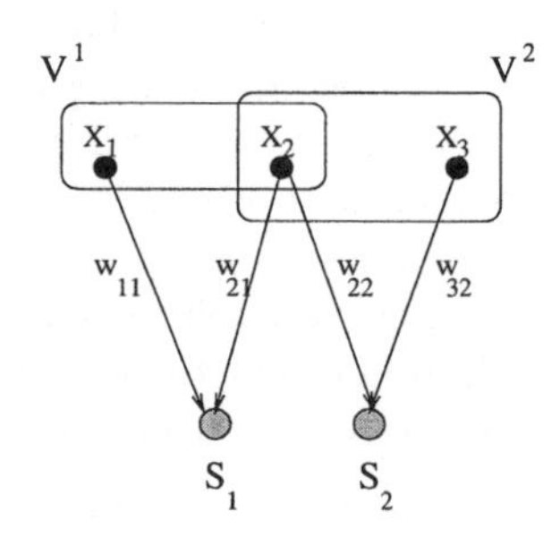

Fig. 4. Two sets of sources transmit their correlated data to two sinks

Suppose the weights are such that $w_{11} < w_{21} + w_{22} < w_{32}$. A decentralized algorithm has to use only local information, that is, information only available in a certain local transmission range or cluster neighborhood). We assume that only local Slepian-Wolf constraints are considered in each set V^j for the rate allocation, and no knowledge about the total weights d_i from nodes in the other subsets is available. Then, it readily follows that the optimal rate allocations in each of the two subsets are:

$$R'_1 = H(X_1)$$
$$R'_2 = H(X_2|X_1), \text{ for set } V^1$$

and respectively,

$$R'_2 = H(X_2)$$
$$R'_3 = H(X_3|X_2), \text{ for set } V^2.$$

If the rate allocation at a source X_i can take different values R_{ij} depending to which sink j source X_i sends its data, then it is straightforward to find this multiple rate assignment. The rate allocation for each cluster will be independent from the rate allocations in the other clusters (see Sect. 4 for the closed-form solution for the rate allocation for each set V^j, and a decentralized algorithm for finding the optimal solution).

However, this involves even more additional complexity in coding, so in some situations it might be desirable to assign a unique rate to each node, regardless of which sink the data is sent to. We can see from this simple example that we cannot assign the correct unique optimal rate R_2 to source X_2, unless node 2 has global knowledge of the whole distance structure from nodes $1, 2, 3$ to the sinks S_1 and S_2. For a general topology, knowledge in a node from the nodes belonging to other different sets is needed at least at nodes that are at the intersection of sets (in this example, source X_2). Even so, it is clear that such global sharing of cluster information over the network is not scalable because the amount of necessary global knowledge grows exponentially.

There are however some important special cases of interest where the problem is tractable, and we treat them in the following two sections.

4 Data Gathering: All Sources ($V^j = V$) Sent to One Sink $S = j$

This case has been studied in the context of the *network correlated data gathering problem* [7], and is a particular case of the problem we consider in this paper. An example is shown in Fig. 5.

In this case, the problem simplifies: if there is a single sink S, then the Steiner tree rooted at i and spanning node S is actually the shortest path, of total weight d_i, between the two nodes. The overall optimal transmission structure is thus the superposition of the shortest paths from each node i to the sink S. This superposition forms the shortest path tree (*SPT*) rooted in S. The *SPT* can be easily found with a distributed algorithm (e.g. Bellman-Ford).

Let us review in Sect. 4.1 the results contained in [7].

4.1 Solution of the LP Problem

The algorithm for finding the optimal rate allocation for this setting is:

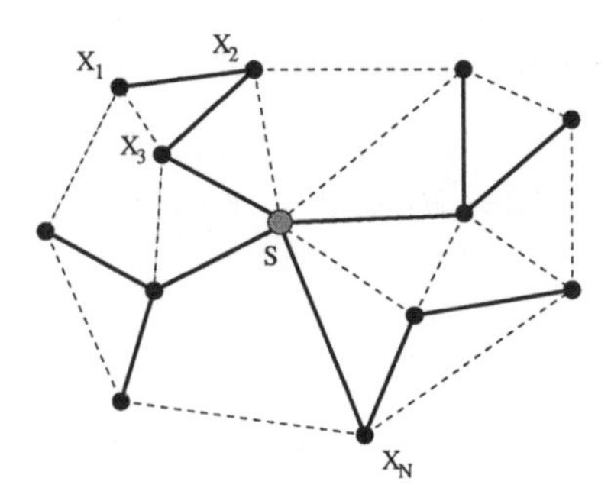

Fig. 5. In this example, data from nodes $X_1, X_2, \ldots, X_N$ need to arrive at sink S.

Algorithm 1. *Data gathering optimal Slepian-Wolf rate allocation:*

- *Find the weights $d_i = d_{SPT}(i, S)$, for each node i, given by the SPT of the graph G, by running e.g. distributed Bellman-Ford.*
- *Solve the constrained LP:*

$$(R_1^*, \ldots, R_N^*) = \arg\min_{\{R_i\}} \sum_i R_i d_{SPT}(i, S),$$

under constraints: (3)

$$\sum_{i \in \mathbf{Y}} R_i \geq H(\mathbf{Y}|\mathbf{Y}^C), (\forall)\mathbf{Y} \subseteq V$$

where $d_{SPT}(i, S)$ is the total length of the path in the SPT from node i to S, and $(R_1^*, \ldots, R_N^*)$ is the optimal rate allocation.

As discussed in Sect. 3, we see that in order to express the rate constraints, centralized knowledge of the correlation structure among *all* nodes in the network is needed. Nevertheless, in this case there is a single set of constraints that involves all the nodes, and because of this the solution for the rate allocation can be expressed in a closed-form.

Suppose without loss of generality that nodes are numbered in increasing order of their distance to the sink on the *SPT*: $(X_1, X_2, \ldots, X_N)$ with $d_{SPT}(X_1, S) \leq d_{SPT}(X_2, S) \leq \cdots \leq d_{SPT}(X_N, S)$.

Proposition 3 (LP solution). *The solution of the optimization problem in (3) is:*

$$\begin{aligned} R_1^* &= H(X_1), \\ R_2^* &= H(X_2|X_1), \\ &\ldots\ldots\ldots \\ R_N^* &= H(X_N|X_{N-1}, X_{N-2}, \ldots, X_1). \end{aligned} \quad (4)$$

That means that each node codes its data with a rate equal to its respective entropy conditioned on all other nodes which are closer to the sink than itself.

4.2 Approximation Algorithm

In the previous subsection, we present the optimal solution of the linear programming rate assignment for the single sink data gathering problem, under Slepian-Wolf constraints. We consider now the problem of designing a distributed approximation algorithm. Even if we can provide the solution in a closed form as (4), the nodes still need local knowledge of the overall structure of the network (distances between nodes and distances to the sink). This local knowledge is needed for:

1. Ordering the distances on the SPT from the nodes to the sink: each node needs its index in the ordered sequence of nodes so as to determine on which other nodes to condition when computing its rate assignment.
2. Computation of the rate assignment:

$$\begin{aligned} R_i &= H(X_i|X_{i-1},\ldots,X_1) \\ &= H(X_1,\ldots,X_i) - H(X_1,\ldots,X_{i-1}) \end{aligned}$$

   Note that *all* distances among nodes $(1,\ldots,i)$ are needed locally at node i for computing this rate assignment.

Such global knowledge might not be available. Thus, we propose a fully distributed approximation algorithm, which avoids the need for a node to have global knowledge of the network, and which provides solutions very close to the optimum.

Suppose each node i has complete information (distances between nodes and distances to the sink) only about a local vicinity $\mathcal{N}(i)$. This information can be computed by running for example a distributed algorithm for finding the SPT (e.g. Bellman-Ford). The approximation algorithm that we propose is based on the observation that nodes that are outside this neighborhood count very little, in terms of rate, in the local entropy conditioning, under the assumption that the correlation decreases with the distance between nodes, which is a natural assumption.

Algorithm 2. *Approximated Slepian-Wolf coding:*

- *Find the SPT.*
- *For each node i:*
  - *Find in the neighborhood $\mathcal{N}(i)$ the set $\mathcal{C}_i$ of nodes that are closer to the sink, on the SPT, than node i.*
  - *Transmit at rate $R_i^\dagger = H(X_i|\mathcal{C}_i)$.*

This means that data are coded locally at the node with a rate equal to the conditional entropy, where the conditioning is performed *only* on the subset formed by the neighbor nodes which are closer to the sink than the respective node.

The proposed algorithm needs only local information, so it is completely distributed. Still, it will give a solution very close to the optimum since the neglected conditioning is small in terms of rate for a correlation function that is sufficiently decaying with distance (see Sect. 6 for some numerical simulations).

Similar techniques can be used to derive decentralized approximation algorithms for some of the other particular cases of interests that we discuss in the next section.

5 Other Particular Cases

5.1 Broadcast of Correlated Data

This case corresponds to the scenario where some sources are sent to all nodes ($S^i = V$). A simple example is shown in Fig. 6. In this example, the traffic matrix has $T_{ij} = 1, (\forall)j$, for some arbitrary L nodes $\{i_1, \ldots, i_L\} \subset V$.

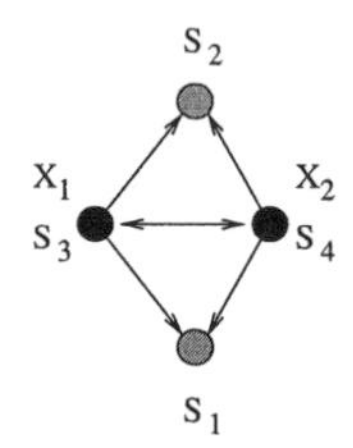

Fig. 6. Data from X_1, X_2 need to be transmitted to all nodes S_1, S_2, S_3, S_4

In this case, for any node i, the value d_i^* in (2) is given by the tree of minimum weight which spans V; this is the minimum spanning tree (MST), and thus, by definition, it does not depend on i.

Note that in this case all weights $\{d_i^*\}_{i=1}^N$ are equal, thus the optimal solution R_i is not unique and therefore we have a degenerate solution case for the LP in (2). There is only one set of constraints, and the cost line is exactly parallel to the diagonal hyper-plane in the Slepian-Wolf region (e.g. in Fig. 3, this happens when the dashed cost line becomes parallel to the diagonal solid line in the boundary). In such a case, not only a corner, but any point on this diagonal hyper-plane of the Slepian-Wolf region is optimal.

Notice that this case includes the typical broadcast scenario where one node transmits its source to all the nodes in the network.

5.2 Multiple Sink Data Gathering

This case corresponds to the scenario where all sources ($V^j = V$) are sent to some set S^a of sinks. In this case (see Fig. 7), finding the optimal weights $\{d_i^*\}_{i=1}^N$ is as difficult as in the arbitrary matrix case, presented in Sect. 3. For every i, the optimal weight d_i^* is equal to the weight of the minimum Steiner tree rooted at i and spanning the nodes in the set S^a.

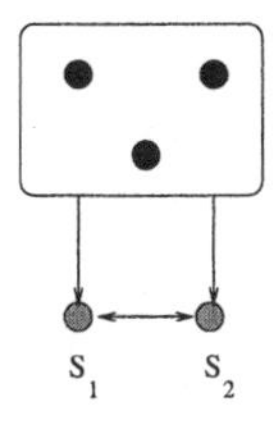

Fig. 7. Data from all nodes has to be transmitted to the set of sinks $S^a = \{S_1, S_2\}$. Each sink has to receive data from *all* the sources

However, given the optimal transmission structure, the optimal rate allocation can be easily found in a similar manner as in Sect. 4. First, we order the nodes in increasing order of increasing distance $d_1^* < d_2^* < \cdots < d_N^*$, and then the optimal rate allocation is given as (4).

5.3 Localized Data Gathering

This case corresponds to the scenario where disjoint sets $\{V^1, V^2, \ldots, V^L\}$ are sent to some sinks $\{S_1, S_2, \ldots, S_L\}$. In this case, for each i, the solution for the optimal weight d_i^* is again the corresponding Steiner tree rooted at i and that spans S^i. If d_i^* can be found, then the rate allocation can be approximated by a decentralized algorithm for each set $\{V^j\}_{j=1}^L$, in the same way as in Sect. 4, that is, we solve L LP programs independently (decentralization up to cluster level).

Algorithm 3. *Disjoint sets.*

- *For each set V^j, order nodes $\{i, i \in V^j\}$ as a function of the total weight d_i.*
- *Assign rates in each V^j as in (4), taking into account this order.*

6 Numerical Simulations

We present numerical simulations that show the performance of the approximation algorithm introduced in Sect. 4, for the case of data gathering. We consider a stochastic data model given by a multi-variate Gaussian random field, and a correlation model where the inter-node correlation decays exponentially with the distance between the nodes. In this case, the joint entropy of the data measured at a set of nodes is essentially given by the logarithm of the determinant of the corresponding covariance matrix.

Then, the performance of our approximation algorithm will be close to optimal even if we consider a small neighborhood $\mathcal{N}(i)$ for each sensor i. We use an exponential model of the covariance $K_{ij} = \exp(-ad_{i,j}^2)$, for varying neighborhood range radius and several values for the correlation exponent a. The weight of an edge (i, j) is $w_{i,j} = d_{i,j}^2$ and the total cost is given by expression (3). Figure 8(a) presents the average ratio between the approximated solution and the optimal one. In Fig. 8(b) we show a comparison of the rate allocations with our different approaches for rate allocation, as a function of the distances from the nodes to the sink.

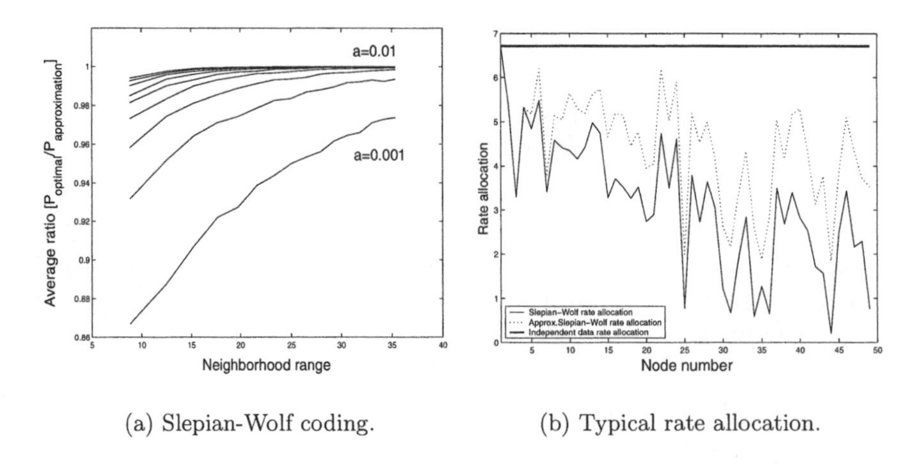

(a) Slepian-Wolf coding. (b) Typical rate allocation.

Fig. 8. (a) Average value of the ratio between the optimal and the approximated solution, in terms of total cost, vs. the neighborhood range. The network instances have 50 nodes uniformly distributed on a square area of size 100×100, and the correlation exponent varies from $a = 0.001$ (high correlation) to $a = 0.01$ (low correlation). The average has been computed over 20 instances for each (a, radius) pair. (b) Typical rate allocation for a network instance of 50 nodes, and correlation exponent $a = 0.0005$. On the x-axis, nodes are numbered in order as the distance from S increases, on the corresponding spanning tree

7 Conclusions and Future Work

We addressed in this paper the problem of joint rate allocation and transmission structure optimization for sensor networks, when the flow cost metric [rate] $\times$ [link weight] is considered. We showed that if the cost function is separable, then the tasks of optimal rate allocation and transmission structure optimization separates. We assess the difficulty of the problem, namely we showed that for an arbitrary transfer matrix the problem of finding the optimal transmission structure is NP-complete. The problem of optimal rate allocation can be posed as a linear programming (LP) problem, but it is difficult in general to find decentralized algorithms that use only local information for this task. We also studied some particular cases of interest where the problem becomes easier and a closed form solution can be found and where efficient approximation algorithms can be derived.

Our future research efforts include the derivation of efficient distributed approximation algorithms for both finding the optimal transmission structure and the optimal distribution of rates among the various subsets of sources for more general cases of transmission matrices. Moreover, an interesting research issue is to find tight bounds for the approximation ratios, in terms of power costs, for these distributed algorithms. Also, we consider more general network problems where for each node i, there is a source vector $\overrightarrow{X}_i = (X_{i1}, \ldots, X_{im})$ and any subvector of this vector has to be transmitted to some set of sinks.

References

1. [2002] Aaron, A., Girod, B.: Compression with side information using turbo codes, Data Compression Conference 2002.
2. [2002], Akyildiz, I.F., Su, W., Sankarasubramaniam, Y., Cayirci, E.: A survey on sensor networks, IEEE Communications Magazine, vol. 40, nr. 8, pp:102–116, 2002.
3. [1996] Arora, S.: Polynomial time approximation scheme for Euclidean TSP and other geometric problems, In Proc. 37th Ann. IEEE Symp. on Foundations of Comput. Sci., IEEE Computer Society, 2–11, 1996.
4. [1989] Bern, M., Plassmann, P.: The Steiner problem with edge lengths 1 and 2, Inform. Process. Lett. 32, 171–176, 1989.
5. [1998] Bertsekas, D.: Network optimization: continuous and discrete models, Athena Scientific, 1998.
6. [1991] Cover, T.M., Thomas, J.A: Elements of information theory, John Wiley and Sons, Inc., 1991.
7. [2003] Cristescu, R., Beferull-Lozano, B., Vetterli, M.: On network correlated data gathering, submitted to Infocom 2004.
8. [2003] Goel, A., Estrin, D.: Simultaneous optimization for concave costs: single sink aggregation or single source buy-at-bulk, ACM-SIAM Symposium on Discrete Algorithms, 2003.
9. [2001] Lindsey, S., Raghavendra, C.S., Sivalingam, K.: Data gathering in sensor networks using the energy*delay metric, Proc. of IPDPS Workshop on Issues in Wireless Networks and Mobile Computing, April 2001.
10. [1984] Luenberger, D.: Linear and nonlinear programming, Addison-Wesley, 1984.
11. [2003] Marco, D., Duarte-Melo, E., Liu, M., Neuhoff, D.L.: On the many-to-one transport capacity of a dense wireless sensor network and the compressibility of its data, IPSN 2003.
12. [2000] Pottie, G.J., Kaiser, W.J.: Wireless integrated sensor networks, Communications of the ACM, nr. 43, pp:51–58, 2000.
13. [2001] Pradhan, S.: Distributed Source Coding Using Syndromes (DISCUS), Ph.D. thesis, U.C. Berkeley, 2001.
14. [1999] Pradhan, S., Ramchandran, K.:Distributed Source Coding Using Syndromes (DISCUS): Design and construction, in Proc. IEEE DCC, March, 1999.
15. [2000] Rabaey, J., Ammer, M.J., da Silva, J.L., Patel, D., Roundy, S.: PicoRadio supports ad-hoc ultra-low power wireless networking, IEEE Computer, vol. 33, nr. 7, pp:42–48.
16. [2000] Rabiner-Heinzelman, W., Chandrakasan, A., Balakrishnan, H.: Energy-efficient communication protocol for wireless microsensor networks, in Proc. of the 33rd International Conference on System Sciences (HICSS '00), January, 2000.
17. [2000] Robins, G., Zelikovsky, A.: Improved steiner tree approximation in graphs, in Proc. 10th Ann. ACM-SIAM Symp. on Discrete Algorithms, 770–779, 2000.
18. [2002] Scaglione, A., Servetto, S.D.: On the interdependence of routing and data compression, in multi-hop sensor networks, in Proc. ACM MOBICOM, 2002.
19. [1973] Slepian, D., Wolf, J.K.: Noiseless coding of correlated information sources. IEEE Trans. Information Theory, IT-19, 1973, pp. 471–480.

WSDP: Efficient, Yet Reliable, Transmission of Real-Time Sensor Data over Wireless Networks

Arasch Honarbacht and Anton Kummert

University of Wuppertal,
Communication Theory,
Rainer-Gruenter-Strasse 21
42119 Wuppertal, Germany,
{honarb,kummert}@uni-wuppertal.de,
http://wetnt7.elektro.uni-wuppertal.de/welcome/index.html

Abstract. In wireless sensor networks, sensor data is usually distributed using a connection-less, multicast-capable transmission service. However, especially in the error-prone wireless environment, packet loss and varying transport delays are common phenomena. In this paper, we present a new way of improving reliability of sensor data distribution without compromising network efficiency. Even better, we show that, when using our technique, transmission interval lengths may be increased, resulting in a gain in network capacity and, as a matter of fact, reduction of overall power-consumption. The proposed solution is applicable to wireless networks in general, and sensor networks in particular.

1 Introduction

When having a look at wireless sensor networks (WSNs), usually a large amount of tiny control systems equipped with low-cost radio transceivers form a self-organizing ad hoc network. Sensor data of different types (temperature, pressure, light, humidity, battery level, acceleration, velocity, ...), originating in different locations, must be distributed among interested groups within the collective, as illustrated in figure 1. This information is required by distributed control algorithms, which often run periodically. Based on this sensor information, output to actuators is calculated. Imagine that actuator output is required every 10ms, for instance. Then this would mean that sensor input is also required every 10ms. However, it might not be feasible – and for sure not very efficient[1] – to transmit the sensor data with the required periodicity. In addition, due to the common switching and multiple access strategies employed in WSNs, transport delays show some variation. At times, even complete packet loss may occur, and retransmission is hardly an option in real-time distributed multipoint-to-multipoint systems.

An efficient way to distribute such information is to use the wireless medium's native broadcast feature. Sadly, broadcasts – and multicasts in general – are

[1] with respect to power-consumption, network traffic load, etc.

H. Karl, A. Willig, A. Wolisz (Eds.): EWSN 2004, LNCS 2920, pp. 60–76, 2004.

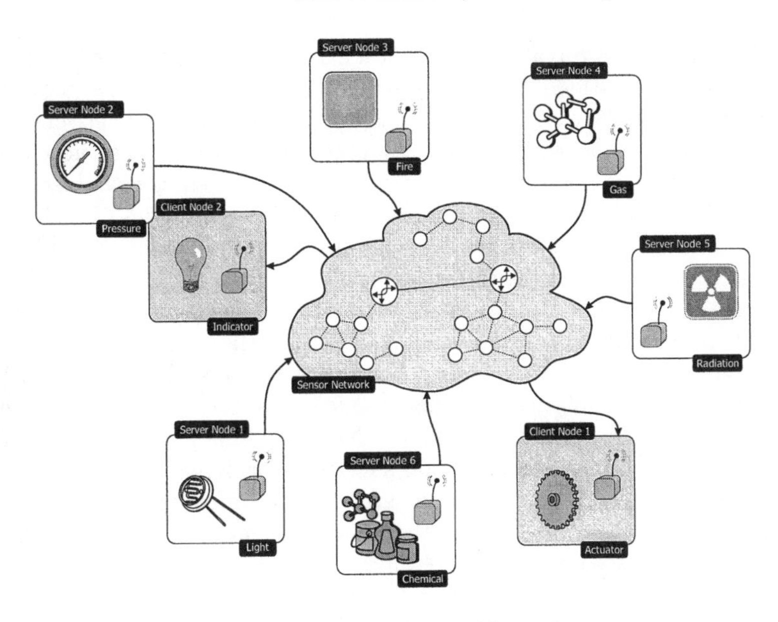

Fig. 1. Wireless Sensor Network

somewhat unreliable compared to unicasts. Let us consider one of the most popular protocols for distributed random access to a shared wireless channel, IEEE 802.11 DCF, as an example. Then, we can identify the following major reasons for the lack of reliability:

- **Acknowledgement**. Because of the well-known ACK-explosion problem, multicast frames are not protected by immediate positive ACKs, while directed frames are.
- **Fragmentation**. In order to reduce loss probability, large unicast frames are split into several smaller portions (fragments). When a fragment happens to be lost, only the short fragment needs retransmission. Multicast frames cannot be fragmented.
- **Virtual Carrier Sense**. *A priori* medium reservation under RTS/CTS regime is not available in multicast operation, resulting in two negative implications. First, packet loss probability is increased, and second, network throughput is degraded in presence of hidden terminals [1].

The source of all of these problems is the point-to-multipoint situation, which also affects network and transport layers. For obvious reasons, TCP is not suitable for multicast. While alternate transmission protocols are in the focus of current research, it is common practice to use UDP instead. Because of its simple nature, UDP is quite an efficient protocol, when packet loss and network congestion are no issues. But for the same reason it is also an unreliable protocol: There is no way for the transmitter to know if any, some, or all of the addressed stations have successfully received a message.

Although UDP on top of 802.11 is not a likely combination for wireless sensor networks, some of the identified problems are inherent to many unacknowledged, connection-less, multi-cast capable transport services and carrier-sense based multiple access protocols. The solution presented in this paper is generic in nature, and does not rely on UDP. What is really required, is a way to transmit data messages with sensor value samples from the source nodes to a sink. UDP has been selected as one possible example, because of its widespread use.

Our solution is based on the combination of KALMAN filter and TAYLOR approximation techniques, continuously estimating relevant sensor values at the sink. Therefore, a brief overview of KALMAN filter theory is given in section 2, and its application to wireless sensor networks is presented in section 3. Based on the theoretical foundation, practical implementations are outlined in section 4, including a generic variant suited for design and simulation stages, and an example of a highly optimized version specifically tailored to the requirements of tiny sensor nodes. A framework in terms of a protocol specification is given in section 5, and compatibility of the proposed solution with existing WSN approaches is discussed in section 6. Numerical results for a real-life sensor signal are presented in section 7, before the paper is finally concluded in section 8.

2 Kalman Filter

2.1 System Model

Considering the state space description,

$$\mathbf{x}(k+1) = \mathbf{A}\mathbf{x}(k) + \mathbf{u}(k) \tag{1}$$

$$\mathbf{y}(k) = \mathbf{C}\mathbf{x}(k) + \mathbf{w}(k) \tag{2}$$

where $\mathbf{u}$ denotes system noise and $\mathbf{w}$ denotes observation noise, respectively, the KALMAN filter [2, 3, 4, 5] is basically responsible for estimating the state vector $\mathbf{x}$ given only the observations $\mathbf{y}$. The filtering operation can be split into two distinct parts: *prediction* (or estimation) and *update*.

2.2 Prediction

During a prediction step, a new value for the state vector $\mathbf{x}$ at time k is estimated based on the value of $\hat{\mathbf{x}}$ at time $k-1$ (3), leading to a new estimate for $\mathbf{y}$ at time k (5). Furthermore, the error covariance matrix $\mathbf{P}$ is adjusted accordingly (4),

$$\hat{\mathbf{x}}(k|k-1) = \mathbf{A}\,\hat{\mathbf{x}}(k-1|k-1) \tag{3}$$

$$\mathbf{P}(k|k-1) = \mathbf{A}\,\mathbf{P}(k-1|k-1)\,\mathbf{A}' + \mathbf{Q} \tag{4}$$

$$\hat{\mathbf{y}}(k|k-1) = \mathbf{C}\hat{\mathbf{x}}(k|k-1) \tag{5}$$

where the superscript prime denotes transposition of a matrix. Matrix $\mathbf{P}$ includes information from two other covariance matrices, namely the system noise covariance matrix $\mathbf{Q}$ and the observation noise covariance matrix $\mathbf{R}$, and thus gives a measure of the current *estimation quality*. In general, these noise covariance matrices may be expressed as

$$\mathbf{Q} = \mathrm{E}\{\mathbf{U}(k)\,\mathbf{U}'(k)\} \tag{6}$$

$$\mathbf{R} = \mathrm{E}\{\mathbf{W}(k)\,\mathbf{W}'(k)\} \tag{7}$$

where $\mathbf{U}$ and $\mathbf{W}$ are stochastic processes describing the two kinds of additive noise mentioned above (1, 2).

2.3 Update

When a new update (or observation) $\mathbf{y}$ is available, this update is used to correct the estimated values of $\hat{\mathbf{x}}$ (9) and, as a result, of $\hat{\mathbf{y}}$. The KALMAN gain matrix $\mathbf{K}$ is adapted (8) and the error covariance matrix $\mathbf{P}$ is recalculated as well (10),

$$\mathbf{K} = \mathbf{P}(k|k-1)\,\mathbf{C}'(\mathbf{C}\mathbf{P}(k|k-1)\,\mathbf{C}' + \mathbf{R})^{-1} \tag{8}$$

$$\hat{\mathbf{x}}(k|k) = \hat{\mathbf{x}}(k|k-1) + \mathbf{K}(\mathbf{y}(k) - \mathbf{C}\hat{\mathbf{x}}(k|k-1)) \tag{9}$$

$$\mathbf{P}(k|k) = \mathbf{P}(k|k-1) - \mathbf{K}\,\mathbf{C}\,\mathbf{P}(k|k-1)\ . \tag{10}$$

2.4 Asynchronous Enhancements

The standard KALMAN filter known from motion tracking applications relies on synchronously and equidistantly available measurement updates. Classical applications, e.g. radar systems used in air traffic control, comply with these requirements. In many other cases, though, the time instants when, (a) input to the filter is available, and (b) output from the filter is desired, may diverge. Obviously, the classical KALMAN filter approach would fail under such conditions. In the past, enhancements to the standard KALMAN filter have been introduced [6, 7], making it accept sporadic, infrequent updates, while still producing periodic output. Since input and output sections are decoupled, the resulting structure is called: Asynchronous KALMAN filter.

In order to be capable of handling time varying transport delays, messages containing sensor values are marked with time stamps at the message generation site (the server site). Hence, update messages consist of a sensor value observation $\mathbf{y}(k_m)$ and its corresponding time stamp k_m. For now, we assume that all processes running on the various embedded controls share the same notion of time, meaning that they share a common time base. Later we will argue that this requirement can be relaxed at the price of an uncompensated delay.

At the client site, messages are received and queued based on their time stamps. At the next synchronous clock instant a message is de-queued and ready to be applied. At this point we can distinguish three cases when comparing the message's time stamp k_m with the client process's local time k_c:

1. The update becomes valid in the **future** ($k_m > k_c$). In this case the update is deferred until the local time reaches the message's time. Then this update belongs to the next case. Note that this case has only been introduced for theoretical comprehensiveness, though it may be useful for a few special applications.
2. The update is valid **right now** ($k_m = k_c$). Then it is applied immediately. This corresponds to the case, when network transmission delay is in the order of time stamp granularity.
3. The update had been valid in the **past** ($k_m < k_c$). Then the update is applied to a previously saved filter state and $M = k_c - k_m$ predictions are performed in a row.

This way, the filter is capable of handling past, present and future updates. Most of the time, the filter will be processing updates that belong to a past clock instant, caused by some transport delay between server and client. This transport delay can be compensated with a number of prediction steps in a row at the cost of an increasing estimation error ϵ.

$$\epsilon = |\mathbf{y}(k) - \hat{\mathbf{y}}(k|k-M)| \tag{11}$$

The overall filter structure is shown in figure 2.

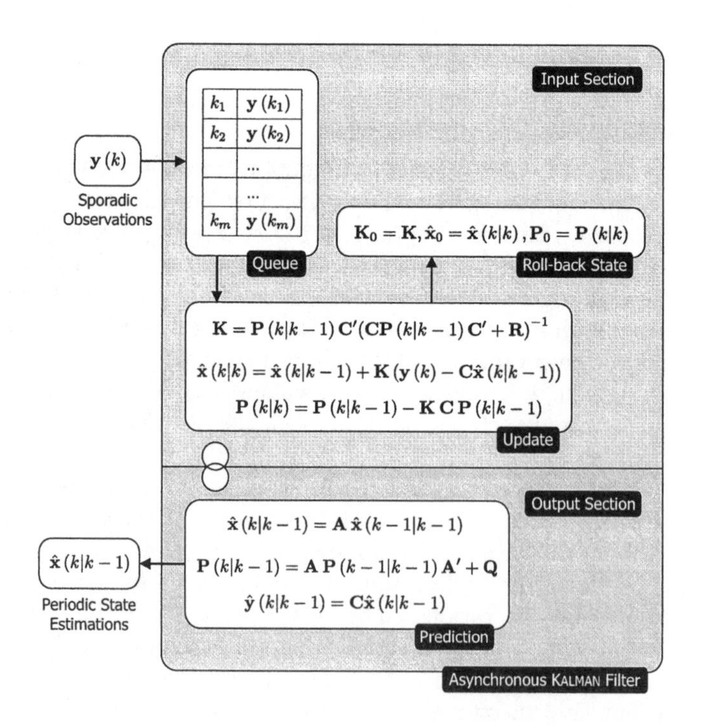

Fig. 2. Asynchronous KALMAN Filter (Structural View)

3 Application to Sensor Values and Wireless Networks

In collision-constrained wireless networks, sensor samples may arrive at random time instants. Nonetheless, the current sensor value may be desired more frequently, than samples are available. Therefore, predictions of the sensor values in question are performed periodically (synchronously) with the desired output rate.

3.1 Generic Deterministic Modelling

When using a KALMAN filter, two deterministic matrices, the system transition matrix **A** and the observation matrix **C**, need to be set up properly. In this section, we provide a generic solution based on a TAYLOR approximation of the desired sensor signal. Thanks to the TAYLOR approximation's generic nature, the solution is applicable to a vast number of entities, for instance, temperatures, pressures, etc. Let $p(t)$ denote the sensor signal in question, then we can approximate its value at time $t + \Delta t$ by

$$p(t+\Delta t) = p(t) + \Delta t \frac{\mathrm{d}p(t)}{\mathrm{d}t} + \frac{(\Delta t)^2}{2}\frac{\mathrm{d}^2 p(t)}{\mathrm{d}t^2} + \cdots . \tag{12}$$

Now, assume a discrete-time signal with sampling period T. This leads to

$$p(kT+T) = \sum_{n=0}^{N-1} \frac{T^n}{n!} \cdot p^{(n)}(kT) . \tag{13}$$

Matrix **A** expresses the transition from one clock instant to the following; therefore equation (13) yields the structure and contents of **A**. For practical reasons, the TAYLOR approximation is aborted at some point, for instance after the 2nd derivative term. In this case we would identify this setup as a 2nd order model, i.e. $n = 2$. Dimensions of state vector $\mathbf{x}$ and system transition matrix **A** depend on n. While the state vector is given by

$$\mathbf{x} = \left(p, p^{(1)}, p^{(2)}, \ldots, p^{(n)}\right)' , \tag{14}$$

the system transition matrix would look like

$$\mathbf{A} = \begin{pmatrix} 1 & T & \frac{T^2}{2} & \cdots & \frac{T^n}{n!} \\ 0 & 1 & T & \cdots & \frac{T^{n-1}}{(n-1)!} \\ \vdots & & & & \\ 0 & 0 & 0 & \cdots & 1 \end{pmatrix} . \tag{15}$$

Finally, the observation matrix **C** is responsible for gathering the filter's output signal $\mathbf{y}$ by collecting the relevant information from the state vector $\mathbf{x}$. For instance, if we were only interested in the sensor value itself, then $\mathbf{C}$ could be expressed as

$$\mathbf{C} = (1, 0, 0, \ldots, 0) , \tag{16}$$

where its dimension equals $1 \times (n+1)$. However, if we were also interested in the first derivative, for instance, such information would be very easy to determine, since it is already given in the state vector. The information is then available to adaptive control algorithms, which can modify some of their internal coefficients accordingly. By the way, this model easily extends to multidimensional sensor input, which may be beneficial when correlational dependencies exist across dimensions. Interested readers should refer to [6] for a 3D example.

3.2 Generic Stochastic Modelling

In addition to the deterministic matrices $\mathbf{A}$ and $\mathbf{C}$ mentioned in the previous section, two other, stochastic, matrices have to be designed: The system noise covariance matrix $\mathbf{Q}$ and the observation noise covariance matrix $\mathbf{R}$, respectively. $\mathbf{Q}$ is of the same dimension as $\mathbf{A}$, while, for one-dimensional sensor data, $\mathbf{R}$ is scalar. In order to have these matrices set up correctly, one has to understand the source of the noise they model.

System noise is caused by the fact that, for practical applications, the TAYLOR approximation (13) is aborted at some point, e.g. for $n = 2$, derivatives $p^{(3)}$, $p^{(4)}$ and so forth are simply discarded (approximation error). The model implies that the highest order derivative included (in the previous example $p^{(2)}$) remains constant for the duration T. Observation noise may be caused by quantization errors, systematic errors associated with the method of measurement, etc.

Again, let us consider the 2nd order case. System noise is introduced by the fact that the highest order derivative, i.e. $p^{(2)}$ is faulty when the sensor signal's 2nd order derivative (acceleration) changes between two clock instants. We take this into account in the following way. First, we determine the covariance σ^2 for some typical sensor characteristic. Second, we set the corresponding matrix element Q_{nn} to the value of σ^2. Note that the covariance is set only in the highest order term, since it will automatically propagate by means of the system transition matrix $\mathbf{A}$ to all lower order terms.

$$\mathbf{Q} = \begin{pmatrix} 0\,0\,0 \cdots & 0 \\ 0\,0\,0 \cdots & 0 \\ \vdots & \\ 0\,0\,0 \cdots & \sigma^2 \end{pmatrix} \tag{17}$$

Observation noise is modelled by its covariance τ^2:

$$\mathbf{R} = \left(\tau^2\right) \tag{18}$$

The last matrix to mention in this context is the error covariance matrix $\mathbf{P}$, which has the same dimension as $\mathbf{Q}$ and $\mathbf{A}$. Since this matrix is adapted automatically at every prediction and update instant, only its initial value $\mathbf{P}(0)$ must be set. We know the sensor's initial value, so we will not make any error there. But because we don't know a second value yet, all derivatives will most probably be faulty in the first place (except for the trivial case, when the value does not change at all), with the highest order derivative being the most faulty. Summarizing these thoughts, we get

$$\mathbf{P} = \begin{pmatrix} 0 & 0 & 0 \cdots & 0 \\ 0 & a_1 & 0 \cdots & 0 \\ \vdots & & & \\ 0 & 0 & 0 \cdots & a_n \end{pmatrix}, \tag{19}$$

where the a_i are some constants with $a_{i+1} \geq a_i$.

4 Practical Filter Implementations

In this section, we provide two practical implementations for the KALMAN filter. One has been developed with emphasis on general applicability; the other is specifically tailored to tiny devices with scarce processing and memory resources.

4.1 Generic Implementation

Efficient, optimized solutions have one drawback: They always rely on certain assumptions and constraints on matrix dimensions, symmetry, etc. Therefore, a generic implementation, which is more flexible but less efficient, has been implemented. It is perfectly suited for simulations, and has been implemented as a set of C++ classes. First, a class `CMatrix` is defined, which encapsulates generic matrix operations, such as addition, multiplication, inversion, transposition, etc. on arbitrarily sized matrices. Then, the class `CKalmanFilter` defines the filtering process in terms of concatenations of these operations, as outlined in listing 1.1. This yields maximum flexibility (ability to use different TAYLOR approximation orders, arbitrarily structured matrices $\mathbf{A}, \mathbf{C}, \mathbf{Q}, \mathbf{R}$, etc.) and readability of the code at the expense of increased memory and processing requirements.

4.2 Efficient Implementation

Different levels of optimization are possible reducing the required computational resources dramatically. These optimizations can be achieved by assuming fixed matrix structures, which are often symmetric and sparse. Furthermore, when performing calculations cleverly, a number of computational results may be reused, while matrix calculation progresses. Using such a progressive technique allows processing of numerous sensor values even with tiny devices offering not much computational capacity.

The results discussed in section 7 are based on a simple, one-dimensional, 1st order model. For this specific case, we provide efficient algorithms to perform

Listing 1.1. Generic C++ Implementation of the KALMAN Filter

```
CMatrix CKalmanFilter::Predict()
{
    m_vectorX = m_matrixA * m_vectorX;
    m_matrixP = m_matrixA * m_matrixP * m_matrixA.Transpose() + m_matrixQ;

    return m_matrixC * m_vectorX;
}

void CKalmanFilter::Update(const CMatrix &vectorY)
{
    m_matrixK = m_matrixP * m_matrixC.Transpose() *
        (m_matrixC * m_matrixP * m_matrixC.Transpose() + m_matrixR).Invert();

    m_vectorX = m_vectorX + m_matrixK * (vectorY - m_matrixC * m_vectorX);
    m_matrixP = m_matrixP - m_matrixK * m_matrixC * m_matrixP;
}
```

the computations associated with KALMAN filter equations. Based on these algorithms, we evaluate computational and memory requirements in practical applications.

Let us consider the equations associated with the prediction step, i.e. equations (3, 4, 5). As already mentioned before, this step is performed quite often, or more precisely, with a frequency equivalent to the desired output rate. When we assume the following,

$$\begin{aligned} &\mathbf{x} = \begin{pmatrix} p \\ \dot{p} \end{pmatrix}, \mathbf{A} = \begin{pmatrix} 1 & T \\ 0 & 1 \end{pmatrix}, \mathbf{C} = \begin{pmatrix} 1 & 0 \end{pmatrix}, \mathbf{P}(0) = \begin{pmatrix} a & b \\ b & c \end{pmatrix}, \\ &\mathbf{Q} = \begin{pmatrix} 0 & 0 \\ 0 & \sigma^2 \end{pmatrix}, \mathbf{R} = \begin{pmatrix} \tau^2 \end{pmatrix} \end{aligned} \tag{20}$$

then equation (4) manifests as:

$$P_{11}(k+1) = P_{11}(k) + TP_{12}(k) + TP_{21}(k) + T^2 P_{22}(k) \tag{21a}$$

$$P_{12}(k+1) = P_{11}(k) + TP_{22}(k) \tag{21b}$$

$$P_{21}(k+1) = P_{21}(k) + TP_{22}(k) \tag{21c}$$

$$P_{22}(k+1) = P_{22}(k) + \sigma^2 \tag{21d}$$

Because of $P_{12}(0) = b = P_{21}(0)$ and (21b, 21c), it is evident that

$$P_{12}(k) = P_{21}(k), \forall k \geq 0. \tag{22}$$

By the way, the same is true for the update step, eliminating the need to calculate and store P_{21} at all. Using a clever interleaving of operations, only three multiply-accumulates (MACs) and one accumulation is required to perform all the calculations stated in equations 21. The next step is to update the state vector according to (3), which in our specific case simply leads to

$$x_1(k+1) = x_1(k) + Tx_2(k). \tag{23}$$

Finally, the output signal needs to be calculated according to (5). However, in our case, the output signal is identical to x_1.

In listing 2, an efficient C++ implementation of prediction and update steps is given, based on the assumptions made in (20). Notice that a plain C or Assembler implementation is readily derived from the C++ representation for small devices where a C++ compiler is not available. The techniques used to obtain the update step's implementation are comparable to those used for the prediction step, hence they are not going to be discussed here in detail.

Listing 2. Efficient C++ Implementation of the 1st order Filter

```
float COptimizedKalmanFilter::Predict()
{
    // P = A * P * A' + Q;
    m_fltP11 += m_fltT * m_fltP12;
    m_fltP12 += m_fltT * m_fltP22;
    m_fltP11 += m_fltT * m_fltP12;
    m_fltP22 += m_fltQ22;

    // x = A * x and
    // y = C * x
    return m_fltX1 += m_fltT * m_fltX2;
}

void COptimizedKalmanFilter::Update(const float fltY)
{
    // Precalculate some intermediate results
    register const float fltM11 = m_fltP11 + m_fltR;
    register const float fltDelta = fltY - m_fltX1;

    // K = P * C' * (C * P * C' + R)^-1
    m_fltK1 = m_fltP11 / fltM11;
    m_fltK2 = m_fltP12 / fltM11;

    // x = x + K * (y - C * x)
    m_fltX1 += m_fltK1 * fltDelta;
    m_fltX2 += m_fltK2 * fltDelta;

    // P = P - K * C * P
    m_fltP22 -= m_fltK2 * m_fltP12;
    m_fltP12 -= m_fltK2 * m_fltP11;
    m_fltP11 *= 1 - m_fltK1;
}
```

4.3 Considerations on Computational Costs

It is worth mentioning that the KALMAN filter is not very costly with respect to computational load when implemented efficiently. Having a look at listing 2, the prediction takes five additions, and four multiplications. The update is somewhat costlier: Three additions, four subtractions, five multiplications, and two divisions are required. Note that computations associated with the update are only performed in response to an incoming update sample – an event that occurs sporadically compared to the periodic predictions. Memory requirements are as follows: Seven 32-bit floating point variables are required, and three floating point constants, for a total of 40 bytes. Memory requirements may be further reduced by hard-coding of the constants, resulting in a memory requirement of 28 bytes per filter.

Asynchronous extensions add a slight overhead, which can be kept very small by limiting the queue size to a single sample, i.e. eliminating the queue. A single sample queue does not degrade performance, if predictions are performed more often than updates, samples arrive with a delay, and update messages are strictly ordered. These conditions are easily met in WSNs.

Moreover, the filtering operation is valuable from a signal processing point of view. In particular, adaptation of **R** and **Q** provides a way of sensor noise reduction. In other words, the computations performed during KALMAN filtering allow to trade-off traffic load reduction against reliability – while at the same time signal quality is improved.

5 Wireless Sensor Data Protocol

Based on the theoretical foundation mentioned in the previous sections, we propose a new protocol for transmission of wireless sensor data: WSDP. Located between application and UDP (or any other datagram service), this middleware component accepts sensor data from applications and utilizes a UDP port for delivery of WSDP protocol data units (WSDP-PDUs), as depicted in figure 3. Again, UDP is just an example, almost any datagram service is suitable on top of almost any kind of network – WSDP is an application layer protocol.

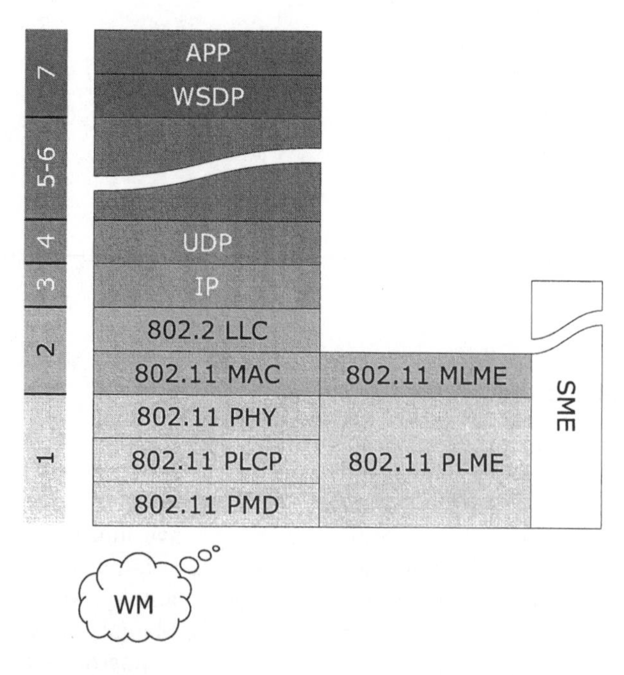

Fig. 3. Example Protocol Stack for WSDP – Also Used in Simulations

The proposed architecture allows a single sensor data server to distribute its information to many clients. And vice versa, a single client may subscribe to a number of data sources. Also, any station may be client and server at the same time, or decide to be either one.

5.1 Sending Sensor Data

Server applications willing to transmit sensor data issue a `WSDP-DATA.request`. This service primitive accepts a time stamp k_m, a sensor value $\mathbf{y}(k_m)$, a unique sensor identifier (SID), and addressing information (destination IP address). The IP address may be a unicast, multicast or broadcast address, while the SID unambiguously identifies the sensor. The request service primitive will construct a datagram and instruct UDP to deliver the message.

5.2 Subscribing to Sensor Data

Clients subscribe to a sensor by invoking a `WSDP-SUBSCRIBE.request`. In addition to a SID, values for the desired approximation order and output period T are specified, too. From this time on, the client will be notified about the sensor's value through periodic `WSDP-DATA.indications`, until the subscription is cancelled by a `WSDP-UNSUBSCRIBE.request`.

Transparent to applications, the client runs an asynchronous KALMAN filter tracking the value of each sensor it subscribes to. Whenever a WSDP-PDU carrying sensor data is received, this data is passed to the filter as an update $\mathbf{y}(k_m)$ – provided the update message has not been corrupted due to transmission errors. Transmission errors are easily identified using UDP's CRC checksum. Analogously, when the next `WSDP-DATA.indication` is scheduled, a KALMAN prediction step is performed and the resulting output $\hat{\mathbf{y}}(k_c|k_c-1)$ is passed to the client. Clients have access to all filter matrices, so they are free to use different tracking models (matrix $\mathbf{A}$), different noise parameters (matrices $\mathbf{R}$, $\mathbf{Q}$), etc.

5.3 Retrieving Sensor Data

Clients will be informed by means of a `WSDP-DATA.indication` about the current state of a sensor value. Previously, they must have subscribed to one or more SIDs. As part of the indication, SID, time stamp k_c, and sensor value estimation $\hat{\mathbf{y}}(k_c|k_c-1)$ are delivered to the client application. In essence, this value is the KALMAN filter's output.

5.4 Data Reduction at the Source

Depending on application and available processing power in each node, an additional server side KALMAN filter may be an option. One can think of many ways of limiting the data rate at the source. One way would be to mimic the same KALMAN filter settings, which are also used at the clients. This yields a predictor/corrector setup, in which a threshold ϵ_c on the tolerable estimation error ϵ may be defined. Then, updates would be sent only if $\epsilon \geq \epsilon_c$, for instance. In other words: When sensor values change quickly, then updates are sent more frequently, while only few updates are required otherwise.

Another straight-forward way to implement data reduction at the source would be a uniform sub-sampling of sensor data streams. Obviously this solution lacks the adaptivity of the previously described solution. However, it is easy to implement and offers reduced computational cost in comparison to a second KALMAN filter. In addition, uniform sub-sampling helps to increase robustness against erroneous bursts.

The decision whether to use an adaptive scheme or not, should be based on the kind of sensor network used: When observing uncorrelated phenomena, an adaptive scheme may improve bandwidth utilization (statistical multiplexing). On the contrary, when a lot of co-located sensor nodes observe the same phenomenon, an adaptive scheme may increase loss probability due to frequent update packet collisions.

6 Compatibility Issues

6.1 Synchronization

It was previously stated that nodes are supposed to share a common time base, which requires some kind of synchronization mechanism. Though, time synchronization in large-scale multi-hop ad hoc networks seems to be a rather complicated task. Fortunately, the asynchronous KALMAN filter can be operated without synchronization at all. This comes at the expense of an uncompensated transmission delay. If a common time base is available, the KALMAN filter will automatically compensate for varying transmission delays using the time stamp information provided in the update messages. In contrast, if source and sink nodes are not synchronized, the transmission delay caused by lower layers cannot be compensated anyway.

In this case, the receiver should use the time stamps k_m only for rejection of duplicate frames and detection of mis-sequenced packets. Then, it simply applies the updates *as soon as they arrive*, effectively bypassing the queue. The filter will output a delayed version of the original sensor data, which is even smoother, compared to the synchronized case with delay compensation [6]. Furthermore, missing synchronization may lead to signal distortions, when jitter (delay uncertainty) is not negligible. Anyhow, these unavoidable distortions are not related to the filtering process, since they would also occur without any filter at all.

6.2 Data-Centric Approaches

Although WSDP has been expressed in terms of node-centric UDP, from the discussions in sections 2, 3 and 4 it should be clear that the asynchronous KALMAN filter may be combined equally well with data-centric approaches [8]. From the application layer point of view, such approaches primarily affect addressing issues. For instance, in lieu of IP addresses and SIDs, attribute names and sensor locations may be used. To take it further: Asynchronous KALMAN filter methodology is easily inserted into almost any data flow.

7 Results

7.1 Simulation Environment

Simulations have been carried out using a proprietary simulator for wireless networks [9]. Designed especially for wireless ad hoc networks, this simulator perfectly matches the requirements inherent to the concepts outlined in this paper. It is entirely written in C++ using the Standard Template Library – no other libraries or third party tools have been used. Besides a powerful and flexible wireless medium model, it also provides sophisticated implementations of the IEEE 802.11 MAC [10], as well as a mating PHY.

7.2 Simulations

We have implemented WSDP in the simulator and run several simulations in order to evaluate different setups. Data used in the simulations is not artificial. In fact, it comes from a magnetic tracking sensor at a sample rate of approximately 180Hz. The entire set contains 2873 samples, corresponding to nearly 16 seconds of recorded sensor data. Besides the particular signal presented here, many other signals have been studied, too – always yielding similar performance.

Three ways of data reduction at the source have been applied in order to reduce the data rate efficiently. Figure 4 shows some representative results. Here, the grey line corresponds to the 180Hz sensor data stream (all samples), while the circles mark those samples that have actually been transmitted by WSDP (updates). The black line represents the results at the receiver, i.e. the value delivered by the periodic `WSDP-DATA.indications`. In all cases, a simple 1st order TAYLOR approximation has been used, $\sigma^2 = 20$, $\tau^2 = 0.05$, and $T = 10$ms. It is remarkable, how quickly start-up transients disappear. This fact gives an idea of the filter's adaptive performance.

In 4(a) data has simply been sub-sampled equidistantly, keeping only 144 of the original 2873 samples. This corresponds to 5% of the original data. For this specific application, obviously 95% of the sensor data can be easily dropped – there is hardly any deviation from the original data. In 4(b), this approach has been carried to extremes, keeping only 2% of the original data stream. This low data rate is not meant to be actually useful; instead differences between the three rate reduction methods become evident. In 4(d) the data has not been sub-sampled uniformly, but updates have been selected randomly, again keeping only 2% of the data. In 4(c), the server-side KALMAN filter approach has been used to adaptively perform update selection. Again, only 2% (56 samples) of the

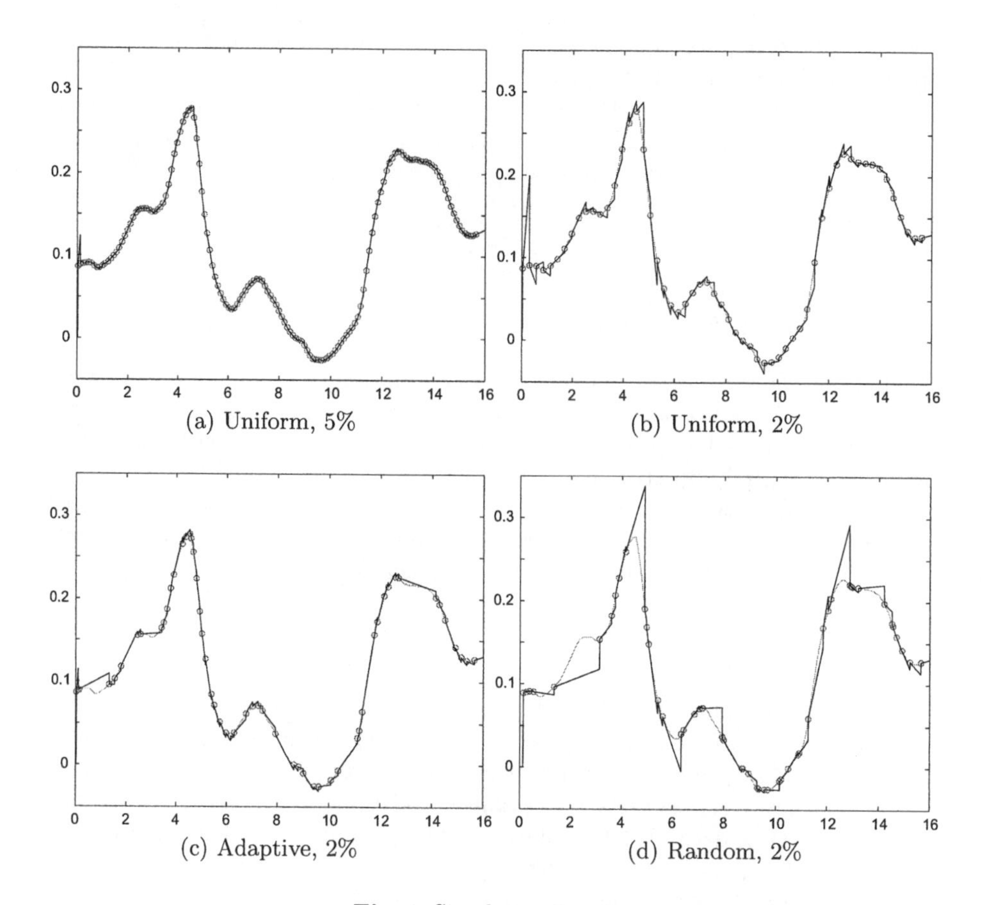

(a) Uniform, 5% (b) Uniform, 2%

(c) Adaptive, 2% (d) Random, 2%

Fig. 4. Simulation Results

available data are used as updates. But here, these samples have been selected adaptively, as outlined in section 5.4. However, the server-side filter runs at a much higher frequency, 1000Hz compared to the 100Hz of the client filter.

The adaptive scheme is best used when data transmission is reliable, since it identifies those updates that are most crucial for reconstruction. However, the threshold provides a means to exchange reliability against traffic reduction. Furthermore, the adaptive scheme could be extended in such a way that it transmits update packets with a minimum periodicity regardless of signal dynamics.

Regarding reliability, we can observe that even complete loss of update messages does hardly affect the quality of sensor value estimations – provided that a sufficient redundancy has been added. For instance, when we choose to reduce the data rate down to 10%, this would still allow us to loose an average 50% of the update messages, while estimation quality would still remain good. The effects of bursty transmission errors may be studied analyzing the results of random reduction presented in figure 4(d). If those updates, which were lost due to a corrupted burst, carry information crucial to successful reconstruction, prediction errors may become non-negligible. Refer to the situation around the turning point at $t = 4\,\text{sec}$. In general, if sensor signal dynamics do not change dramatically during erroneous bursts, this will do only marginal harm to signal reconstruction. Furthermore, when receiving a valid update, the filter adapts quite fast, usually within a single cycle.

8 Conclusion

Having identified some major problems associated with the transmission of sensor data over wireless networks, a new application layer protocol mitigating these deficiencies has been presented. It features ready-to-use, reliable and efficient data distribution across meshy ad hoc networks, which are exposed to hostile wireless conditions. Since WSDP works on top of almost any transport layer protocol, a huge number of potential applications may be thought of. The proposed framework may be combined with additional measures in lower layers, which aim at more reliable multicasts. In addition, the aforementioned benefits come at relatively low cost. Bearing in mind that computational costs are generally negligible compared to transmission costs in wireless sensor networks, adaptive filtering at transmitters and reveivers takes full advantage of this notion. Furthermore, the underlying mathematics, i.e. a KALMAN filter combined with a low order TAYLOR approximation, behaves very well with respect to numerical stability and robustness. The proposed solution is real-time capable, since the filter adds almost no latency.

References

1. Kleinrock, L., Tobagi, F.A.: Packet switching in radio channels: Part II – The hidden terminal problem in carrier sense multiple-access and the busy-tone solution. IEEE Transactions on Communications **23** (1975) 1417–1433
2. Kalman, R.E.: A new approach to linear filtering and prediction problems. Transactions of the ASME: Journal of Basic Engineering **82(1) D** (1960) 35–45
3. Petersen, I.R., Savkin, A.V.: Robust Kalman Filtering for Signals and Systems with Large Uncertainties. Birkhäuser (1999)
4. Bozic, S.M.: Digital and Kalman Filtering. 2nd edn. Arnorld (1994)
5. Chui, C.K., Chen, G.: Kalman Filtering with Real-Time Applications. 2nd edn. Springer (1991)
6. Honarbacht, A., Boschen, F., Kummert, A., Härle, N.: Synchronization of distributed simulations - a Kalman filter approach. In: Proceedings of the IEEE International Symposium on Circuits and Systems (ISCAS). Volume IV., Phoenix, Arizona, USA (2002) 469–472
7. Kummert, A., Honarbacht, A.: State estimation of motion models based on asynchronous and non-equidistant measurement updates. In: Proceedings of the 8th IEEE International Conference on Methods and Models in Automation and Robotics (MMAR), Szczecin, Poland (2002) 55–60

8. Akyildiz, I.F., Su, W., Sankarasubramaniam, Y., Cayirci, E.: Wireless sensor networks: A survey. The International Journal of Computer and Telecommunications Networking **38** (2002) 393–422
9. Honarbacht, A., Kummert, A.: WWANS: The wireless wide area network simulator. In: Proceedings of the International Conference on Wireless and Optical Communications (WOC), Banff, Alberta, Canada (2002) 657–662
10. Honarbacht, A., Kummert, A.: A new simulation model for the 802.11 DCF. In: Proceedings of the International Workshop on Multimedia Communications and Services (MCS), Kielce, Poland (2003) 71–76

Context-Aware Sensors

Eiman Elnahrawy and Badri Nath

Department of Computer Science, Rutgers University
Piscataway, NJ 08854, USA
eiman@paul.rutgers.edu,
badri@cs.rutgers.edu

Abstract. Wireless sensor networks typically consist of a large number of sensor nodes embedded in a physical space. Such sensors are low-power devices that are primarily used for monitoring several physical phenomena, potentially in remote harsh environments. Spatial and temporal dependencies between the readings at these nodes highly exist in such scenarios. Statistical contextual information encodes these spatio-temporal dependencies. It enables the sensors to locally predict their current readings based on their own past readings and the current readings of their neighbors. In this paper, we introduce context-aware sensors. Specifically, we propose a technique for modeling and learning statistical contextual information in sensor networks. Our approach is based on Bayesian classifiers; we map the problem of learning and utilizing contextual information to the problem of learning the parameters of a Bayes classifier, and then making inferences, respectively. We propose a scalable and energy-efficient procedure for online learning of these parameters in-network, in a distributed fashion. We discuss applications of our approach in discovering outliers and detection of faulty sensors, approximation of missing values, and in-network sampling. We experimentally analyze our approach in two applications, tracking and monitoring.

1 Introduction

The availability of various sensors that can be networked using the wireless technology allows for large-scale sensing of physical spaces. These sensing applications are in various domains. For example, environment and habitat monitoring [1], tracking [2], travel and transportation, inventory management and supply chain [3], monitoring of building structure [4,5], creating a highly efficient temperature-controlled environment by monitoring the readings of temperature/humidity/pollen sensors in buildings, etc. With the deployment of such sensor networks it is not only possible to obtain a fine grain real-time information about the physical world but also to act upon that information.

The accuracy and the timeliness of this sensed information are extremely important in all these applications; since detrimental actions are usually taken based upon these sensed values. In general, the quality and the reliability of this data are important issues that have received attention recently [6,3]. In this paper, we consider several problems that affect the accuracy of sensor data: missing sensor values, and outliers/anomalies or malicious sensors. Our approach is based on exploiting the spatio-temporal relationships that exist among sensors in wireless sensor networks that are used in monitoring temperature, humidity, etc., or in tracking [2]. Such networks are dense with redundant and

H. Karl, A. Willig, A. Wolisz (Eds.): EWSN 2004, LNCS 2920, pp. 77–93, 2004.

correlated readings for coverage and connectivity purposes, robustness against occlusion, and for tolerating network failures [1,4,2]. In particular, we propose an energy-efficient technique for learning these relationships and then using them for prediction. We call such relationships the Contextual Information of the network (CI). Specifically, CI defines the spatial dependencies between spatially adjacent nodes as well as the temporal dependencies between history readings of the same node. It therefore enables the sensors to locally predict their current readings knowing both their own past readings and the current readings of their neighbors.

The loss and the transience in wireless sensor networks, at least in their current form, as well as the current technology and the quality of wireless sensors contribute to the existence of missing information/outliers in sensor data and imply that these networks must operate with imperfect or incomplete information [7,8]. Furthermore, energy-efficient modes imply that some sensors may be in a doze mode and will not report readings all the time. In order to solve the problem of missing readings we can either use sensor fusion or predict these readings as we propose. However, fusion is not usually practical; knowledge of a particular missing reading may be critical, readings from entire clusters of sensors may be missing due to communication failures, etc. Even if one argues that sensor networks will become more robust against missing readings over the time due to advances in the wireless communication or in the technology, learning of CI will still be very advantageous for detecting anomalies and discovering malicious sensors. Nevertheless, CI provides a tool for reducing the overall energy spent in the network by sampling, especially that energy is a scarce commodity of sensor networks [4, 5,7]. We shall expand on these applications in Section 2.

To the best of our knowledge, there is no work so far neither on online learning of spatio-temporal correlations in wireless sensor networks nor on utilizing them in the missing values/outliers/sampling problems. This is probably due to several energy and communication limitations of these networks that make this problem a challenge. Therefore, the main focus of this paper is to introduce an approach for "modeling" and online in-network "learning" of CI in sensor networks statistically. Our approach is based on Bayesian classifiers; we map the problem of learning and utilizing statistical CI to the problem of learning the parameters of a Bayes classifier, and then making inferences, respectively. Our approach is both scalable and energy-efficient; we utilize the concept of in-network aggregation [5,7] in our learning procedure, which has been shown to be energy-efficient theoretically and experimentally [9,10,5].

In general, we aim at systems of context-aware sensors where sensors can infer their own readings from knowledge of readings in the neighborhood and their own history. In other words, sensors are aware of their context (the neighborhood and history). We argue that such sensors are a step toward autonomous reliable networks where nodes not only sense the environment and report their readings, but also perform more sophisticated tasks such as detection of possible outliers, and so forth. Our contributions are threefold: First, we introduce a novel Bayes-based approach for modeling and learning of CI in sensor networks that is scalable and energy-efficient. Second, we apply our approach to outlier detection, approximation of missing values, and in-network sampling. Finally, we analyze the applicability and the efficiency of our proposed approach experimentally.

We are aware that other factors and dynamics such as noise and environmental effects (weather, sensitivity, etc.) or calibration errors (e.g., sensor ageing) may have an impact on our learning process. In this paper, however, we assume that these effects have been suppressed using, e.g., [6,3]. In fact, based on our experiments, our approach is readily tolerant to random noise and environmental effects since it is based on learning over the time and works in a distributed fashion. In the future, we plan to integrate the research on calibration and noise with our approach for further improvements.

The rest of this paper is organized as follows. We introduce applications of our approach in Section 2. In Section 3, we present our approach for modeling CI statistically. We present our proposed online learning procedure in Section 4. We then revisit the applications of our work in Section 5, where we show how they are achieved. A performance analysis is given in Section 6. Section 7 discusses related work. Section 8 concludes this paper and highlights our future work directions.

2 Applications

In this section, we briefly discuss the causes of missing values and outliers in sensor networks and show how they can be solved using CI. We also introduce an important application, called in-network sampling.

2.1 Outlier Detection

Similar to traditional data sets, collected for the purpose of machine learning or data mining, false outliers exist in data sampled from sensor networks. This problem is even more severe in sensor networks due to various factors that affect their readings such as the environmental effects, the technology and the quality of the sensors (motes), nonsensical readings arising when sensors are about to run out of power (malicious sensors), etc. False outliers are critical in applications that involve decision making or actuator triggers since they may fire false alarms. Detecting real anomalies (true outliers) in sensor data, which are contrary to the hypothesis, is also crucial. Specifically, it draws attention to any abnormality in the environment being monitored.

Detection and reasoning about outliers in traditional databases involve lengthy offline techniques such as visualization [11]. These techniques however are not suitable for sensors due to the real time nature of their applications. CI provides a tool for real-time detection of outliers. Consider the scenario shown in Figure 1. Frame (a) shows random observations that are not expected in dense sensor networks. On the other hand, frames (b), (c) show two consecutive data samples obtained from a temperature-monitoring network. The reading of sensor i in frame (c) looks suspicious given the readings of its neighbors and its last reading. Intuitively, it is very unusual that its reading will jump from 58 to 40, from one sample to another. This suspicion is further strengthened via knowledge of the readings in its neighborhood. In order for sensor i to decide whether its reading is an outlier, it has to know its most likely reading in this scenario. Therefore, based on the CI that defines the spatio-temporal correlation at i, the last reading of i, and the readings of sensors j, n, sensor i can predict that its reading should be somewhere between 55, 65. It then compares this prediction with its sensed value. If the two differ

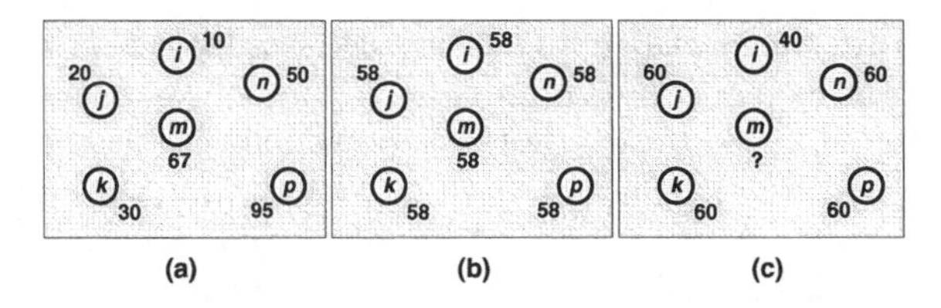

Fig. 1. Frame (a) shows random observations that are not expected in dense sensor networks. Frames (b), (c) show two consecutive data samples obtained from a dense sensor network.

significantly, and if the sensed value is very unlikely, then i will decide that its reading is indeed an outlier. It may then react accordingly, e.g., it may choose to report the predicted value and then sends a separate "alert" message to its base-station.

Distinguishing anomalies from malicious sensors is somewhat tricky. One approach is to examine the neighborhood of the sensor at the base-station. In particular, if many correlated sensors in the same neighborhood reported alert messages then this is most likely a serious event. We are strictly considering one-dimensional outliers in this paper. Multidimensional outliers are part of our future work.

2.2 Missing Values

Missing values are missing readings from sensor data sampled at a specific time instance. They highly exist in sensor networks and generally arise due to node failures and packet losses [7,8]. Heavy packet loss and asymmetric links are prevalent in sensor networks due to signal strength fading, sporadic or constant environmental interference (e.g., wind or rain), hardware crash, packet collision, etc [7,1]. Zhao *et al.* showed experimentally that more than 10% of the network links suffer loss rate $> 50\%$, and that the loss of most of the links fluctuates over the time with variance $9\% - 17\%$ [7]. Node failure is also quite common. It can be isolated or clustered [8].

Missing values impact the result of any query over sensor data. The resultant errors can be very significant as in in-network processing and aggregations [9,5,7]. Several solutions have been suggested to tolerate this error such as link quality profiling [7], and child caching [5]. However, the problem still persists, though less severely.

CI can be used to predict any missing reading and hence recover missing values. Our intention is not to replace the previously mentioned approaches, but rather to use ours in conjunction for additional tolerance. The specific recovery steps vary from application to another. If the network is sampled for a detailed data set then the recovery can be performed at the base-station, assuming that the geographic location of each sensor is known. E.g., consider Figure 1, the missing reading of sensor m in frame (c) is predicted using its CI, knowing the readings of i, j, k, n, p, and m's last reading. Moreover, if all k, j, m are missing, the recovery proceeds progressively from m, where the neighborhood is defined, to j then k, and so on. If, on the other hand, an in-network aggregation is performed using routing trees then only the leaf nodes can be recovered. This is due to the fact that individual readings are aggregated as we climb up the tree [5,7]. In this case, every parent should cache the past readings of its leaf children. It then recover the missing reading using the CI of that child, its past reading, and the readings of its neighbors, which include that parent and its neighbors as well.

2.3 Super-Resolution and In-Network Sampling

Our discussion of the missing values problem was only a motivation for another important application that we call "super-resolution" or in-network sampling. We borrow the former notion from the image processing and computer vision communities. It is defined as the estimation of missing details from a low resolution image [12,13]. In our context, we define this problem as the estimation of missing readings from the set of all sensor readings at a specific time instance. The analogy is that the missing readings are scattered in the network as if we have a low resolution image of all the readings. E.g., consider Figure 1. If the readings of n, j, p are missing from frame (b) then the readings of the rest of the sensors constitute a low resolution form of all the readings.

This application is advantageous when summaries or detailed sets are collected. Specifically, if the missing values in the low resolution readings are predicted within an acceptable error margin, then a lot of energy will be saved. In the former (i.e., summaries), we save the energy spent on continuously sensing unnecessarily, especially that the sensing task is usually expensive [14]. While in the latter, the energy spent on the sensing and on the communication of those missing readings is saved since the prediction step is performed at the base-station. The benefits in this latter scenario are clearly significant, since the communication cost in sensor networks is very high. Furthermore, collisions are effectively reduced by decreasing the number of forwarded packets.

Similar to the recovery of missing values discussed above, we can *strategically* select a subset of the sensors to sense the environment at a specific time while predicting the missing readings. The selection criteria should be based on the geographical locations, remaining energy, etc., and on the CI. In fact, a complete re-tasking of the entire network can be performed; e.g., if some sensors are about to run out of battery then they should not be sampled very often, and so on. Information about the remaining energy is obtained using, e.g., [15]. This re-tasking is controlled at the base-station or via the use of a proxy [1]. A basic algorithm is to control the nodes such that they alternate sensing the environment by adjusting their sampling rate appropriately.

3 Modeling Contextual Information

Statistical modeling of the spatio-temporal correlations in sensor networks is directly translated to modeling of the CI. In this section we discuss related work in literature for modeling of spatial and temporal dependencies in data. We briefly cite why precise statistical models for such a data are usually difficult to work with, and are therefore not used in practice. We will introduce several assumptions used in literature and in our approach to make these models feasible to work with. Finally, we formally present our approach for modeling of CI.

3.1 Modeling Structured Data

Structured data refers to temporal, spatial, and spatio-temporal data. The individual observations (measurements) in this data are correlated; in time-series temporal data the measurements are correlated in time while in spatial data the measurements are correlated in space. Hence, knowledge of one observation provides information about the others. A general statistical modeling of the correlations between all the observations is very complicated; since too many probability distributions (i.e., parameters of the model) need to be learned or explicitly quantified. Consequently, a huge learning data need to be available to estimate these parameters. Moreover, some combination of data may rarely occur, and therefore, its corresponding probability cannot be computed by learning [16,19,13]. As a simple example, consider a spatial data consists of p points, each point can take on discrete values from 1 to n. Let us assume that we will model this data "assuming" that every point is correlated with each other point. In order to fully define this complex model we need to compute, for each point, the probability that it takes on different values (n such values) given all the possible values of each other point. The resultant number of combinations and probabilities that need to be computed are prohibitive, even for small n, p. In conclusion, even if it is possible to build a model that involves each possible correlation, it is not feasible to compute its parameters.

Therefore, Markov-based models that assume "short-range dependencies" have been used in literature to solve this problem [16,19]. Specifically, for modelling of time-series data, e.g., first order Markov models [16], and for spatial data, e.g., Markov Random Fields (MRF) in computer vision [17,12,13]. The Markov assumption is very practical. It drastically simplifies the modeling and makes the parameter-learning process feasible. Specifically, the size of the combinations and the learning data needed to define Markov-based models are significantly reduced. E.g., in temporal data, it states that the influence of the past is completely summarized by the last observation, while in spatial data, the observation at a specific location depends only on its immediate neighbors. Sensors can be thought of as pixels of an image, where the neighbors are the spatially nearby sensors, while the readings of a specific sensor follow a time-series. Hence, we apply Markov assumption to our model. We assume that the influence of all neighboring sensors and the entire history of a specific sensor on its current reading is completely summarized by the readings of its immediate neighbors and its last reading.

3.2 A Probabilistic Approach

We now present our Markov-based model of CI. We map the problem of learning and utilizing the CI to the problem of "learning parameters" of a Bayes classifier, and then "making inferences", respectively. The different values of sensor readings represent the different classes. The features of classification are: (1) the current readings of the immediate neighbors (spatial), and (2) the last reading of the sensor (temporal). The parameters of the classifier is the CI, while the inference problem is to compute the most likely reading (class) of the sensor given the parameters and the observed spatio-temporal information. Figure 2 shows the structure of this Bayes-based model, where N is a feature that represents readings of neighboring nodes, H is a feature that represents the last reading of the sensor, and S is the current reading of the sensor.

A continuously changing topology and common node failures prohibit any assumption about a specific spatial neighborhood, i.e., the immediate neighbors may change over the time. To account for this problem, we model the spatial information using "some" neighbors only. The exact number varies with the network's characteristics and application. Our model below is fairly generalizable to any number of neighbors as desired.

As part of our future work, we plan to study criteria for choosing neighbors. In particular, we will allow the sensors to learn correlations with each available neighbor and weight them according to their degree of correlation. To make inference, the sensor can choose those that yield the highest confidence. However, in the following discussion and experimental evaluations we assume a neighborhood consists of two randomly-chosen indistinguishable neighbors.

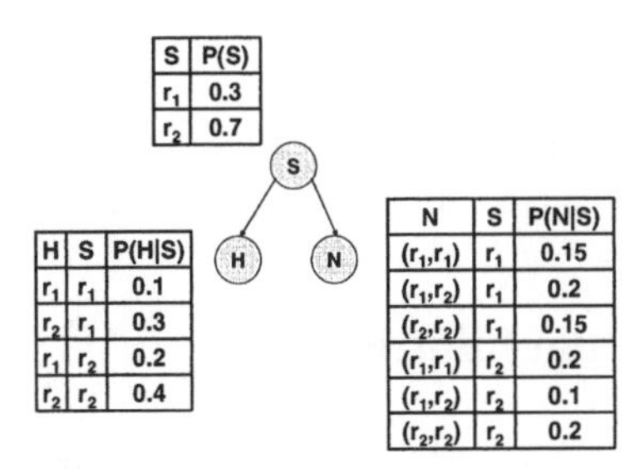

Fig. 2. Our Bayes-based model of spatio-temporal correlations.

To formally describe our model, without loss of generality, assume that sensor readings represent a continuous variable that takes on values from the interval $[l, u]$. We divide this range, $(u - l)$, into a finite set of m non-overlapping subintervals, not necessarily of equal length, $R = \{r_1, \ldots, r_m\}$. Each subinterval is considered a class. These classes are *mutually exclusive* (i.e., non-overlapping) and *exhaustive* (i.e., cover the entire range). R can be quantized in different ways to achieve the best accuracy and for optimization purposes, e.g., by making frequent or important intervals shorter than infrequent ones. Entropy of the readings may guide this decision. The features of the classifier are the history H, and the neighborhood N that represents the reading of any two nearby sensors. H takes on values from the set R, while N takes on values from $\{(r_i, r_j) \in R \times R, i \leq j\}$, and the target function (class value) takes on values from R. To define the parameters of this model, we first show how the inference is performed.

Bayes classifier is a model for probabilistic inference; the target class, r_{NB}, output by the classifier is inferred using maximum a posteriori (MAP) [18,16,11].

$$r_{MAP} = argmax_{r_i \in R} P(r_i|h, n) \tag{1}$$

Where h, n are the values of H, N, respectively. Rewritten using Bayes rule yields,

$$r_{MAP} = argmax_{r_i \in R} \frac{P(h, n|r_i)P(r_i)}{P(h, n)} = argmax_{r_i \in R} P(h, n|r_i)P(r_i) \tag{2}$$

Since the denominator is always constant, it does not affect the maximization and is omitted. From Equation 2, the terms $P(h, n|r_i)$, $P(r_i)$ should be computed for every h, n, r_i, i.e., they constitute the parameters of our model. The size of training data needed to reliably learn them, especially the term $P(h, n|r_i)$ is expected to be very large, for the same reasons we discussed earlier. Therefore, in order to cut down the size of training data we utilize the "Naive Bayes" assumption which states that the feature values are conditionally independent given the target class [18,16,11]. That is, we assume that the spatial and temporal information are conditionally independent given the reading of the sensor. Although the Naive Bayes assumption is not true in general, "Naive" Bayes classifiers have been shown to be efficient in several domains where this assumption does not hold, even competing with other more sophisticated classifiers, e.g., in text-mining [18,16,11]. Based on this assumption we obtain the following.

$$r_{NB} = argmax_{r_i \in R} P(h|r_i)P(n|r_i)P(r_i) \quad (3)$$

The parameters (CI) become (a) the two conditional probabilities tables (CPT) for $P(h|r_i)$, $P(n|r_i)$, and (b) the prior probability of each class $P(r_i)$.

4 Learning Contextual Information

In this section, we present a procedure for online learning of the model parameters in-network, in a distributed fashion. We also discuss some applicability issues.

4.1 Training Data

The training data used to estimate the CI consists of the triples (h, n, r_t), where r_t represents the current reading of the sensor, h, its last reading is cached at the sensor, and n as defined above. Information about n is available at each sensor node when sampling the network; since the shared channel can be utilized for snooping on neighbors, broadcasting their readings. Snooping, if done correctly should incorporate a very little cost with no communication complications. The neighbors can be, e.g., the parent of the node and one of its children in case of a routing tree. To account for lack of synchronization, the nodes quantize the time, cache the readings of their neighbors over each time slot, and use them for learning at the end of the slot. If the training instance is incomplete (some information is missing), it is discarded and not used for the learning.

4.2 An Online Distributed Learning Algorithm

Non-Stationary Phenomena. If the sensed phenomenon is completely non-stationary in space, i.e., if the dependencies are coupled with a specific location, the parameters are learned as follows. Each node estimates $P(r_i), i = 1 \dots m$ simply by counting the frequency with which each class r_i appears in its training data. That is, its sensed value belongs to r_i. The node does not need to store any training instance; it just keeps a counter for each r_i, and an overall counter of the number of instances observed so far, all initialized to 0. It increments the appropriate counter whenever it observes a new instance. The CPT of H, N are estimated similarly. Notice that $P(H = h|r_i)$ is the number of times ($H = h$ *and* the sensor reading belongs to r_i), divided by the number of times the class is r_i. Since the node already keeps counters for the latter, it only needs m^2 counters for the former. To estimate the CPT of $P(n|r_i)$, in case of two indistinguishable neighbors, the node keeps $\frac{m^2(m+1)}{2}$ counters for each ($n = (r_i, r_j), i \leq j$ *and* the reading belongs to r_i) since (r_i, r_j) is indistinguishable from (r_j, r_i). That is a total of $1 + m + \frac{3}{2}m^2 + \frac{m^3}{2}$ counters are needed.

After a pre-defined time interval, a testing phase begins where the sensor starts testing the accuracy of its classifier. It predicts its reading using the learned parameters and compares it with its sensed reading at each time slot. It keeps two counters for the number of true and false predictions. At the end of this phase, the sensor judges the accuracy by computing the percentage of correctly classified test instances. If it is not acceptable according to a user-defined threshold, the learning resumes. The sensor repeats until the accuracy is reached or the procedure is terminated by the base-station.

If the phenomenon is not stationary over time, the sensors re-learn the parameters dynamically at each change. This change can be detected, e.g., when the error rate of the previously learned parameters increases significantly. The old learned correlations can be stored at the base-station and re-used if the changes are periodic. The sensors periodically send their parameters to the base-station to be recovered if they fail. A hybrid approach for non-stationarity in time and space is formed in a fairly similar way.

Stationary Phenomena. Sensor networks may have thousands of nodes. If the sensed phenomenon is stationary over space, then the above learning procedure can be modified to scale with the size of the network in an energy-efficient fashion. Notice that stationarity in space does not imply that we assume a static or a fixed topology. In fact, the following procedure adapts to topology changes. Space-stationarity however implies that training data from all the nodes can be used for learning. It therefore enables collecting of a large number of training instances in a relatively short time.

Specifically, the locally learned parameters at each node can be combined together. Hence, the distributed learning reduces to "summing" the individually learned counters over all the nodes. To do that, we build on existing techniques for in-network aggregation that has been proven to be efficient with respect to communication and energy costs. Particularly, it yields an order of magnitude reduction in communication over centralized approaches [5,7]. We benefit from their routing mechanism, topology maintenance, etc. Such a functionality is readily provided in several sensor applications. However, designing of efficient routing protocols for Bayes learning and our specific tasks is an interesting research direction that we plan to investigate.

In particular, after a predefined epoch duration that depends on the sampling rate and the sensed phenomenon, the sensors perform an in-network SUM aggregate. The local counters are sent from the children to their parents. The parents sum the counters of all their children and add their own, and so on, in a hierarchical fashion. Eventually, the overall counters reach the base-station. The sensors reset their counters to 0 after propagating their values to their parents and continue the learning process. The base-station accumulates the received counters over epochs, estimate the parameters after a number of epochs, then flooded them into the network to reach every sensor.

A testing phase begins in which each sensor tests the classifier using two local counters as before. After a testing epoch, the overall counters are collected using in-network SUM aggregation. The base-station judges the accuracy of the learned parameters by computing the overall percentage of the correctly classified test instances. If it is not acceptable, the learning phase resumes. The procedure then continues until the accuracy becomes acceptable. The overall parameters are then used by every sensor for inference.

A refresh message containing the parameters is flooded every while to account for node failure. When new nodes join the network, they eventually get engaged in the process.

This approach adapts to less perfect situations when the space-stationarity holds inside clusters of sensors over the geographic space. A separate model is learned and used by the sensors in every region, each is computed separately following the above procedure by using a GROUP BY (region) operator. Nevertheless, we implicitly assumed that the sensors are homogeneous to simplify the discussion. However, this assumption is not necessary. The dynamics of the network, the characteristics of each sensor such as its energy budget, sensing cost, importance of its reading, are all factors that control how the sensors are involved in training/testing/sensing. This control is done locally at the sensor, via the base-station, or both. Due to the limited space we do not expand on these details. One final issue is that our Bayes-based model converges using a small training set [18]. The convergence also makes it insensitive to common problems such as outliers and noise given that they are usually random and infrequent, and duplicates since the final probabilities are ratios of counters.

4.3 In-Network Distributed versus Centralized Learning

Learning the parameters can generally be performed either in-network, as we propose, or centrally by analyzing detailed sensor data sets at the base-station. Our overall approach for modeling and utilizing the CI is applicable under both procedures. For completeness however we briefly discuss some of the tradeoffs. We argue that our approach does not introduce any significant overhead, especially given its important applications. Most networks are readily used for collecting the learning data, e.g., for monitoring, and hence this data is considered "free". The size of learning data, and consequently the overall communication cost, clearly vary from application to another. They also depend on other factors such as the desired accuracy and the routing mechanism. They therefore require experimentations and cannot be quantified explicitly. The evaluations in Section 6 are focused on the performance of our model in prediction and not on the energy cost. We plan to evaluate the latter experimentally once we finish our prototype.

The cost of the learning will generally be dominated by the communication since it is the major source of power consumption [5,4]. For complete non-stationarity (in space), where learning is performed at each node, a centralized approach is inferior due to obvious communication cost. For stationary or imperfectly stationary networks the tradeoff is not that clear. Assume in-network learning is performed via aggregation and a fairly regular routing tree. Notice that in-network learning involves computing of a distributive summary aggregate while centralized learning can be viewed as computing of a centralized aggregate or as collecting of individual readings from each node [5]. Therefore, the communication cost of in-network learning is roughly $k \times O(m^3) \times O(n)$, where k is the number of epochs, m is the number of classes, and n is the number of nodes involved in learning, which can be as large as the size of the network. This is equivalent to the cost of computing $O(m^3)$ summary aggregates k times. The cost of a centralized learning is roughly $p \times O(n^2)$, where p is the size of training data at each sensor which is application-dependent. This is equivalent to the cost of computing p centralized aggregates[1]. For a realistic situation where $p = 1000, k = 2, m = 5, n = 10$, the cost of the centralized learning is an order of magnitude higher. This difference further increases for perfectly stationary situations since n becomes very large. Even when m increases the difference remains significant. Nevertheless, in-network learning effectively reduces the number of forwarded packets which is a serious disadvantage of centralized learning. The analysis fairly extends to non-stationarity in time.

5 Utilizing Contextual Information

Once the parameters are learned, they can be used for prediction at any time. Applications of our model then become an inference problem. In particular, the probability of the sensor reading being in different classes $r_i, i = 1 \ldots m$, is computed for every r_i from Equation 4 The class with highest probability is then output as the prediction.

$$P(r_i|h,n) \sim P(h|r_i)P(n|r_i)P(r_i) \tag{4}$$

Example 1. Consider the scenario shown in Figure 1. Assume that the sensor readings can take values in the range $[30, 60]$, and that we divided this range into two classes, $r_1 = [30, 45]$, $r_2 = [45, 60]$. Assume that we have already learned the parameters (CPTs) shown in Figure 2. To infer the missing reading of sensor m in frame (c), we use the readings of j, k in this frame, and the history of m, $H = r_2$. We compute $P(r_1|h = r_2, n = (r_2, r_2)) \sim 0.3 \times 0.3 \times 0.15 = 0.0135$, while $P(r_2|h = r_2, n = (r_2, r_2)) \sim 0.7 \times 0.4 \times 0.2 = 0.056$. The second class is therefore more likely. This indicates that the reading of m is expected to be somewhere in the range $[45, 60]$.

For outlier detection, the sensor locally computes the probability of its reading being in different classes using Equation 4. It then compares the probability of its most likely reading, i.e., highest probability class, with the probability of its actual sensed reading. If the two differ significantly then the sensor may decide that its reading is indeed an outlier. For example, we follow the steps of Example 1 to compute the probability of the reading of i being in $[30, 45]$ and $[45, 60]$. We find that its reported reading, i.e., 40, in Figure 1(c), is indeed an outlier since the probability of its reading being in $[30, 45] \sim 0.0135$ is small compared to $[45, 60] \sim 0.056$.

To approximate the missing readings and in-network sampling, the objective is to predict the missing reading of a specific sensor. This is performed by inferring its class using Equation 3, as in Example 1. The predicted class represents a set of readings (a subinterval) and not a single value. We can, e.g., choose the median of this subinterval as the predicted reading, and therefore the error margin in prediction becomes less than half the class width. We may think of this approach as significantly reducing the uncertainty associated with the missing reading from $[l, u]$ to r_i. As the width of each class becomes smaller, the uncertainty further decreases. There is a tradeoff between the complexity of the model and the uncertainty. Smaller subintervals translate to more classes, and consequently, to bigger CPTs that are hard to work with and to locally store at the sensor. This indicates that the width of each class should be chosen wisely, e.g., assign

[1] A detailed analysis of the cost of in-network aggregation can be found in [5]; it has been shown that it yields an order of magnitude reduction in communication over centralized approaches.

short classes to important readings that require tight error margins. This decision is very application-dependent, it depends on the data and its overall range, sensor's storage and its energy resources, required accuracy, significance of knowledge of a specific reading, etc. Entropy metrics may guide this decision as we mentioned earlier.

6 Experimental Evaluation

This section experimentally evaluates the performance of our approach in prediction. We are still working on a prototype on 50 Rene motes and experimenting with different data sets. More evaluations on this prototype will be reported in the future. The following are some initial results from an off-line analysis of two data sets that fairly represent two real applications, tracking and monitoring. The first set is a synthetic spatio-temporal data with sharp boundaries; a typical data in tracking [2]. We simulated the readings of 10000 sensors over a grid of 100×100 using MATLAB. The data was generated in the form of a shockwave propagating around a center based on Euclidean distance. The data exploits a strong spatio-temporal behavior with sharp boundaries as shown in Figure 3. We divided the range of sensor readings into 10 bins, each is represented by a class. The learning proceeded using 10 frames in each learning epoch. The error rate of the classifier is then computed using 20 test frames. It represents the percentage of the incorrectly predicted readings in the test data. The learning procedure continued until the error rate stabilized.

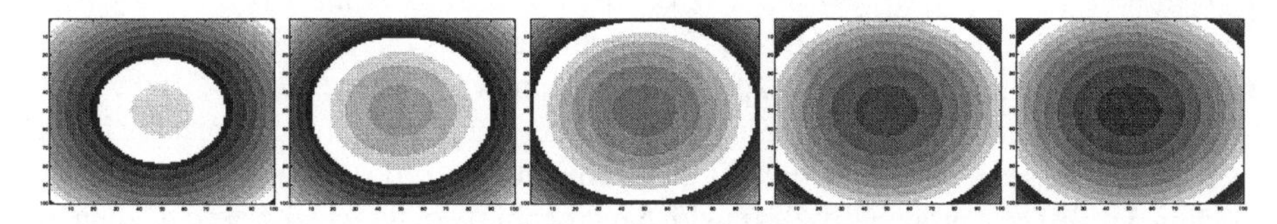

Fig. 3. Consecutive frames of the synthetic data.

To simulate a realistic behavior we added random outliers to each frame of the generated data. We varied the percentage of outliers in each frame of the learning data to test the robustness of our classifier. As shown in Figure 4(a), as the percentage of outliers increases, the classifier takes more iterations (frames) to learn the parameters. By iteration we mean one new frame of data. The curves from the top to the bottom represent 90% to 10% outliers with a decrement of 10%. Figure 4(b) shows the error percentage in prediction using a set of 20 test frames after the learning phase. As the number of outliers increases the error increases and then remains constant at 7%.

To evaluate the efficiency of our approach in detecting outliers, we computed the percentage of detected outliers in the test data as shown in Figure 5(a). Figures 4(b), 5(a) show an interesting behavior; our classifier was able to learn the correlations even with the existence of a high percentage of outliers. When we investigated this behavior, we found that the effect of the spatial feature diminished as the number of outliers increases, i.e., its CPT for feature N is almost uniformly distributed. Therefore, the classifier learned the temporal correlation, H, instead and used it as the major classification feature. We also measured the mean absolute error of the incorrectly predicted readings and the actual

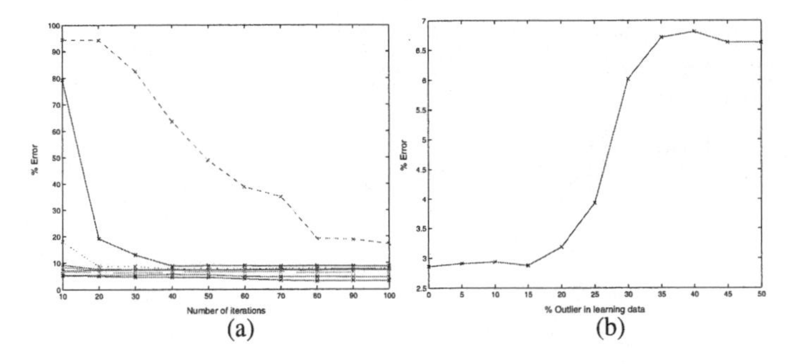

Fig. 4. (a) Effect of outliers on the learning rate. (b) Effect of outliers on the prediction accuracy.

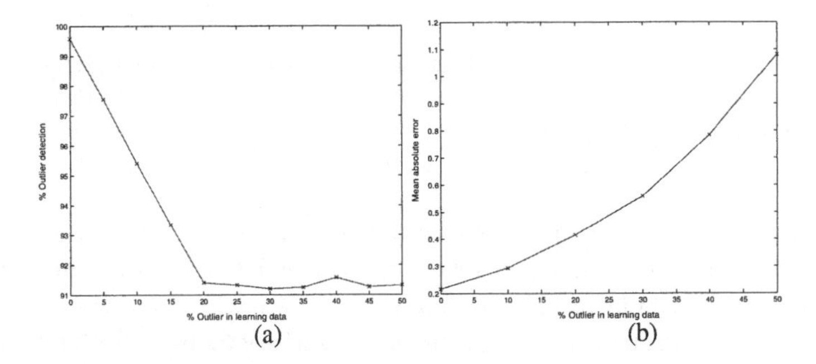

Fig. 5. (a) Effect of outliers in learning data on outlier detection. (b) Mean absolute error of the incorrectly predicted readings versus percentage of outliers in the learning data.

reading versus the percentage of outliers in the learning data as shown in Figure 5(b). As expected the error increases as the number of outliers increases, however, the difference is not significant.

The second analyzed data set is Intel's monitoring data, collected from an entire summer of habitat monitoring in the Great Duck Island, located off the shore of Maine. [1] contains a detailed description of this data. It is publicly available at (www.greatduckisland.net). We experimented with a subset of the nodes, specifically nodes 2, 12, 13, 15, 18, 24, 32, 46, 55, and 57. They are spatially adjacent and clustered in the geographic bounding box (560282,4888140),(560291,4888149). Each node includes light, temperature, thermistor, humidity, and thermopile sensors. We considered their readings from August 6 to September 9, 2002 which is about 140,000 readings for each sensor. We assumed that space-stationarity holds in this region and applied our approach. Figure 6 shows strong correlations in the analyzed nodes over the first 4 sensors. To better illustrate them, the figure also shows a snapshot of the data over a 48 hours period. Figure 7 shows the results of our experiment when the learning and testing were repeated periodically over the entire data set. The range was divided into 15 equally spaced classes in each sensor except the temperature sensor where we include the effect of changing the number of classes on learning. The other sensors showed a similar effect as well. We noticed a saw teeth effect which is due to random imprecision, noise, and outliers in the testing phase. Notice that the error becomes small enough in a relatively

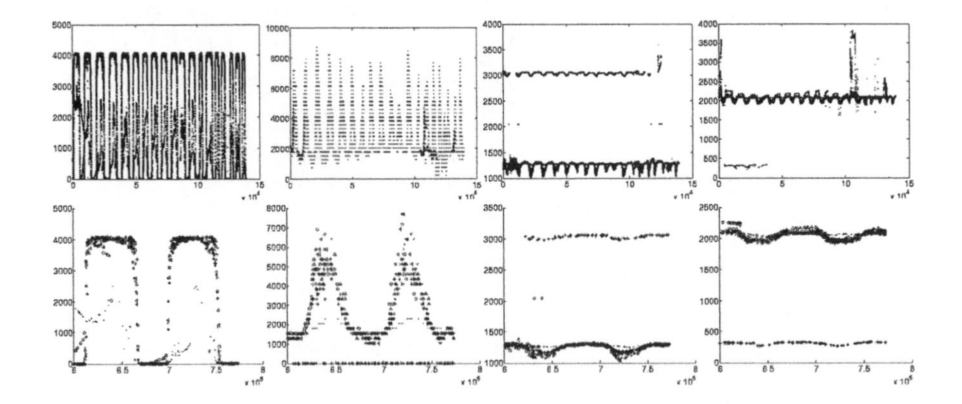

Fig. 6. From left to right, overall reported data (top), 48 hours data (bottom), from all nodes for light, temperature, thermistor, and humidity sensors. The y-axis represents the readings while the x-axis represents samples over time.

short time ($> 90\%$ accuracy in most of the cases). The thermopile sensor showed the same behavior and results, however we did not include it due to space limitations.

In general, these initial results show that our approach works well in sensor data with spatio-temporal characteristics, even with the existence of outliers (first and second set) and random noise (second set). The small prediction error $\approx 10\%$ also demonstrates the efficiency of our approach. As apparent from the figures the learning converges after a relatively few number of learning frames.

7 Related Work

In deriving our model of CI, we drew from research in spatial and temporal data [19,16, 20], graphical models in computer vision and image processing [17,12,13], and machine learning [18,16,11]. We addressed the specific relations to each of these areas throughout this paper. In general, our work focused on deriving a model that adapts to the limitations of sensor networks such as their resource, topology changes, etc.

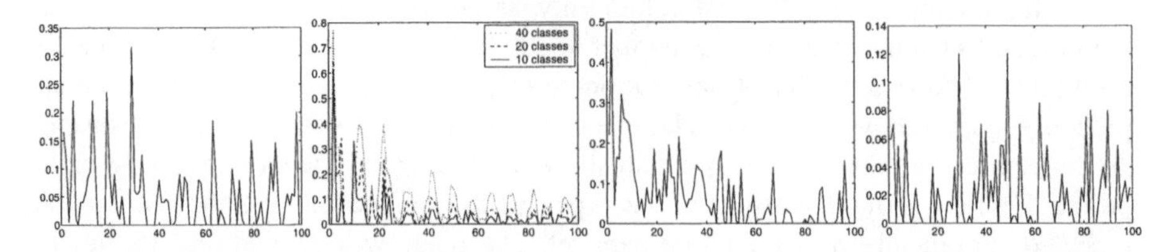

Fig. 7. From left to right, the accuracy of the leaned classifier over epochs for light, temperature, thermistor, and humidity sensors. The y-axis represents the error fraction while the x-axis represents the percentage of data used for learning.

Spatio-temporal correlations in sensor data have been addressed in [4,21,22]. However, these approaches assume the existence of such correlations without attempting to explicitly quantify them. E.g., Ganesan *et al.* utilize them in storage of sensor data, in a compressed form, using wavelet coefficients [4]. The benefits of utilizing spatial correlations in monitoring were also discussed in [22]. Our focus however is to explicitly model and quantify these correlations by learning them. We proposed an algorithm for achieving this task and introduced several applications such as outlier detection.

Goel *et al.* also utilize the spatio-temporal correlations to save communication energy [21], however, they do not explicitly learn them as we do. Their work is focused on continuously sending prediction models to the sensors, which are computed based on similarities between sensor data and MPEG. Their approach to saving communication energy, by instructing the sensors to not report their readings if the predictions are within a threshold, bears similarities to our sampling application. However, they do not save the sensing energy in this case as we do. Their predictions are continuously sent to the sensors which is energy-inefficient. Furthermore, they are not necessarily accurate since the CI is not learned, but rather the predictions are computed based on matching algorithms borrowed from MPEG. However, the dynamics of the monitored phenomenon over successive frames are not usually identical to the dynamics of MPEG. Many predictions may therefore turn out to be wrong. In contrast, our approach to prediction is based on learning the correlations, i.e., accurate. Once the correlations are learned they are reused over the time, without the need to continuously send any prediction models from the base-station. Nevertheless, since they do not learn the correlations, it is not clear how outliers and abnormalities can be detected in their model.

Recent research in sensor networks showed that in-network processing is more energy-efficient theoretically and experimentally [5,7,23], since valuable communication energy is saved. This motivated the recent work on computing aggregates by processing the query in-network [5,7], and on designing and implementing database functionalities [14]. This work also motivated our online in-network learning of CI; we adopted an in-network procedure that is similar to computing distributed aggregates, hierarchically, in a distributed fashion. Learning the CI is very advantageous. We have already introduced some of its important applications. We argue that it should be supported as part of the system deployment similar to aggregations.

8 Conclusions and Future Work

Spatio-temporal correlations highly exist in sensor data and are modeled using the CI. It enables the sensors to locally predict their readings at a specific time. We presented an approach for modeling such correlations that is based on Bayesian classifiers. We proposed an energy-efficient algorithm for online learning of CI in-network, in a distributed fashion. We also briefly compared it with centralized learning. We presented several practical assumptions in our model such as the markovianity and the conditional independence. Although these assumptions are not usually true it has been shown in literature, and in our evaluations, that they work well in practice. We introduced applications of our model in outlier detection, approximation of missing values, and sampling.

We showed that once the CI is learned these applications reduce to an inference problem. Our evaluations showed the applicability and a good performance of our approach.

Our future work directions include finishing our prototype and using it for further experimentations and evaluations of the overall cost and accuracy. We also plan to apply our approach to different sensor applications and to further investigate several design and deployment issues such as working with heterogeneous sensors, the number of neighbors, selecting the neighbors intelligently versus randomly, designing more efficient routing techniques, dealing with rare events, etc. We plan to consider the different tradeoffs and the cost of the different decisions that we presented in this paper, especially those that were not possible to quantify. Finally, extending our approach to the multi-dimensional case is far more complicated and is still an open problem.

Acknowledgement. We are thankful to the anonymous reviewers for their valuable comments to improve the quality of the paper. This work was supported in part by DARPA under contract number N-666001-00-1-8953 and NSF grant ANI-0240383.

References

1. A. Mainwaring, J. Polastre, R. Szewczyk, D. Culler, and J. Anderson, "Wireless sensor networks for habitat monitoring," in *Proceedings of ACM WSNA'02*, September 2002.
2. J. Liu, P. Cheung, L. Guibas, and F. Zhao, "A dual-space approach to tracking and sensor management in wireless sensor networks," in *Proceedings of ACM WSNA'02*, 2002.
3. E. Elnahrawy and B. Nath, "Cleaning and querying noisy sensors," in *Proceedings of ACM WSNA'03*, September 2003.
4. D. Ganesan and D. Estrin, "DIMENSIONS: Why do we need a new data handling architecture for sensor networks?," in *Proceedings of Hotnets-I*, October 2002.
5. S. Madden, M. J. Franklin, and J. M. Hellerstein, "TAG: a Tiny AGgregation Service for Ad-Hoc Sensor Networks," in *Proceedings of 5th OSDI*, December 2002.
6. V. Bychkovskiy, S. Megerian, D. Estrin, and M. Potkonjak, "A collaborative approach to in-place sensor calibration," in *Proceedings of IPSN'03*, April 2003.
7. J. Zhao, R. Govindan, and D. Estrin, "Computing aggregates for monitoring wireless sensor networks," in *IEEE SNPA*, April 2003.
8. D. Ganesan, R. Govindan, S. Shenker, and D. Estrin, "Highly-resilient, energy-efficient multipath routing in wireless sensor networks," in *MC2R*, vol. 1, no. 2, 2002.
9. B. Krishanamachari, D. Estrin, and S. Wicker, "The impact of data aggregation in wireless sensor networks," in *DEBS'02*, July 2002.
10. J. Heidemann, F. Silva, C. Intanagonwiwat, R. Govindan, D. Estrin, and D. Ganesan, "Building efficient wireless sensor networks with low-level naming," in *Proceedings of SOSP*, 2001.
11. I. H. Witten and E. Frank, *Data Mining: Practical Machine Learning Tools and Techniques with JAVA Implementations.* Morgan Kaufmann, 2000.
12. W. T. Freeman and E. C. Pasztor, "Learning low-level vision," in *Proceedings of International Conference on Computer Vision*, 1999.
13. J. S. Yedidia, W. T. Freeman, and Y. Weiss, "Understanding belief propagation and its generalizations," Tech. Rep. TR2001-22, MERL, November 2001.
14. S. R. Madden, M. J. Franklin, J. M. Hellerstein, and W. Hong, "The design of an acqusitional query processor for sensor networks," in *ACM SIGMOD*, June 2003.
15. Y. J. Zhao, R. Govindan, and D. Estrin, "Residual energy scan for monitoring sensor networks," in *IEEE WCNC'02*, March 2002.

16. D. Hand, H. Mannila, and P. Smyth, *Principles Of Data Mining*. The MIT Press, 2001.
17. P. Smyth, "Belief networks, hidden markov models, and markov random fields: a unifying view," *Pattern Recognition Letters*, 1998.
18. T. Mitchell, *Machine Learning*. New York: McGraw Hill, 1997.
19. S. Shekhar and S. Chawla, *Spatial Databases: A Tour*. Prentice Hall, 2003.
20. Q. Jackson and D. Landgrebe, "Adaptive bayesian contextual classification based on markov random fields," in *Proceedings of IGARSS'02*, June 2002.
21. S. Goel and T. Imielinski, "Prediction-based monitoring in sensor networks: Taking lessons from MPEG," *ACM Computer Communication Review*, vol. 31, no. 5, 2001.
22. J. Heidemann and N. Bulusu, "Using geospatial information in sensor networks," in *Proceedings of Workshop on Intersections between Geospatial Information and Info. Tech.*, 2001.
23. C. Intanagonwiwat, D. Estrin, R. Govindan, and J. Heidemann, "Impact of network density on data aggregation in wireless sensor networks," in *Proceedings of ICDCS '02*, July 2002.

Analysis of Node Energy Consumption in Sensor Networks

Katja Schwieger, Heinrich Nuszkowski, and Gerhard Fettweis

Vodafone Chair Mobile Communications Systems
Dresden University of Technology, Mommsenstr. 18, D-01062 Dresden, Germany
Phone: +49 351 463 33919, Fax: +49 351 463 37255
{schwieg,nuszkows,fettweis}@ifn.et.tu-dresden.de

Abstract. The energy consumption of network nodes in low data rate networks is analyzed using Markov chains and signal flow graphs. It allows for a comprehensive OSI-layer analysis, taking into account important system characteristics like the transmission channel, the basic receiver structure, the network traffic, the error control coding (ECC) and the kind of channel access. The results show the energy necessary to transmit one bit, depending on the data rate, the ECC, and the number of retransmissions allowed. Thus, it is possible to determine, which design parameters are important and what values are to be preferred. Another performance measure, the probability of a non-successful transmission of a burst, is evaluated as well. Furthermore, the shares of energy in transmit (TX)/receive (RX)/sleep-states are determined.

1 Introduction

Energy efficient low data rate networks are a challenging topic emerging in wireless communications. Sensor networks are the number one application cited in that context. As the sensors are usually battery-driven, each node has to get by with very little energy. By now, it has been understood that the energy consumption of a system can not be minimized by optimizing the components independently. Instead, the design process has to consider the interactions between many layers [1], taking into account many parameters describing the physical layer, error control codes (ECC), protocols, network topology etc. Energy-efficient protocols are widely explored, e.g. [2], [3]. Also, some investigations towards energy-efficient modulation schemes have been made. Nevertheless, only few papers manage to couple important parameters of various layers in order to trade-off performance vs. energy [4], [5], [6]. There are also some interesting prototyping projects, which bring together the experts from different fields in order to do layer-comprehensive optimization, e.g. [7], or trying to miniaturize the node itself like Smart Dust, just to mention some big research approaches.

This paper analyzes the energy consumption of a sensor network using Markov chains and Signal Flow Graphs (SFG). Although we only consider a single-hop network, the analysis is extendable to multi-hop networks as well. We believe, that this should be the first step since one can easily include parameters from different layers. Thus, the main focus is firstly to investigate the influence of physical layer parameters while keeping in mind protocol and network issues. In contrast to the common approach of tailoring the

H. Karl, A. Willig, A. Wolisz (Eds.): EWSN 2004, LNCS 2920, pp. 94–105, 2004.

protocol according to the desired application we believe that that protocol design has to taken into account the envisioned hardware specifications as well. It will be shown that each hardware has different shares of energy in TX/RX/sleep mode.

This paper explains the analysis itself as well as selected results using a given system model. The analysis could be used to evaluate the performance capabilities of other parameters as well, e.g. the influence of the modulation or channel access scheme.

Before describing the general analysis method in Section 3, which can be used to derive the energy needed to transmit one bit (or one data packet, respectively), we roughly explain the main concept of our transmission system in Section 2 in order to understand Section 3. Details about the system parameters investigated and results pointing to important design criteria will be finally given in Section 4.

2 System Description

The energy analysis presented here was part of an industry driven project. Thus, many parameters and system requirements like modulation scheme, ECC, network structure, transmit/receive power consumption of the radio module etc. were taken over from the target specifications. Other parameters like duty cycle, packet length etc. were taken from the recently approved low data rate standard IEEE 802.15.4 [8]. Nevertheless, the developed analysis tool can be adapted and then employed for a wide range of other systems as well.

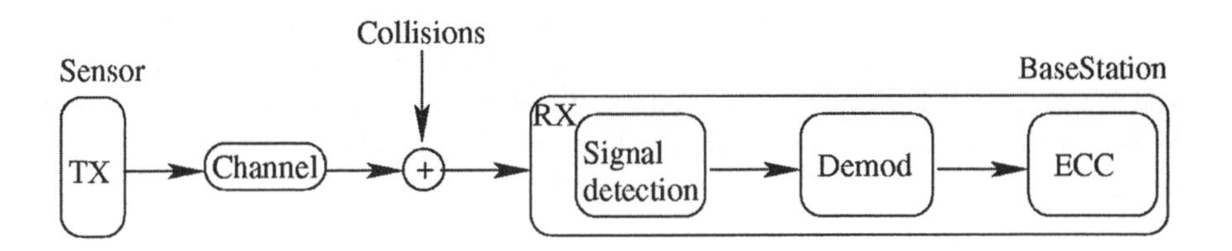

Fig. 1. System Model

Fig.1 shows the signal flow when data is transmitted from a battery-powered sensor to the base station: The packet is sent over a channel, which is assumed to be Additive White Gaussian Noise (AWGN). Of course, depending on the channel access scheme, collisions with other data may occur. At the base station, firstly a signal detection process is performed, which is based on a known synchronization sequence. Then the demodulation of the data takes place. At the end, the error decoding proves whether a packet arrived successfully, i.e. the number of error should not exceed the number of corrigible errors. Detailed parameter specifications will be found in 4.1.

After reception of data, the base station sends out an acknowledgment (ack). This ack is assumed to arrive properly as the base station possesses unlimited energy resources and can thus transmit with appropriate power. If the sensor does not receive an ack, it assumes a transmission failure. Then automatic repeat request (ARQ) is applied and its packet may be retransmitted up to L times.

3 Analysis Method

3.1 State Machine and Signal Flow Graph (SFG)

Now, that we know the functioning of the system, we can derive its state diagram, which is shown in Fig. 2. Herein, the vector above a branch denotes the probability of the respective transition and the energy needed for taking that step. The calculation of these probabilities will be shown in 3.2-3.4. Whenever a sensor sends out data, it has to spend energy E_T for it and then waits for an ack to arrive, which costs energy E_R. Going from the left to the right in the upper rows denotes a successful transmission cycle, covering all steps described in 2: collision-free transmission, detection and acceptance. If one of these steps fails, a retransmission takes place, i.e. the graph continues one row further down until the number of allowed retransmissions L is reached. If the Lth retransmission fails, the packet is lost forever. This will be considered an unsuccessful transmission. The energy consumption is a random variable. Thus, the theory of Markov chains can be applied. From the state machine a Signal Flow Graph (SFG) can be derived. Herein, the contribution of one distinct brach to the overall energy consumption is shown. For it the vector $[p_i, E_i]$ will be replaced by:

$$[p_1, E_1] \Rightarrow p_1 \delta(e - E_1) . \tag{1}$$

Eq. 1 represents the contribution of a particular branch to the probability density function (pdf) of the total energy needed. Transition probabilities between non-neighboring nodes could be calculated by convolution. To avoid this rather complex operation, Fourier-transform can be applied to 1 with respect to e, yielding to simple multiplications:

$$\mathcal{F}\{p_1\, \delta(e - E_1)\} \;=\; p_1\, \exp(-juE_1) . \tag{2}$$

Now, we would like to consider the probability of a (non)-successful transmission of data $P_{(non)succ}$, more precisely the Fourier transformed pdf of the transition from $Start$ to $End_{s/n}$. It denotes the probability for a (non)successful transmission of a sensor packet, which is weighted by the energy spent. Inputs to that equation are the single transition probabilities and the energy spent for those. L is the maximum number of allowed retransmissions. Applying Mason's rule we find:

$$P_{succ}(z) \;=\; p_1\, p_2\, p_3 \sum_{l=1}^{L+1} (1 - p_1\, p_2\, p_3)^{l-1}\, z^{-1}, \text{ where } z = \exp\left(ju(E_T + E_R)\right) \tag{3}$$

$$P_{nonsucc}(z) \;=\; (1 - p_1\, p_2\, p_3)^{L+1} z^{-(L+1)} . \tag{4}$$

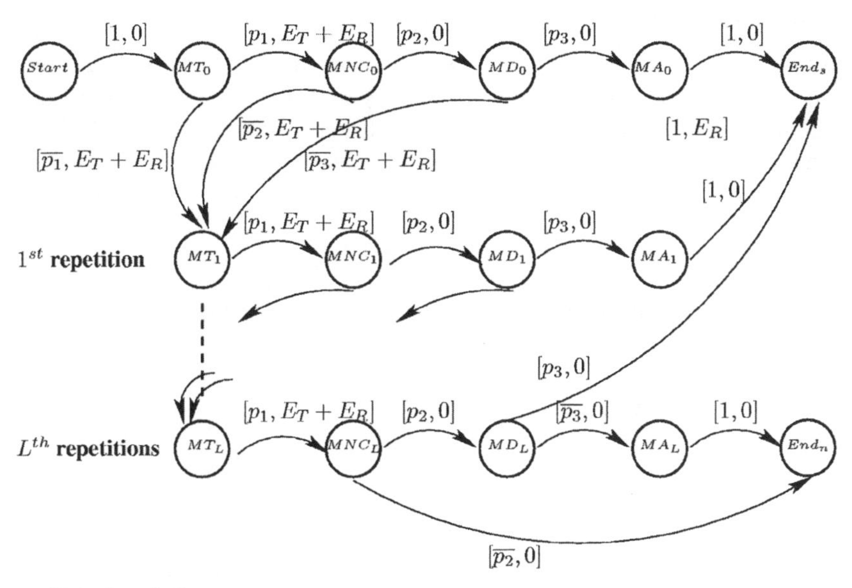

State description
MT_i: Message transmit during $i^{th} repetion$
MNC_i: Message not collided during i^{th} repetition
MD_i: Message detected during i^{th} repetition
MA_i: Message accepted during i^{th} repetition
End_s: Successful transmission of data
End_n: No successful transmission of data

Probabilities
p_1: Probability of no message collision
p_2: Probability of message detection
p_3: Probability of message acceptance

Fig. 2. System State Machine

Then, the characteristic function $\Phi_e(u)$ for the energy is:

$$\Phi_e(u) = [P_{succ}(z) + P_{nonsucc}(z)]|_{z=\exp(ju(E_T+E_R))} . \tag{5}$$

The mean value of the overall energy consumption $< E >$ for the transmission of a single packet can be calculated using the first moment of the characteristic function [9]:

$$< E > = \frac{d\,\Phi_e(u)}{-j\,du}|_{u=0} \tag{6}$$

$$= (E_T + E_R)\frac{1 - (1 - p_1\,p_2\,p_3)^{L+1}}{p_1\,p_2\,p_3} . \tag{7}$$

Now, we have to find the appropriate probabilities p_1, p_2, p_3.

3.2 Probability of No Message Collision p_1

For a first analysis we would like to consider a slotted ALOHA [10] channel access scheme. There will be a collision, if two or more sensors use the the same time slot. The overall arrival rate in the network λ_{net} is modeled as a Poisson process; if n denotes the number of packets arriving in the same time slot of duration T then $p(n)$ is the probability of n arrivals in one time slot :

$$p(n) = \frac{(\lambda_{net}\, T)^n}{n!} \exp(-\lambda_{net}\, T)\,. \tag{8}$$

A message does not collide if there are no other packets in the channel:

$$p_1 = \exp(-\lambda_{net}\, T)\,. \tag{9}$$

The overall traffic λ_{net} is a compound of the original traffic λ caused by the sensors and the traffic due to retransmissions. Thus, we get:

$$\lambda_{net}\, T = \lambda T \sum_{l=0}^{L} (1 - p_1\, p_2\, p_3)^l\,. \tag{10}$$

Combining (9) and (10) gives a non-linear system, which has to be solved for obtaining p_1. Other channel access schemes are under investigation.

3.3 Message Detection Probability p_2

As mentioned in Section 2 burst detection is accomplished by a signature detection algorithm, which uses a preamble of length N. The demodulated symbols are fed into a matched filter (MF). Whenever the output of the MF exceeds a given threshold $thresh$ the message is detected. In order to obtain the message detection probability p_2 the pdf $p_Y(y)$ of the MF has to be known. As we are dealing with non-coherent 2-FSK demodulation scheme, this pdf can be derived by considering the squared amplitudes Z_n of the incoming preamble:

$$Y = \sum_{n=1}^{N} |Z_n|^2 = \sum_{i=1}^{2N} X_i^2\,. \tag{11}$$

Herein, the X_i's denote the Inphase/Quadrature components of the preamble, which are corrupted by white Gaussian noise. Thus, it yields, that Y underlies a non-central chi-square distribution with $2N$ degrees of freedom:

$$p_Y(y) = \frac{1}{2\sigma^2} \left(\frac{y}{s}\right)^{(N-1)/2} \exp\left(-\frac{s^2 + y}{2\sigma^2}\right) I_{N-1}\left(\sqrt{y}\frac{s}{\sigma^2}\right), \quad y \geq 0, \tag{12}$$

whereas s is the sum of the squared mean values of the random variables Z_n and σ^2 is their variance. I_{N-1} indicates a Bessel-function of order $(N-1)$. Then, the message detection probability can be simply expressed by:

$$p_2 = \int_{tresh}^{\infty} p_Y(y)dy\,. \tag{13}$$

For the case of a coherent detection scheme, p_2 can be derived similarly, yielding a non-central χ^2 distribution as well, but being of order 2 and having different s, katbob20 and σ^2 parameters.

3.4 Message Acceptance Probability p_3

Two important parameters are inputs to the message acceptance probability p_3: The applied ECC and the modulation scheme used. Suppose, an error correcting code is applied, which is able to correct up to m errors and detects errors almost surely. Then, the packet is accepted as long as the number of false bits does not exceed the number of corrigible bits. The probability of a single bit error p_{err} depends on the modulation scheme, and for non-coherent 2-FSK it can be calculated by:

$$p_{err} = 0.5 \exp\left(-0.5\frac{E_b}{N_0}\right) . \tag{14}$$

Herein, E_b/N_0 denotes the bit energy-to-noise ratio. When W denotes the code word length, then p_3 is given by a Binomial distribution:

$$p_3 = \sum_{i=0}^{m} p_{err}^i (1 - p_{err})^{W-i} \binom{W}{i} . \tag{15}$$

4 Results

4.1 Parameters

In the following we give details about the parameters used in the analysis. We assume a network with 100 sensors, which send data packets of 128 bits (uncoded) every 5 minutes, what is a realistic scenario for home surveillance purposes. Since the data length is very short it is reasonable to assume equally short acks. As already mentioned non-coherent 2-FSK demodulation is used. Complying with 802.15.4 the carrier frequencies of 900 MHz and 2.4 GHz are investigated while applying a data rate of 20 kbps and 250 kbps, respectively. The number of retransmissions in case of packet delivery failure was chosen to be either 0 or 3. Two different coding schemes are compared: A BCH(152, 128, 7) code ('short code') and a BCH(252,128,37) code ('long code'), which are able to correct 1 or 10 errors, respectively. The current consumption of sensor was assumed to be 15 mA in receive mode, 6 μA in sleep mode and around 28 mA in transmit mode, whereas this last value heavily depends on the operating point of the power amplifier (PA) and therefore varies with the SNR.

As the data rate is low (and thus the required bandwidth) and multi-path spread is small (indoor scenario), the channel can be considered frequency flat. Moreover, we will assume it to be AWGN. A maximum distance between node and base station of 30 m and an additional loss of 50 dB due to walls and other obstacles was assumed. For our purposes, this is a rather pessimistic scenario, so that the results can be considered as an upper limit for the consumed energy.

4.2 Analysis Results

Theoretically, energy of only a few pJ is necessary to transmit a bit. As Fig. 3 shows, in a real system energy at the order of tens of μJ is required for the given scenario. Obviously, the most energy is needed when transmitting in the 900 MHz band (i.e. at a rate of 20 kbps) with the long error code and allowing 3 retransmissions. Since the number of retransmissions decreases when the SNR increases, a steep drop in the energy consumption can be recognized towards higher SNRs in the case of 3 allowed retransmissions. At good SNRs channel degradation is low and thus the number of retransmissions depends on the number of collisions only (and thus on the number of retransmissions, the number of nodes and their activity). Obviously, there is a saturation of energy consumption. This might not be true for all scenarios but depends on the operating point of the PA.
Another important performance measure is the reliability of the sensor's data transmission. Thus, in Fig. 4 the probability of packet transmission failure is depicted.
Four major observations can be made:

- The transmission at high rates (i.e. 250 kbps) outperforms the low rate (i.e. 20 kbps) transmission even though the free path loss will be higher at 2.4 GHz. The reason is the decreased collision probability due to (temporally) shorter packets.Thus retransmissions are avoided. Furthermore, the sensor will remain in the high power consuming TX/RX states for much shorter periods of time.
- ECC with high error correcting capabilities are not preferable most of the time. At very low E_b/N_0 the correcting capabilities are not sufficient, and the packet has to be retransmitted anyways. At high E_b/N_0 a long code is simply not necessary and causes long packets and therefore a higher collision probability. Only for a small range (around 7,8,9 dB) at high data rates and for 3 retransmissions the long code yields a lower energy consumption than the short code. This happens, when the number of collisions is low and the performance only depends on the channel degradation. But at high E_b/N_0 the long code does not bring any performance improvement over the short code anymore.
- From Fig. 3 it is evident that from an energy point of view it does not make sense to allow retransmissions. Nevertheless, the reliability for a successful reception of sensor data at the base station is important as well. Thus, Fig. 4 shows the probability for a non-successful transmission of sensor packets. Obviously, three possible retransmissions can make the whole system much more reliable by spending only very little more energy: with it, the loss of packet can be reduced to 10^{-4} and less, whereas without retransmissions this rate does not fall below $4 * 10^{-3}$. Thus, a limited number of retransmissions should always be allowed.
- In our case it is advantageous to work at high SNRs as this guarantees high reliability by spending little power. This is due to the fact, that high SNRs avoid retransmissions, and the PA (which finally generates the transmit power) has only little influence on the overall energy consumption. Again, if the share of the energy of the PA increases, E_b would have a remarkable rise towards higher SNR. Then an optimum SNR at the minimum E_b could be determined.

Another interesting measure is the probability of messages colliding in the network, which is shown in Fig. 5 for the aforementioned conditions. Clearly, for only 100 sensors

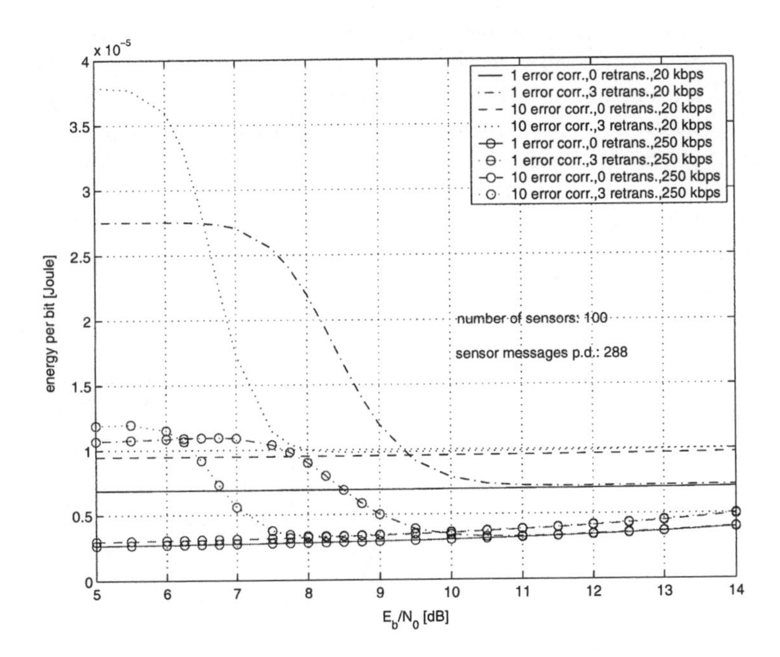

Fig. 3. Energy consumption per bit

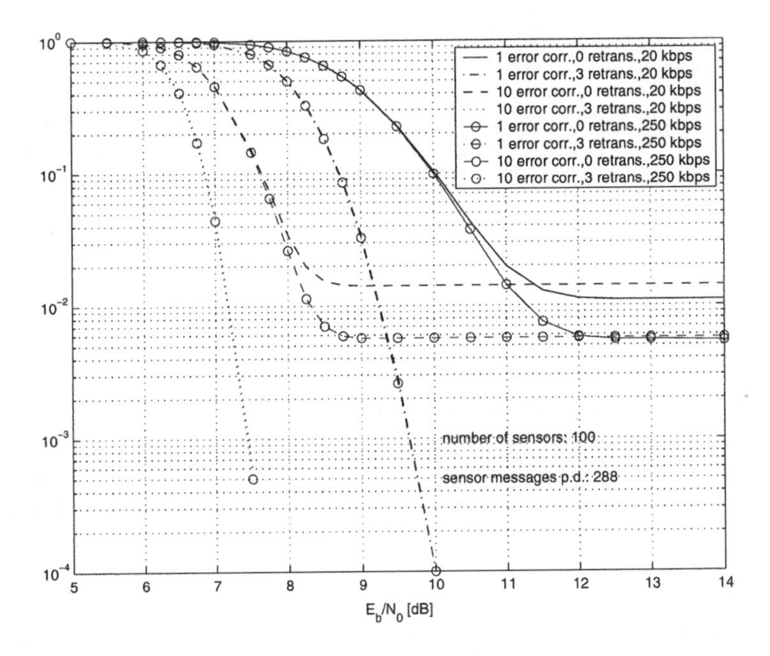

Fig. 4. Probability of non-successful transmission of sensor packets (curves 2/5 and 4/8 overlap)

in a network sending every 5 minutes the collision probability is very low; long packets cause more collisions as well as several allowed retransmissions. It is interesting to note, that there is a saturation for E_b/N_0 greater than 10-11 dB. This means, that at high E_b/N_0 the number of collisions remains constant. Further investigations show, that a significant number of collisions occurs for network sizes beyond 1000 sensors. To be more precise, a network larger than 2500 nodes exceeds a collision probability of 0.5. More than 5000 sensors cause a collision almost sure (more than 90%). Nevertheless, the characteristics (the general devolution of the curves as well as the saturation) remains the same as shown in Fig. 5.

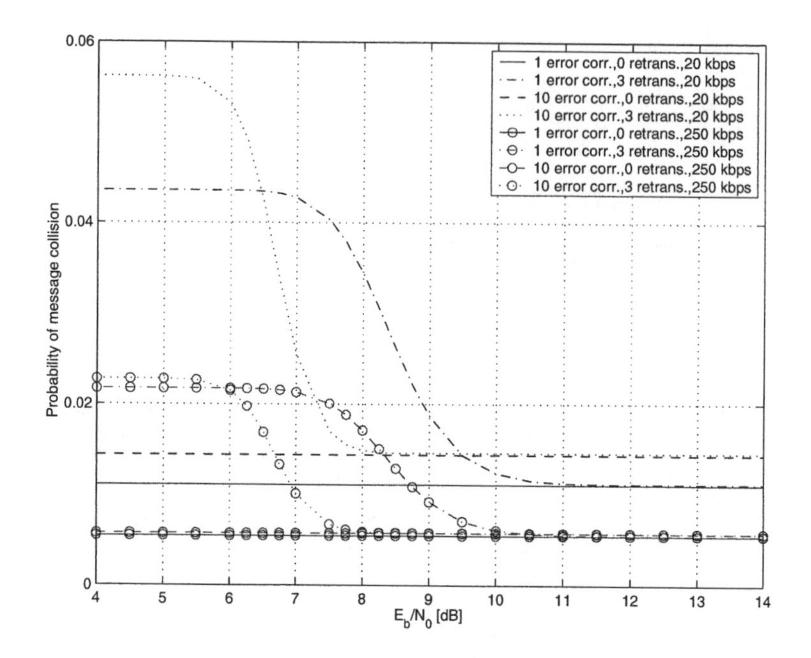

Fig. 5. Probability of message collision

In order to estimate the maximum energy per bit needed for a transmission, the energy consumption in the case of unlimited allowed retransmissions is shown in Fig. 6. Obviously, for low E_b/N_0 the energy needed for each packet transmission converges to infinity since there are too many errors and thus all packets have to be retransmitted. At good E_b/N_0 (more than 9 dB) the energy approaches the one needed if a limited number of retransmissions applies.

Turning to a slightly different scenario, we will now look for the energy distribution in the node in the three possible modes: transmit, receive, sleep. Switching times between the modes are neglected. For both, HF-engineers and protocol designers this is important to know as it shows where there are the most potentials in energy savings.

Fig. 7 shows the shares of the three modes, when considering transmission as above but in the 433 MHz ISM-band at a rate of 2.4 kbps. A different node hardware is used,

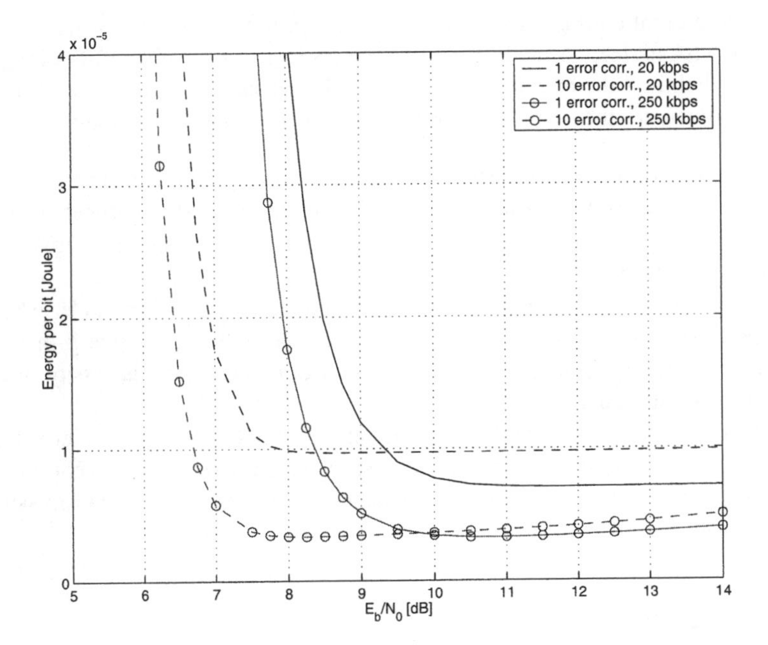

Fig. 6. Energy consumption for unlimited retransmissions

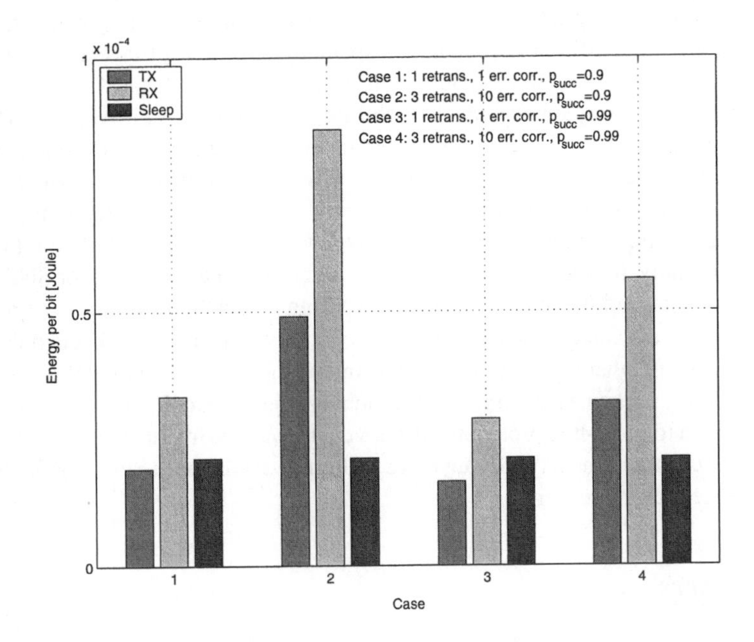

Fig. 7. Energy Consumption of different modes

so that the current consumption is around 8 mA in transmit mode, 14 mA in receive mode and 6 μA in sleep mode. Either 1 or 3 retransmissions are allowed. Looking at two levels of reliability (the probability of a successful packet arrival is either 0.9 or 0.99), the figure illustrates the shares of energy in all modes. It can be seen that:

- All modes consume a significant amount of energy. Although energy consumption in receive state is dominant, other scenarios might have a different behavior.
- As sensors spend most of their time in sleep mode, a remarkable energy is needed for that mode as well.
- In case 1 and 3 less energy is spent in TX-mode than in sleep mode, whereas is vice versa in case 2 and 4. Thus the outstanding influence of the protocol/ECC on the energy consumption is shown. This again verifies the necessity of a layer comprehensive analysis.
- Again, it is clearly visible that working at high SNR is energy efficient. That can be verified by comparing case 2 and case 4, which basically represent 2 different SNRs. Obviously, more energy is consumed in case 2 as more retransmissions are necessary.

5 Conclusions

Using the powerful capabilities of the theory of Markov chains and SFGs, a single-hop wireless low data rate network was analyzed concerning the energy consumption of a single node. Many different parameters spanning several OSI-layers were taken into account. It is shown, how the interaction between those influence the overall energy consumption of a single link and its reliability. The analysis also investigates the energy consumption for the case of unlimited retransmissions and covers the collision probabilities in different networks. Furthermore, it was shown, that each mode consumes a significant share of energy. Thus the necessity of designing a protocol not just based on the application but on the hardware as well was verified. An important influence has the PA. Usually, for indoor applications, the additional energy for higher SNR is only a small percentage of the fixed cost. Hence, working at high SNRs seems to be preferable.

Nevertheless, many problems still need to be solved. Current work is dedicated to the investigation of other modulation schemes (including higher order modulation), different channel access schemes, the impact of switching energies etc. In future, the analysis will be extended to multi-hop protocols. Moreover, it would be interesting to see the analysis when the channel is not AWGN anymore, but time varying due to moving scatters in the environment of the sensors.

References

1. A. Goldsmith and S.B. Wicker. Design challenges for energy-constrained ad hoc wireless networks. *IEEE Commun. Mag.*, 9:8–27, August 2002.
2. K. Sohrabi, J. Gao, V. Ailawadhi, and G.J. Pottie. Protocols for self-organization of a wireless sensor network. *IEEE Personal Communications*, pages 16–27, October 2000.

3. W.R. Heinzelmann, A. Chandrakasan, and H. Balakrishnan. Energy-efficient communication protocol for wireless microsensor networks. *International Conference on System Sciences*, January 2000.
4. W. Stark, H. Wang, S. Lafortune, and D. Teneketzis. Low-energy wireless communication network design. *IEEE Wireless Communications*, pages 60–72, August 2002.
5. V. Raghunathan, C. Schurgers, S. Park, and M.B. Srivastava. Energy-aware wireless microsensor networks. *IEEE Signal Processing Magazine*, 19:40–50, March 2002.
6. R. Min et al. Energy-centric enabling technolgies for wireless sensor networks. *IEEE Commun. Mag.*, 9:28–39, August 2002.
7. J. Rabaey et al. Picoradio supports ad hoc ultra-low power wireless networking. *Comp.*, 33:42–48, July 2000.
8. E. Callaway et al. Home networking with IEEE 802.15.4: A developing standard for low-rate wireless personal area networks. *IEEE Commun. Mag.*, pages 70–77, aug 2002.
9. G.R. Grimmett and D.R. Stirzaker. *Probability and Random Processes*. Oxford University Press, 1992.
10. N. Abramson. The ALOHA system-another alternative for computer communications. *AFIPS Conference Proceedings*, 36:295–298, 1970.

Silence Is Golden with High Probability: Maintaining a Connected Backbone in Wireless Sensor Networks

Paolo Santi[1] and Janos Simon[2]

[1] Istituto di Informatica e Telematica, Pisa, 56124, Italy,
paolo.santi@iit.cnr.it
[2] Dept. of Computer Science, Univ. of Chicago, IL 60637, USA ,
simon@cs.uchicago.edu

Abstract. Reducing node energy consumption to extend network lifetime is a vital requirement in wireless sensor networks. In this paper, we present and analyze the energy consumption of a class of cell-based energy conservation protocols. The goal of our protocols is to alternately turn off/on the transceivers of the nodes, while maintaining a connected backbone of active nodes. The protocols presented in this paper are shown to be optimal, in the sense that they extend the network lifetime by a factor which is proportional to the node density.

1 Introduction

Wireless sensor networks (WSN for short) are composed of battery-operated microsensors, each of which is integrated in a single package with low-power signal processing, computation, and a wireless transceiver. Sensor nodes collect the data of interest (e.g., temperature, pressure, soil makeup, and so on), and transmit them, possibly compressed and/or aggregated with those of neighboring nodes, to the other nodes. This way, every node in the network acquires a global view of the monitored area, which can be accessed by the external user connected to the WSN. Examples of scenarios where WSN can be used are described in [6, 9,10,12].

Perhaps the most important cost in WSN is energy, since it determines battery use, and therefore the lifetime of the network. Since the major source of energy consumption in the sensor node is the wireless interface, considerable energy can be saved if the transceivers are completely shut down for a period of time. Of course, these sleeping times must be carefully scheduled, or network functionality could be compromised. A *cooperative strategy* is a distributed protocol that maintains a (minimal) connected backbone of active nodes, and turns into sleeping state the transceivers of non-backbone nodes. Periodically, the set of active nodes is changed to achieve a more uniform energy consumption in the network.

In this paper, we consider cell-based cooperative strategies for WSN. In the cell-based approach [1,13], the network deployment area is partitioned into iden-

H. Karl, A. Willig, A. Wolisz (Eds.): EWSN 2004, LNCS 2920, pp. 106–121, 2004.

tical non-overlapping cells. Each cell has an active representative. We focus on the process that for each cell updates its active representative.

We found and analyzed two interesting and simple strategies. The first assumes (possibly with unwarranted optimism) that all information about the nodes of the cell is available to the algorithm. The second is a probabilistic algorithm. If the number of nodes in each cell is known to the nodes within the cell, the algorithm is shown to elect a representative in constant expected time. The same behavior can be shown for the model in which the number is not known, but the nodes are assigned to cells uniformly and identically at random. The algorithms are energy-optimal, in the sense that they extend the network lifetime by a factor which is proportional to the node density. This is the best that any cell-based strategy can do [1]. We also find precise bounds for the running time of these algorithms.

Besides the potential usefulness of our algorithms, we believe that the ideas are interesting. They show how relatively inaccurate location information (the cell to which a node belongs) and loose synchronization can be exploited to elect leaders in a very simple and efficient way. Furthermore, we show how different amounts of "local" knowledge (deterministic vs. probabilistic knowledge of the number of nodes in a cell) result in different performance (worst-case vs. average case energy optimal). As discussed in Section 9, the techniques presented in this paper techniques could be useful in other settings.

2 A WSN Application Scenario

A possible sensor networks application scenario in which the algorithms presented in this paper could be applied is the following.

A very large set of sensors is used to monitor a remote geographical region, for example to promptly detect fires in a vast forest. The (thermal) sensors are dispersed from a moving vehicle (e.g., an airplane or helicopter), hence their positions in the deployment region cannot be known in advance (but we have information about their probability distribution.) Once the sensors are deployed, their positions are mostly fixed.

The network should operate for a long period of time (e.g., the whole summer season), and should guarantee that, wherever the fire occurs, the alarm message generated by the sensor that detects the event quickly reaches the external supervisor, who is connected to some of the nodes in the network. Since the supervisor could be mobile (e.g., the supervisor is a ranger equipped with a PDA who wanders in the forest), the actual sensor nodes to which the supervisor is connected cannot be known in advance. This means that the alarm message should quickly propagate in the whole network, so that the supervisor can get the message independently of his physical location.

Since the batteries that power the sensors cannot be replaced once the unit is deployed, energy is a vital resource. With the current technology (see Section 4), the sensor node component that consumes most energy is the radio interface, even when the radio is in idle mode. Hence, considerable energy savings can be achieved if the radio interface of sensors can be turned off most of the time. In principle, a sensor does not need to have its radio on while it is sensing: it could turn the radio on only when an event (fire) is detected, and the alarm message must be sent. However, some of the sensor nodes must keep their radio on also when they are sensing, so that the alarm message generated somewhere in the network could quickly reach the supervisor. I.e., *a connected backbone of awake nodes must exist at any time.*

The class of energy saving protocols presented in this paper aims at extending network lifetime by alternately turning on/off sensor transceivers, in such a way that a connected backbone of awake nodes always exists. The emphasis will be on minimizing the *coordination cost*, i.e., the cost due to the message exchange needed to assign roles (sleep/awake) to sensors.

3 Related Work

Several cooperative strategies aimed at extending network lifetime have been recently introduced in the literature. All of these strategies are related to some extent to the class of energy saving protocols presented in this paper.

In [2], Chen et al. present SPAN, a cooperative strategy aimed at reducing power consumption while preserving both the network capacity and connectivity. SPAN adaptively elects coordinators from all nodes in the network, which are left active, while non-coordinator nodes are shut down. The coordination algorithm is transparent to the routing protocol, and can be integrated into the IEEE 802.11 MAC layer. An inconvenient displayed by SPAN is that, contrary to what is expected, the energy savings achieved by the protocol do not scale with node density. This is due to the fact that the coordination cost with SPAN tends to "explode" with node density, and counterbalances the potential savings achieved by the increased density.

The CPC strategy presented in [11] is based on the construction of a connected dominating set: the nodes in the dominating set schedule the sleeping times of the other units. The authors show through simulation that CPC achieves energy savings with respect to the case where no cooperative strategy is used. However, the analysis of how CPC performance scales with node density is lacking.

The work that is most related to our are the papers by Xu et al. [13] and by Blough and Santi [1]. In [13], the authors introduce the GAF protocol for cell-based cooperation. The protocol is based on a subdivision of the deployment region into an appropriate number of non-overlapping square cells, with the property that if at least one node is awake for every cell, then the network is connected. Nodes in the same cell elect a representative, which is left active, while the transceivers of non-representative nodes are shut down. Periodically, the representative election phase is repeated to balance power consumption and to deal with mobility.

In [1], Blough and Santi investigates the best possible energy savings achievable by any cell-based cooperative strategy, i.e., the energy savings that are achievable under the assumption that the coordination cost is 0. They show that cell-based cooperative strategies have the potential to increase the network lifetime of a factor which is proportional to the node density. They also show that GAF achieves good energy savings, but that it leaves room for further improvement.

In this paper, we show that, if relatively inaccurate location information and loose synchronization are available, the theoretical limit of [1] can actually be achieved using very simple leader election algorithms. With respect to GAF, our algorithms are relatively more demanding (we require loose synchronization), but considerably simpler. Concerning the energy savings, our algorithms are proved to be optimal (in the worst case, or in expectation), while GAF is evaluated only through simulation.

A final comment concerns the synchronization assumption. In many WSN application scenarios, sensors must synchronized anyway (for instance, to report the events detected by the network in a consistent temporal order). So, we believe that our assumption is not particularly demanding.

4 The Energy Model

We assume that a sensor can be in the following states:

1. **sleeping:** node can sense and compute, but radio is switched off;
2. **idle:** radio is on, but not used;
3. **receiving:** node is receiving a radio message;
4. **transmitting:** node is transmitting a radio message.

The well known point is that while currents are minuscule in a chip, even a small radio requires energy levels that are orders of magnitude greater than those inside a processor. Measurements on a Medusa II sensor node [7] have shown sleep:idle:receiving:transmitting ratios of 0.25:1:1.006:1.244. It is also known that the energy cost of sensing is very small as compared to the energy consumption of the wireless transceiver, and is comparable to the cost of the radio interface in sleeping state [3].

As discussed above, we assume that nodes are able to determine the cell to which they belong through some form of location estimation. Similar assumptions were made in [1,13]. Furthermore, we assume that sensors are loosely synchronized. In the following we will assume that nodes are equipped with a GPS receiver. We remark that this assumption is done only for clarifying the presentation, and that any other localization and synchronization mechanisms could be used in practice. When the GPS receiver is used only to estimate node positions (at the beginning of the node operational time) and for loose synchronization purposes, its power consumption is about 0.033W [13], which is comparable to the 0.025W consumed by the wireless interface in sleep mode.

Given the discussion above, in this paper we use the following energy model:

- the energy consumption per time unit of a sensor with the radio on is modeled as a constant C, independently of the actual radio activity (idle, receiving, or sending);

- the energy consumption per time unit of a sensor with the radio off is modeled as a constant c, with $c \ll C$ (given the numbers above, we have $c \approx C/100$), independently of the actual sensor activity (sensing, or receiving the GPS clock).

5 Cell-Based Cooperative Strategies

We assume that the some initial information (such as node ID, node position, and so on) is available to the nodes by an initialization protocol. Since nodes are not coordinated in the initialization phase, conflicts to access the wireless channel could arise. These can be solved using traditional backoff-based techniques. The details of the initialization protocol implementation are not discussed in this paper. Rather, we are concerned with minimizing the energy consumed for node coordination *after* the initialization phase.

We assume that all the nodes are equipped with batteries with the same characteristics, i.e., that all the nodes have the same amount of energy E_{init} available at the beginning of the network operational time, which for simplicity, we define as immediately after initialization[1].

Our goal is to turn the radio off as long as possible, while maintaining a connected backbone of active nodes. To further simplify the analysis, we assume here that $c=0$, i.e., the energy consumed by a node for sensing and receiving the GPS signal with the radio off is negligible with respect to the energy consumed when the radio is on.

Observe that the energy consumption of a node in general is influenced by external factors, such as environmental conditions and, more notably, by "alarm" events (e.g., a fire in the WSN scenario of Section 2). During the alarm detection phase, energy consumption is no longer an issue: sensors must use all the energy that is needed to promptly propagate the alarm (e.g., the radio is turned on for a relatively long period of time). However, external factors are unpredictable by nature, and their influence on the nodes' energy consumption cannot be formally analyzed. For this reason, in the energy analysis of the protocols presented in this paper we will assume that there is no external influence on energy consumption during the network operational time. Thus, we will analyze the intrinsic energy cost of maintaining a connected backbone of active nodes (*the cost of silence*).

With the assumptions above, if no cooperative strategy is used to alternately turn the node transceivers on/off, the network operates for $T_b = E_{init}/C$ time units after initialization. At time T_b, all the nodes die simultaneously, and the network is no longer operative. Time T_b is called the *baseline time*; the objective of our analysis will be to quantify the network lifetime extension yielded by our protocols with respect to this baseline time.

[1] This is likely not to be true even if nodes are equipped with batteries with the same characteristics, since the energy cost for initialization of a node is expected to be proportional to the number of nodes in its neighborhood. So after initialization different processors in different cells will not have the same charge. This does not affect our protocol, but would make the presentation more cumbersome.

Similarly to the GAF protocol of [13], our protocols are based on a subdivision of the deployment region into non overlapping cells of the same size. More specifically, we make the following assumptions:

- n sensors are deployed (according to some distribution) in a square region of side l; i.e., the deployment region is of the form $R=[0,l]^2$, for some $l>0$;
- all sensors have the same transmitting range r, with $r<l$; i.e., any two sensors can communicate directly if they are at distance at most r from each other;
- R is divided into $N=8\frac{l^2}{r^2}$ non-overlapping square cells of equal side $\frac{r}{2\sqrt{2}}$.[2]

Given the subdivision above, the following properties are immediate:

P1. any two sensors in the same cell can communicate directly with each other; in other words, the subgraph induced by all the nodes belonging to the same cell is a clique;

P2. any two sensors in adjacent cells (horizontal, vertical, and diagonal adjacency) can communicate directly with each other.

A straightforward consequence of properties P1 and P2 is that leaving an active node for every cell is sufficient to provide a connected backbone of active nodes (under the assumption that the underlying communication graph is connected). Similarly to GAF, the goal of our protocols will be to elect a representative (also called *leader*) node for each cell, which is the only node that remains active in the cell. After a given *sleep period*, non-active nodes awake (i.e., turn their radio on) and start a new negotiation phase, with the purpose of (possibly) electing a new representative. Periodic leader re-election is needed to balance power consumption over the cell. If leaders are not re-elected periodically, nodes will die one after the other, at very distant times: when the first leader dies (approximately at time T_b), a new leader is elected; then the second leader dies, and so, until the last node in the cell is dead. With this scheme, network coverage, i.e., the ability of the network to "sense" any point of the deployment region R, could be compromised. Also, sleeping nodes would have to periodically check that their leader is still alive, which costs energy. A preferable solution would be to let the sensors consume energy more homogeneously, in such a way that they will all die approximately at the same time. This way, network coverage is not impaired: as long as the cell is operational (i.e., there is at least one alive node), almost all the nodes in the cell are still alive. This point will be further discussed below.

Before presenting the algorithms, we need to define the concept of network lifetime more formally. As discussed in [1], the definition of network lifetime depends on the application scenario considered. In this paper, we will use the concept of *cell lifetime*, which is defined as follows:

[2] We use here the cell subdivision used in [1], which is slightly different from that used in [13] (see [1] for details). This choice is due to the fact that we will want to compare the energy consumption of our protocols to the best possible energy savings achievable by any cell-based strategy, which have been analyzed in [1].

Definition 1 (Cell lifetime). *The lifetime of a cell is defined as the interval of time between the instant in which the cell becomes operative and the death of the last node in the cell.*

The time needed for network initialization is not considered in the analysis of lifetime, i.e., we assume that the cells are operative only when the initialization phase is over. To simplify the analysis, we further assume that all the cells become operative at the same time.

To conclude this section, we observe that nodes in adjacent cells could cause contention to access the wireless channel also after the initialization phase. To avoid this, we assume that an overall checkerboard pattern is used to allocate transmissions in adjacent cells at different time slots.

6 The FULL Protocol

In this section, we assume that the following information is initially available to the nodes:

- its location (in particular, its cell ID);
- the ID of every other node in its cell;
- access to global time (for synchronization purposes).

To obtain such information, every node sends a message containing its node and cell ID (which can be easily obtained from the GPS signal) during the initialization phase. We recall that at this stage, since intra-cell synchronization is not yet established, conflicts to access the wireless channel must be solved using traditional backoff-based techniques.

Given the assumptions on the initial information available to each node, the following coordination protocol for leader re-election can be easily implemented.

ALGORITHM FULL (Algorithm for cell i)
Assumptions:

- assume there are n_i nodes in the cell (this is known by hypothesis);
- assume also that each node p knows its ordering in the set of all IDs of nodes belonging to i. For simplicity, we can equivalently assume that nodes in cell i have IDs ranging from 0 to $n_i - 1$, with $0 \leq p \leq n_i - 1$ denoting the ID of node p (the p-th in the ordering);
- assume the re-election process starts at time T_r. Each step below will take time T_s, assumed sufficient to turn the radio on, send a message to a node in the same cell, and to perform the simple calculation needed by the algorithm.

Protocol for generic Node p ($p \neq 0, n_i - 1$):

- at time $T_r + (p-1) \cdot T_s$:
  - turn radio on and receive message $M = (E_{max}, m)$ from node $p - 1$
  - let E_p be the estimated energy remaining in the battery of node p at the end of the protocol execution
  - $E_{max} = \max(E_{max}, E_p)$
  - if $E_{max} = E_p$, then $m \leftarrow p$

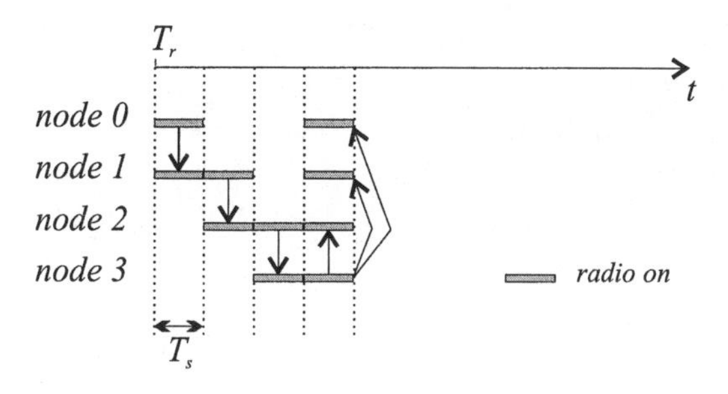

Fig. 1. Time diagram of the FULL protocol execution when $n_i = 4$.

- at time $T_r + p \cdot T_s$:
  - send message (E_{max}, m)
  - turn radio off
- at time $T_r + n_i \cdot T_s$:
  - turn the radio on and receive message $M = (E_{max}, m)$; m is the leader for the next sleep period, and E_{max} it the remaining energy of the leader
  - if $p \neq m$ then turn the radio off
  - END ; the negotiation phase ends: a new leader is elected

The protocol for nodes 0 and $n_i - 1$ is slightly different: node 0 simply sends message $(E_0, 0)$ at time T_r, and wakes up at time $T_r + n_i \cdot T_s$ to know the leader identity; node $n_i - 1$ ends the protocol after sending the message $M = (m, E_{max})$ at time $T_r + n_i \cdot T_s$ (in case it is not the leader, it turns its radio off before ending the protocol).

The time diagram of the protocol execution is depicted in Figure 1. It is immediate that each node other than the first and the last has its radio on for exactly $3T_s$ time, while node 0 and node $n_i - 1$ have their radio on for only $2T_s$ time. Assuming without loss of generality that T_s corresponds to one time unit, we have that the overall energy consumption of the FULL protocol is $(3(n_i - 2) + 4)C \in \Theta(n_i)$, i.e., the per node energy consumption is $\Theta(1)$. Thus, the energy spent by nodes to coordinate themselves is asymptotically optimal (note that every node must at least turn the radio on for one time unit to know the identity of the next leader, which require $\Omega(1)$ energy cost per node). In contrast, a naive re-election algorithm would have all the nodes with the radio turned on in the whole negotiation phase, with a $\Theta(n_i)$ energy cost per node ($\Theta(n_i^2)$ overall).

When the protocol ends, all the nodes in the cell (except the leader) have their radio off. At some time, the nodes must turn their radio on again and start a new negotiation phase, possibly electing a new leader. The duration of the sleep period should be chosen in a such a way that the energy consumption is balanced in the cell, and nodes die approximately at the same time. In what follows, we will assume that the duration of the sleep period depends on the energy E_{max} still available in the battery of the leader. More specifically, we set

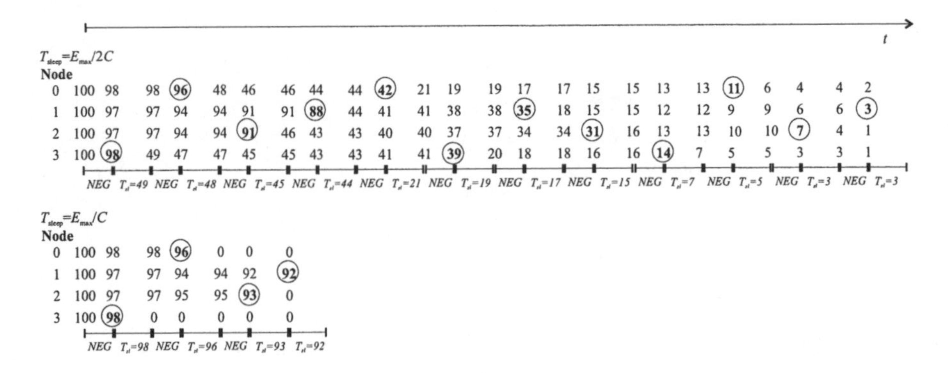

Fig. 2. Time diagram of the FULL protocol execution for different choices of the sleep period. Every node has 100 units of energy available after initialization. The leader is highlighted with a circle.

$T_{sleep} = \lfloor \frac{E_{max}}{2C} \rfloor$; i.e., we set T_{sleep} to half of the expected leader lifetime[3]. The example shown in Figure 2 motivates our choice.

In the figure, we report the time diagram of the FULL protocol execution for different choices of the sleep period. If the sleep period is set to E_{max}/C, only three negotiation phases (each during 4 units of time) are needed. Node 3 dies at time 102, node 0 at time 202, node 2 at time 299, and node 1 at time 391, causing the cell death. Thus, the cell lifetime is extended from the baseline time of 100 to 391, with an increment of 391%. This is the 97.75% of the best possible cell lifetime achievable, which is 400. However, the energy consumption is very unbalanced, and this is detrimental for coverage. In fact, the average number of alive nodes during the cell operational time is 2.54, which is only 63.5% of the best possible coverage (obtained when all the nodes in the cell are alive). Further, the average lifetime per node is only 248.5 units of time.

The situation is very different when we set T_{sleep} to $\lfloor \frac{E_{max}}{2C} \rfloor$. In this case, the energy spent for cooperation is higher (12 negotiation phases are needed), and the overall cell lifetime is decreased with respect to the previous case (324 instead of 391, corresponding to 81% of the best possible lifetime). However, energy consumption is very well balanced among nodes: the average number of alive nodes is 3.97 (99.25% of the best possible coverage), and the average lifetime per node is 321.75. Thus, we have traded a relatively small decrease in the cell lifetime with a significantly larger increase in the average lifetime per node, which induce a much better cell coverage. We remark that also in this scenario the energy spent during the 12 negotiation phases is limited (only 30% of the overall energy).

[3] We recall that we are assuming that the energy needed to sense and to receive the GPS signal is negligible with respect to that needed to power the radio. Further, we are assuming that no external factor is influencing energy consumption. Recalling that the leader has its radio on during the sleep period, with these hypotheses the expected leader lifetime is $\frac{E_{max}}{C}$.

Theorem 1. *Assume that cell i contains n_i nodes. If nodes coordinate their sleep periods using the* FULL *protocol, the cell lifetime is at least $\Theta(n_i T_b)$, where T_b is the baseline cell lifetime that corresponds to the situation where no cooperative strategy is used.*

Proof. Let us consider a given node p in the cell. We recall that a node spends $\Theta(1)$ energy during each negotiation phase, while during the sleep period it consumes $E_{max}/2$ energy if it is the leader, and 0 energy otherwise. Under the hypothesis that $E_{init} - 3n_iC > \frac{E_{init}}{2}$ (which holds in many realistic scenarios, where $E_{init} \gg n_iC$), p will be elected as leader before any other node is elected leader for the second time (see Figure 2). In fact, when a node is first elected as leader it consumes at least $E_{init}/2$ energy, and the condition above ensures that, at most during the n_i-th negotiation phase, the energy remaining in node p (which is at least $E_{init} - 3n_iC$) is greater than the energy remaining in every other node (which is at most $E_{init}/2$). Assuming w.l.o.g. of generality that p is the last node to be elected as leader for the first time, and observing that each negotiation phase lasts exactly n_i time units, we have that the cell lifetime is at least $n_i\left(n_i + \frac{T_b - 3n_i}{2}\right) \in \Theta(n_i T_b)$.

A straightforward consequence of Theorem 1 is that the FULL protocol (when it is used in every cell) consumes the minimal energy required to maintain a connected backbone of active nodes, i.e., it is optimal from the point of view of energy consumption (see also [1]). Another consequence of Theorem 1 is that a sensor network that uses FULL to coordinate node activity takes full advantage of a high node density: the lifetime of a cell is extended of a factor asymptotically equal to the number of nodes in the cell.

The FULL protocol described above works only if all the n_i nodes that were in the cell at the beginning of the cell operational time are still alive. Otherwise, the message containing the current value of E_{max} is lost when it is supposed to be received by a node which is actually dead. However, a simple modification of the protocol is sufficient to deal with this situation. A n_i-bits mask is included in the message propagated during the negotiation phase. The purpose of this mask is to keep trace of the nodes that are still alive. Initially, all the bits in the mask are set to 1 (the number of alive nodes is n_i). When node p receives the message, it compares the energy E_p remaining in its battery with a "minimum threshold" value E_{min}. If $E_p \leq E_{min}$, the node sets the p-th bit of the mask to 0, and propagate the message to the next node. When the last node in the ordering sends the message containing the leader for the next step, it includes in the message also the bit mask. This way, all the nodes in the cells know the number of nodes still alive, and they can turn their radio on at the appropriate instant of time (which depends on the position of the node ID in the ordering of the nodes which are still alive).

The value of E_{min} must be chosen carefully. In principle, E_{min} could be set to 3, which is the minimum value that guarantees that the node can remain alive for the entire next negotiation phase[4]. However, a more conservative value for E_{min} that accounts for the possible occurrence of external factors (that induce an increased energy consumption in the node) is preferable in practice.

7 The Random Protocol

In this section, we present and analyze a simple randomized coordination protocol, which accounts for the fact that, in general, a node might not know the actual number of nodes in its cell.

Algorithm Random (Algorithm for cell i)
Assumptions:

- each node knows its cell ID (this information can be easily obtained via GPS);
- each node has access to global time;
- nodes can detect conflicts on the wireless channel;
- the re-election process starts at time T_r. Step 2 below will take time T_s, assumed sufficient to flip a coin, turn the radio on, possibly send a message to the nodes in the same cell, and detect conflicts on the wireless channel.

Protocol for generic Node j:

- at time T_r:
  1. END = **False**
  2. repeat until END=**True**
     a) flip a coin, with probability of success p
     b) if SUCCESS, send message (E_j, j) (E_j is the estimated energy remaining in the battery); otherwise, set the radio in receiving mode;
     c) if nobody sent a message or COLLISION, goto step 2
     d) END=**True**
     e) if not SUCCESS, turn the radio off (if node j is not the leader, it turns its radio off; since exactly one node $k \neq j$ sent a message, node j knows the current leader k and its remaining energy E_k)
     f) goto step 2

Let $\sharp S$ be the random variable denoting the number of times that step 2 must be repeated before a unique leader is elected. Clearly, the energy consumption of the Random protocol depends on $\sharp S$, which, in turn, depends on the probability p of obtaining a success in the coin flip. It is easy to see that $\sharp S$ has geometric distribution of parameter $(1-q)$, where $q = 1 - n_i p(1-p)^{(n_i-1)}$ and n_i is the number of nodes in cell i. In fact, step 2 is repeated if the number of successes in the n_i independent coin flip experiments is 0 or at least 2, which occurs with probability q. The value of $Prob(\sharp S \leq k) = 1 - q^k$ converges to 1 as k increases for any $q < 1$, and the speed of convergence to 1 is higher for smaller value of q. Thus, the minimum value for q is desirable. This value is obtained setting

[4] Observe that a node with $E_p = E_{min}$ is elected as the leader for the next sleep period only if the energy remaining in the other nodes still alive is at most E_{min}. If this is the case, the cell is at the end of its operational time, and no further negotiation phase will take place.

$p=\frac{1}{n_i}$, for which we get $q=1-(1-\frac{1}{n_i})^{(n_i-1)}$. For large enough values of n_i, we have $q\approx 1-\frac{1}{e}$. Thus, with the best possible setting for p, RANDOM is expected to execute $E[\sharp S]=\frac{1}{1-q}\approx e$ times step 2 before a unique leader is elected. If we assume w.l.o.g. that T_s equals the time unit, and observing that all the nodes in the cell i have their radio on during the execution of RANDOM, we have that the average energy cost of the protocol is $en_iC\in\Theta(n_i)$, with a per node average energy consumption of $eC\in\Theta(1)$, which is optimal. Note that the deterministic FULL protocol is energy optimal in the worst case, while RANDOM is energy optimal on the average.

The analysis of the average case behavior of RANDOM in the entire network (under the hypothesis that the value of p in every cell is set to the actual number of nodes in the cell) is quite straightforward. Recalling that N denotes the total number of cells, we have that, on the average, a fraction of $\frac{1}{e}$ cells select the leader in the first step, a fraction of $\frac{1}{e}$ of the remaining $N\left(1-\left(\frac{1}{e}\right)\right)$ cells select the leader in the second step, and so on. Thus, the average number of cells that have selected the leader in at most k steps is $N_k=N\left(1-\left(1-\frac{1}{e}\right)^k\right)$. Note that N_k converges to N quite quickly; for instance, when $k=10$, we have $N_k=0.9898N$. We can conclude that, on the average, about ten steps are sufficient to elect the leader in every cell.

Note that setting p to the optimal value $\frac{1}{n_i}$ implies that the number of nodes in every cell must be known, which is the same assumption as in the deterministic FULL protocol. Assume now that we only have *probabilistic* information about node placement. In particular, assume that the n nodes are distributed uniformly and independently at random in the deployment region R. In this case, we know that the expected number of nodes in a cell is $\alpha=\frac{n}{N}=\frac{nr^2}{8l^2}$. If we set $p=\frac{1}{\alpha}$, we have the following result.

Theorem 2. *Assume that n nodes are distributed uniformly and independently at random in $R=[0,l]^2$, and set $p=\frac{1}{\alpha}$ in the* RANDOM *protocol. If n, r and l are chosen in such a way that $r^2n=kl^2\ln l$ for some constant $k\geq 16$, then:*

a. the communication graph obtained when all the nodes are awake is connected w.h.p.
b. $\lim_{n,l\to\infty} E[\sharp S]=e$.

Proof. Part *a.* of the Theorem is proved in [8] (Th. 8). To prove part *b.*, let $Max(n,N)$ and $Min(n,N)$ be the random variables denoting the maximum and minimum number of nodes in a cell when n nodes are distributed into N cells, respectively. Setting $n=\frac{kl^2\ln l}{r^2}$ for some constant $k\geq 16$, and $N=8\frac{l^2}{r^2}$, we have that $Max(n,N)\in\Theta(\alpha)$ and $Min(n,N)\in\Theta(\alpha)$ [1,5].[5] Assume w.l.o.g. that RANDOM is executed in cell i with minimum occupancy (i.e., $n_i=Min(n,N)$), and set $p=\frac{1}{\alpha}$. The probability $q(n,N)$ of repeating step 2 of the protocol when n nodes are distributed uniformly into N cells is:

$$q(n,N)=1-\frac{Min(n,N)}{\alpha}\left(1-\frac{1}{\alpha}\right)^{Min(n,N)-1}$$

[5] A similar and somewhat stronger property has been proved in [14].

We have:

$$\lim_{n,N\to\infty} q(n,N) = 1 - \lim_{n,N\to\infty} \frac{Min(n,N)}{\alpha}\left(1-\frac{1}{\alpha}\right)^{Min(n,N)-1}$$

Taking the logarithm, and using the first term of the Taylor expansion of $\ln(1-x)$ for $x \to 0$, we obtain:

$$\lim_{n,N\to\infty} \ln\frac{Min(n,N)}{\alpha} - \frac{1}{\alpha}\left(Min(n,N)-1\right) \tag{1}$$

Rewriting the first term of (1) as $\ln\left(1-\frac{\alpha-Min(n,N)}{\alpha}\right)$, and observing that $Min(n,N) \in \Theta(\alpha)$ implies that $\lim_{n,N\to\infty}\frac{\alpha-Min(n,N)}{\alpha}=0$, we can use again the Taylor expansion, obtaining:

$$(1) = \lim_{n,N\to\infty} -\frac{\alpha-Min(n,N)}{\alpha} - \frac{Min(n,N)-1}{\alpha} = -1\ .$$

It follows that $\lim_{n,N\to\infty} q(n,N) = \lim_{n,l\to\infty} q(n,N) = 1-\frac{1}{e}$, and the theorem follows by observing that $E[\sharp S] = \frac{1}{1-q}$, since $\sharp S$ has geometric distribution of parameter q.

Theorem 2 is very interesting, since it states that, even in the case when only probabilistic information about the number of nodes in a cell is known, the energy consumption of RANDOM is optimal on the average.

8 Time Requirements

While our main goal is to analyze battery life, it is interesting to analyze more precisely the time spent in the coordination routines. Assume that the number n of deployed sensors is sufficient to ensure that the communication graph obtained when all the nodes are awake is connected w.h.p.. If the nodes are distributed uniformly at random, the minimum and maximum cell occupancy is $\Theta(\log N)$ with overwhelming probability (see [1,5]). Thus, $n_i \in \Theta(\log N)$ for every i, and, since in FULL we have n_i rounds of $O(1)$ messages in every cell, we have that the total time spent by the algorithm to coordinate all the nodes is $\Theta(\log N)$.

The analysis for algorithm RANDOM is more involved. If we assume that in each cell n_i is known, and the probability p is set to $\frac{1}{n_i}$, then it is not hard to see that with high probability there will be a cell where the protocol will require at least $\Omega(\log N)$ steps. We sketch the proof of our time estimate for the case n_i is known. In fact, we have N cells, and in each of them we independently run a process that has probability $\frac{1}{e}$ of success (a unique representative for the cell is chosen). The probability that a leader has not been elected in t steps is $q_t = (1-\frac{1}{e})^t$. Since the random processes are independent, the probability that all N cells have leaders after t steps is $(1-q_t)^N$. Fix some parameter ϵ (we want to succeed with probability $1-\epsilon$). So we need $(1-q_t)^N = (1-(1-\frac{1}{e})^t)^N > 1-\epsilon$, which holds only if t is $\Omega(\log N)$.

The observation above may be surprising, as the energy consumption is $O(1)$ in every cell. The difference is due to bounding the expected value of the time (which is a constant) and bounds on the tail of the distribution, which are needed to analyze the time needed to terminate the protocol in all the cells (which is $\Theta(\log N)$).

The same observation as above is true even if n_i is not known. More precisely:

Theorem 3. *Under the hypothesis of Theorem 2, there are constants a and b such that*

1. $\lim_{n,l\to\infty} Pr[\textit{all cells have a leader by time } a\log n]=0$.
2. $\lim_{n,l\to\infty} Pr[\textit{all cells have a leader by time } b\log n]=1$.

Proof. [SKECTH] We use the fact that if we distribute $N\log N$ balls into N bins at random, there are constants a and b such that the probability that there exists a bin with a number of balls outside the interval $[a\log N, b\log N]$ tends to 0 as N goes to infinity [14]. Furthermore, the expected time bound for algorithm RANDOM holds also for non-optimal choice of the parameter α.

9 Discussion and Further Work

The protocols presented in this paper are designed to work in WSN where the node density is quite high, and sufficient to ensure the connectivity of the underlying communication graph. We have demonstrated that our protocols take full advantage of the node density, increasing the cell lifetime of a factor which is proportional to the number of nodes in the cell, and thus is optimal. However, when the node density is particularly high, the energy spent during the initialization phase by the FULL protocol (which is needed to know the number of nodes in every cell) could become an issue. In this setting, the adoption of the RANDOM protocol might be preferable. We remark that, when the node density is very high, the performance of RANDOM with $p=\frac{1}{\alpha}$ (i.e., without precise knowledge of the number of nodes in the cell) is very likely to be close to that of the average case.

As we claimed, these techniques have wider applicability. For example, in the case of a sensor network with high density, it is likely that an event will be reported by a large number of sensors. This can potentially yield communication delays due to contention, paradoxically making the system less reliable. Our techniques may be used to regulate access. For example, if units have a good estimate of the number of other sensors s that are observing the event, they may choose, in a first phase, to communicate it with probability $\Theta(\frac{1}{s})$. The analysis of our protocol (both energy consumption and time) applies to this scenario also.

Another issue for our leader election algorithms is the choice of the time for the next negotiation. If the only criterion is to maximize battery life – and this paper is an attempt to quantify how much one can gain by these techniques – then our algorithms are optimal. In practice, one might not have good estimates of actual battery use, so one might want to run the protocols much more frequently than $\frac{E_{max}}{2C}$. It makes sense to make each of these as efficient as possible, so our efficient algorithms are useful, but a quantitative estimate depends on the specifics of the protocol.

The choice of the value of p in the RANDOM protocol leaves space for several optimizations. For example, the value of p could account for the amount of energy still available at the node; this way, nodes with more energy are more likely to be elected as the next leader, and energy consumption is likely to be well balanced. Another possibility would be to change the value of p depending on the duration of the previous negotiation phase. We make this argument clearer. If p would be set to the optimal value, the expected duration of the negotiation phase is e time units. If the negotiation phase lasts much more than this, it is likely that p is far from the optimal value. For instance, if the previous negotiation phase required many steps because nobody transmitted, this could indicate the the value of p is too low. Conversely, if the excessive duration of the negotiation phase was due to conflicts on the wireless channel, it is likely that the value of p is too high. Since the information about the duration and the collision/absence of transmission during negotiation is available to all the nodes in the cell, a common policy for modifying the value of p can be easily implemented. The definition of such a policy is matter of ongoing work.

Acknowledgements. The work of Paolo Santi was partially funded by European Union under COMBSTRU, Research and Training Network (HPRN-CT-2002-00278).
Portions of this work were done during Janos Simon's stay at Istituto di Informatica e Telematica. Support of the CNR is gratefully acknowledged.

References

1. D.M. Blough, P. Santi, "Investigating Upper Bounds on Network Lifetime Extension for Cell-Based Energy Conservation Techniques in Stationary Ad Hoc Networks", *Proc. ACM Mobicom 02*, pp. 183–192, 2002.
2. B. Chen, K. Jamieson, H. Balakrishnan, R. Morris, "Span: An Energy-Efficient Coordination Algorithm for Topology Maintainance in Ad Hoc Wireless Networks", *Proc. ACM Mobicom 01*, pp. 85–96, 2001.
3. S. Coleri, M. Ergen, J. Koo, "Lifetime Analysis of a Sensor Network with Hybrid Automata Modelling", *Proc. ACM WSNA 02*, Atlanta, pp. 98–104, 2002.
4. L.M. Feeney, M. Nilson, "Investigating the Energy Consumption of a Wireless Network Interface in an Ad Hoc Networking Environment", *Proc. IEEE INFOCOM 2001*, pp. 1548–1557, 2001.
5. V.F. Kolchin, B.A. Sevast'yanov, V.P. Chistyakov, *Random Allocations*, V.H. Winston and Sons, Washington D.C., 1978.
6. A. Mainwaring, J. Polastre, R. Szewczyk, D. Culler, J. Anderson, "Wireless Sensor Networks for Habitat Monitoring", *Proc. ACM WSNA 02*, pp. 88–97, 2002.
7. V. Raghunathan, C. Schurgers, S. Park, M. Srivastava, "Energy-Aware Wireless Microsensor Networks", *IEEE Signal Processing Magazine*, Vol 19, n. 2, pp. 40–50, 2002.

8. P. Santi, D.M. Blough, "The Critical Transmitting Range for Connectivity in Sparse Wireless Ad Hoc Networks", *IEEE Trans. on Mobile Computing*, Vol. 2, n. 1, pp. 25–39, 2003.
9. L. Schwiebert, S.K.S. Gupta, J. Weinmann, "Research Challenges in Wireless Networks of Biomedical Sensors", *Proc. ACM Mobicom 01*, pp. 151–165, 2001.
10. M.B. Srivastava, R. Muntz, M. Potkonjak, "Smart Kindergarten: Sensor-based Wireless Networks for Smart Developmental Problem-solving Environments", *Proc. ACM Mobicom 01*, pp. 132–138, 2001.
11. C. Srisathapornphat, C. Shen, "Coordinated Power Conservation for Ad Hoc Networks", *Proc. IEEE ICC 2002*, pp. 3330–3335, 2002.
12. D.C. Steere, A. Baptista, D. McNamee, C. Pu, J. Walpole, "Research Challenges in Environmental Observation and Forecasting Systems", *Proc. ACM Mobicom 00*, pp. 292–299, 2000.
13. Y. Xu, J. Heidemann, D. Estrin, "Geography-Informed Energy Conservation for Ad Hoc Routing", *Proc. ACM Mobicom 01*, pp. 70–84, 2001.
14. F. Xue, P.R. Kumar, "The Number of Neighbors Needed for Connectivity of Wireless Networks", *internet draft*, available at `http://decision.csl.uiuc.edu/~prkumar/postscript_files.html#Wireless%20Networks`.

Topology Transparent Support for Sensor Networks*

Robert Simon and Emerson Farrugia

Department of Computer Science
George Mason University
Fairfax, VA USA 22030
{simon,efarrugi}@cs.gmu.edu

Abstract. Due to their energy and performance limitations and unpredictable deployment strategies, sensor networks pose difficult engineering challenges for link configuration and topology coordination. This paper presents a novel, topology transparent technique for automatic time slotted link management in sensor networks. Our approach relies on the use of mutually orthogonal latin squares to determine transmission schedules between sensor nodes. We present three different protocols which, when compared to other approaches, significantly reduce or completely eliminate control message overhead. Our approach is probabilistic in the sense that it does not entirely guarantee either complete transmission collision freedom or sensor area coverage. However, we show through analysis and extensive simulation that by proper choice of system parameters we obtain excellent performance and coverage results.

1 Introduction

Emerging wireless sensor networks offer extremely flexible and low cost possibilities for monitoring and information collection for a wide range of environments and applications [1]. In such a network, each node consists of one or more sensors, an embedded processor, a low powered radio, and generally speaking a battery operated power supply. These nodes cooperate to perform monitoring tasks, and typically report their readings and are under the control of external stations.

Sensor networks can be deployed over large areas and in unpredictable ways. This makes the need for low overhead and decentralized procedures for network configuration and management of paramount importance. Further, it may not be possible to replace a node's battery when it dies, and because of this all protocols must be energy efficient. These unique challenges have lead to a recent flurry in current research into the design and performance analysis of network and data link management protocols for sensor networks.

The performance of a sensor network is best judged in terms of meeting the sensing requirements of the users. Other performance metrics include maximizing

* This work was supported in part by the NSA under grant number MDA904-98-C-A081-TTO55

H. Karl, A. Willig, A. Wolisz (Eds.): EWSN 2004, LNCS 2920, pp. 122–137, 2004.

the lifetime of the network, generally by promoting energy efficient policies, and adopting scalable and fault tolerant deployment and management approaches. It has been noted that simply increasing the number of sensors often does not lead to better performance from the users point of view, and in fact may actually degrade the sensing results [12]. The reason for this is that an environment densely populated with sensors leads to a larger number of collisions and congestion in the network. This in turn increases reporting latency and rapidly degrades power reserves. Moreover, large numbers of samples may exceed the accuracy requirements of the observer.

Based on the above observations, we note that a fundamental sensor network design principle is to strike a balance between adequate and accurate sensor coverage for the sensed environment, transmission collision reduction, and extending sensor network lifetime. This paper presents several novel, topology transparent time slotted link configuration mechanisms that directly address this design principle. Specifically, our technique provides probabilistic guarantees for sensor network connectivity, collision reduction, and sensed area coverage. The protocols described here are designed to minimize or eliminate the setup cost for link configuration, and completely eliminate control overhead during data exchange. Congestion control is achieved by using a scheduling technique based on mutually orthogonal latin squares [5]. Each node schedules its transmissions according to a particular symbol assignment, in such a way as to reduce or completely eliminate transmission collisions. Our approach is topology transparent in the sense that it requires little or no knowledge of the underlying network topology.

Our technique does not guarantee absolute collision freedom, and may also result in some sensors intentionally disabling themselves under certain circumstances. However, by protocol design and in conjunction with results in scaling and dimensioning wireless systems [6], we show how to balance the system so that these effects are minimized or eliminated.

We judge the effectiveness of our approach based on the number of correctly configured (e.g., collision reduced) links and the coverage area of the sensor network after configuration. We have run a large number of simulation experiments to evaluate the frequency of collisions in different traffic scenarios, sensor densities, and sensor transmission ranges. These results show that our protocols can achieve a high level of communication performance while simultaneously reducing or eliminating link configuration setup and maintenance cost.

2 Background and Related Work

Our work targets large scale sensor networks that need to be configured and reconfigured as rapidly and cheaply as possible. We assume that the sensor network is under the control of one of more sinks. Further, even though the network may require sensor to sensor communication for purposes of data aggregation, the basic communication pattern is sink to sensor(s) or sensor(s) to sink.

2.1 Link Configuration and Routing

It has been observed that a major source of inefficiency and energy waste in sensor networks is due to congestion and link level collisions[12][13]. Because of this there has been a recent interest in designing link configuration and maintenance protocols specifically for sensor networks. As compared to other wireless network protocols, for instance those proposed for general purpose mobile ad hoc systems, sensor link configuration protocols are severely limited by the need to avoid control overhead and reduce retransmissions.

Due to above considerations, pure contention based link access schemes, such as 802.11[10], are not considered appropriate for use in sensor networks. Most work in sensor network link configuration can be characterized as either a hybrid (contention plus scheduling) or a contention free approach. For example, Shorabi et. al. [11] proposed S-MACS, a self-organizing protocol that uses different time slots and channels to arrange link configuration. In [4], Woo and Culler proposed an adaptive rate control algorithm under the assumption of different types of CSMA mechanisms. A protocol that address energy efficiency in a hybrid contention based approach is presented in [13]. In contrast to this work, we focus on control messaging elimination in a simple, single channel architecture.

Bao and Garcia-Luna-Aceves also describe work in collision free scheduling for ad hoc networks[2]. Although this work can provide collision free scheduling over a two hop range, thus eliminating hidden terminal problems, it requires a higher level of control information than we present here, and is not directly designed for ad hoc sensor networks.

Similar to our work are a class of topology transparent link scheduling methods [3],[8],[9]. The underlying idea is to avoid the need for passing control information. These schemes work by assigned nodes codes or symbols that correspond to specific time slot assignments. For instance, in [9] Ju and Li consider TDMA scheduling based on latin squares in multi-channel mobile network. Our work differs from this earlier work in a number of important ways. We are applying topology transparency directly to sensor networks. Further, in the above a primary concern was scheduling optimality, while our concern is to ensure adequate sensed environmental coverage, at the expense of disabling sensors. In addition, we present several new types of initialization protocols. Unlike this earlier work, we also suggest and evaluate associated routing mechanisms in the context of sensor and sink systems. Finally, for two of our protocols we allow a minimal learning phase.

2.2 Mutually Orthogonal Latin Squares

This section provides a brief introduction to mutually orthogonal latin squares [5]. They are explained as follows: first, assume that there is a set of Y symbols, $1, 2, \ldots, y$. Now consider a $p \times q$ rectangular array, with $y \geq p, y \geq q$. Then we have the following definition:

Definition 1. *A rectangular array is called a* latin rectangle *if every symbol from the symbol set appears at most once in each column and once in each row. If* $p = q$*, then the array is called a* latin square.

We next introduce the concept of mutually orthogonal latin squares. Assume that we have two distinct $p \times p$ latin squares A and B. Write the symbol in the ith row and the jth column from A (respectively, B) as $a_{i,j}$ (respectively, $b_{i,j}$). By definition we have $a_{i,j}, b_{i,j} \in Y$. Squares A and B are orthogonal if the p^2 ordered pairs $\langle a_{i,j}, b_{i,j} \rangle$ are all different. This concept can be generalized to entire families of latin squares of the same order. A set of $\mathcal{A}$ latin squares of order p are a mutually orthogonal family of latin squares if every pair in the set $\mathcal{A}$ is orthogonal. Figure 1 shows two mutually orthogonal latin squares of order 5.

$$\mathbf{A} = \begin{vmatrix} 2&3&4&0&1 \\ 3&4&0&1&2 \\ 4&0&1&2&3 \\ 0&1&2&3&4 \\ 1&2&3&4&0 \end{vmatrix} \qquad \mathbf{B} = \begin{vmatrix} 3&0&2&4&1 \\ 4&1&3&0&2 \\ 0&2&4&1&3 \\ 1&3&0&2&4 \\ 2&4&1&3&0 \end{vmatrix}$$

Fig. 1. Examples of mutually orthogonal latin squares.

3 Sensor Scheduling with Mutually Orthogonal Latin Squares

We assume that the sensor network can be time synchronized. We do not address how nodes are time synchronized in this paper. This can be achieved by one of several techniques, including listening to data frame transmissions and aligning time slots to the latest starting point of a complete packet transmission, or by using one of a number of proposed time synchronization protocols for sensor networks [7].

For simplicity assume that each sensor has only one channel, and that the size of each sensor packet is fixed (or upper bounded) at d bits. Notice, however, that our technique can immediately be applied to multiple channels if available. Transmission scheduling is achieved by assigning specific sensors to transmit during specific slots, e.g, TDMA. Each TDMA frame F consists of p subframes, each of which has p time slots. Each slot is s seconds long. Therefore, each TDMA frame consists of p^2 slots. Suppose that each sensor has a link layer bandwidth of L bits per second. By design, $s \geq \frac{d}{L}$.

Scheduling works as follows. We have a family of M mutually orthogonal latin squares (MOLS). Each sensor i is assigned a particular symbol from a particular latin square. We write this assignment function as $c(i)$, with $c(i) : i \rightarrow p \times M$. For instance, consider the latin squares shown in Figure 1. If we assign sensor

i to symbol 2 from latin square A, then sensor i transmits in the first time slot in the first subframe, the fifth time slot in the second subframe, the fourth time slot in the third subframe, the third time slot in the fourth subframe, and the second time slot in the fifth subframe. All transmission schedules are thus repeated every p^2 seconds.

Assume that each sensor has a radio range of R meters. We say two sensors have a *direct conflict* if they transmit packets during the same time slot and if they are no more then R meters away from each other.

Definition 2. *Two sensors are* Conflict-Free *if they do not have a direct conflict.*

An objective of the scheduling algorithm is to maximize the number of *Conflict-Free* sensors in radio range. To achieve this goal we first present a simple Theorem:

Theorem 1. *Two sensors are* Conflict-Free *if they are assigned different symbols from the same latin square.*

Proof. Two sensors can be in *direct conflict* only if they are in radio range and transmit during the same time slot. However, since both sensors determine their transmission slots based on the occurrence of their specific symbols from the same latin square, then by Definition 1 their symbols and therefore their transmissions will be *Conflict-Free.*

Sensors that are not *Conflict-Free* can suffer from direct collisions. We define two types of conflicts. The worst case, in the sense of maximizing the chances for collision, occurs when in-range sensors have the same symbol assignment from the same square. For this case we use the following terminology:

Definition 3. *Two sensors are* Strongly-Conflicted *if they are in radio range and are assigned the same symbol from the same latin square.*

Notice that if a sensor is *Strongly-Conflicted* it still will be able to have collision free transmissions if none of the other sensors that it conflicts with have a packet to transmit at the same time. Finally we have,

Definition 4. *Two sensors are* Weakly-Conflicted *if they are in radio range and are assigned symbols from different latin squares.*

Two sensors that are *Weakly-Conflicted* will have their transmissions collide at most 1 out of F times over each time interval F. This follows directly from the MOLS definition, since each ordered symbol pair from two different squares occurs once and only once.

4 Assignment and Routing

This section describes a family of MOLS symbol-square assignment protocols and routing techniques. At the start of the protocol there is a family of M MOLS of order p. We have defined three different assignment methods, given below:

- **STATIC** – Sensors are assigned random symbols and random squares at deployment time.
- **GREEDY** – One sink initiates a broadcast protocol in which after a passive learning period each sensor ultimately broadcasts its symbol and square assignment. Sensors choose an unused symbol in the *most used square* based on the broadcasts they receive from their neighbors during the learning period. A sensor declares itself *DARK*, i.e. disabled, if the broadcasts it receives show that any assignment would result in it being *Strongly-Conflicted.*
- **RANDOM** – One sink initiates a broadcast protocol in which after a passive learning period each sensor ultimately broadcasts its symbol and square assignment. Sensors choose an unused symbol in a *random square* based on the broadcasts they receive from their neighbors during the learning period. A sensor declares itself *DARK*, i.e. disabled, if the broadcasts it receives show that any assignment would result in it being *Strongly-Conflicted.*

The **STATIC** method completely eliminates a configuration cost, while the other two methods require a neighbor exchange, although of a lower cost than the protocols described in Section 2.1. Once a sensor assigns itself a symbol-square combination it never needs to change it. The intuition behind the **GREEDY** method is to maximize the conditions that make Theorem 1 true.

There are several factors involved in choosing the MOLS family. We would like to pick a family of M squares such that for any order p the value of M is maximized, thus minimizing the number of *Strongly-Conflicted* sensors. To do this we make use of the following results. First, note that $M \leq (p-1)$ [5]. Next, define an MOLS family of size $M = p-1$ as *complete.* Now, suppose p is a prime number. We give the following Theorem without proof:

Theorem 2. *If $n \geq 1$ and $n = p^k$, where k is a positive integer, then there is an MOLS family of order n. [5]*

In order to minimize the number of *Strongly-Conflicted* sensors we chose a MOLS family of size M such that $M = p^k$, where p is a prime and $k \geq 1$. By Theorem 2 we are guaranteed that such a family exists, when p is prime.

4.1 Initialization Protocols

Both the **GREEDY** and the **RANDOM** approaches require that an external sink run an initialization protocol. Sensors are in different states at different times. Initially, all sensors are in an *IDLE* state. For both the **GREEDY** and the **RANDOM** protocols, there is a learning phase where the sensors are in an

UNASSIGNED state. Once the sensors have assigned symbols they are in the *ASSIGNED* state. If a sensor determines that it will be *Strongly-Conflicted* with another sensor then it goes into a *DARK* state. The protocol works as follows. Each sensor initializes itself with the procedure shown in Figure 2. The *Assignment_bag* is used to keep the collection of received symbol-square assignments.

my_state ⇐ IDLE;
Current_order ⇐ NULL;
my_symbol ⇐ NULL;
Assignment_bag ⇐ NULL;

Fig. 2. Initialization procedure for **GREEDY** and **RANDOM** protocol

A designated sink initiates the assignment phase by sending out a *NET_INIT* message. This message is then rebroadcast through the system. During this state sensors contend for access to the transmission media, and there is no guarantee that messages will be received correctly. We evaluate this effect in Section 5. The message is quite simple:

NET_INIT(order, square, symbol)

where the value of *order* is the MOLS family size, *square* is the particular square picked by the sender, and *symbol* is the symbol picked by the sender.

Sensors receive this message and enter a learning phase, during which time they listen for their neighbors to send out other *NET_INIT* messages. At some point each sensor leaves the learning state and, if it could successfully assign itself a symbol-square pair, transmits out its own initialization message. This works by defining a function $\mathcal{F}$ that examines the symbol assignments in the assignment bag and picks a new symbol-square combination, if possible, based upon either the **GREEDY** or **RANDOM** approach. The procedure followed by sensors running either of the two protocols is shown in Figure 3.

One of the effects of both **GREEDY** and **RANDOM** is that part of the environment may be left uncovered due to sensors declaring themselves *DARK*. This is best illustrated by an example. Figure 4 shows a sensor node field, with the sensing range for each node shown as a dotted circle centered around that node. The figure shows an uncovered area, no longer able to be sensed, because several adjacent sensors went *DARK*.

4.2 Routing

As noted in [2] and elsewhere, one of the drawbacks of previous topology transparent link configuration schemes can be the difficulty in integrating these schemes into higher level routing and reliability protocols. However, this problem can be addressed in sensor networks whose primary form of communication

```
On receipt of the message NET_INIT(order, square, symbol)

Current_order ⇐ order;
Assignment_bag ⇐ ⟨square, symbol⟩
my_state ⇐ UNASSIGNED;
time_wait ⇐ random_time();

WHILE (waited less then time_wait seconds)
{
        On receipt of New Message NET_INIT(order, square, symbol)
                Assignment_bag ⇐ ⟨square, symbol⟩ ⋃ Assignment_bag;
}

IF ⟨ my_symbol, my_square ⟩ ⇐ F(Assignment_bag) ≠ NULL
{
        my_state ⇐ ASSIGNED;
        TRANSMIT NET_INIT( order, my_square, my_symbol);
}
ELSE
        my_state ⇐ DARK;
```

Fig. 3. Assignment phase **GREEDY** and **RANDOM** protocol

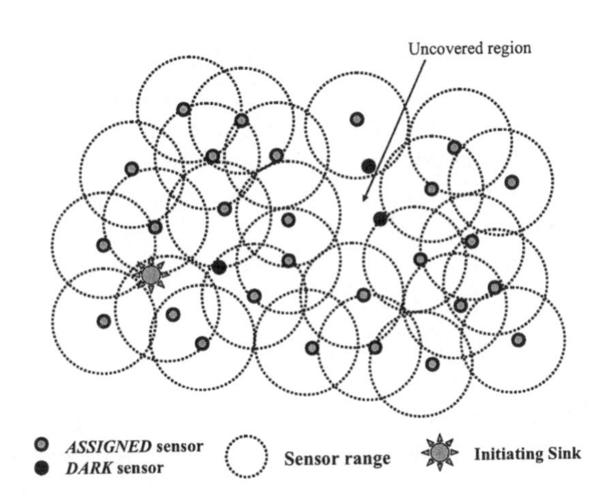

Fig. 4. Uncovered sensor region after assignment phase.

consists of sink to sensor and sensor to sink. The simplest and lowest cost solution is to form spanning trees from each sink to all the sensors in the network. The only state that each sensor needs to maintain is a parent pointer.

Spanning trees can be formed when a sink broadcasts a message, containing the sink ID, a hop count, and a tree ID. Initially, the sender ID is the same as the tree ID, and the hop count is 0. When a sensor receives a beacon, it checks to see if the sender would yield a better hop count than its current parent for the specified tree. If that is the case, it sets its parent pointer for the specified tree to the sender, and makes its hop count "1 + the received hop count". It then schedules its own message with this new information for its next available transmission slot, and the process is repeated. Sensors are likely to receive the best available parent and hop count message first.

5 Analysis

Section 4 presented three different assignment techniques which radically reduce or even completely eliminate the cost of link configuration and maintenance. However, as already noted, these protocols do not guarantee that sensors will avoid all one hop collisions. Further, the problem of two hop, "hidden terminal" collisions is not addressed by our protocol. [1] Finally, as seen in Section 4.1, it is possible that part of the sensed environment may be left uncovered. We address these issues by analysis and performance simulation.

5.1 Coverage and Collision Analysis

Coverage and collision issues are determined by the number of *DARK* or *Strongly-Conflicted* sensors. We present here a way to estimate for a given system architecture the expected number of *Strongly-Conflicted* sensors. Assume that $N \geq 2$ sensors are deployed over a $D \times D$ grid, and that sensors are distributed within the grid according to a uniform random distribution. Without loss of generality assume that the sensing range is the same as the radio range, which we denote by R. For the environment with parameters R and D, let $H(N)$ be the average direct link degree of the network. In other words, $H(N)$ represents the average number of neighbors each sensor has. An approximation of $H(N)$ is

$$H(N) = \frac{\pi R^2}{D^2}(N-1)$$

Suppose each sensor i is assigned a symbol-square assignment given by $c(i)$. Let p be the order of the latin square. Notice that this is the same as the number of symbols. Let M be the number of MOLS. Define a random variable $X_{i,j}$ as

$$X_{i,j} = \begin{cases} 1 & \text{if c(i) == c(j) \&\& } i \neq j \\ 0 & \text{otherwise} \end{cases}$$

[1] We note that both the **GREEDY** and **RANDOM** approach can easily be extended to two hops to address the hidden terminal problem. Since two hop information requires a higher level of protocol overhead, we will not evaluate those approaches here.

In other words, the value of $X_{i,j}$ is 1 if i and j have the same symbol assignment. Let

$$\Gamma = p \times M$$

For any arbitrary pair of sensors i, j, $i \neq j$, the probability p that $X_{i,j}$ is 1 is

$$p[X_{i,j} = 1] = \frac{1}{\Gamma}$$

By definition, sensors can only be *Strongly-Conflicted* if they are in radio range of each other. For each disk with radius R within the grid, let $E[S_n]$ be the expected number of *Strongly-Conflicted* sensors. Then

$$\begin{aligned} E[S_n] &= \sum_{i=2}^{\lfloor H(n) \rfloor} \sum_{j=1}^{i-1} 1 \cdot (p[X_{i,j}]) \\ &= \binom{\lfloor H(n) \rfloor}{2} \frac{1}{\Gamma} \\ &= \frac{\pi^2 R^4}{2D^4} \frac{(N-1)(N-2)}{\Gamma} \end{aligned} \tag{1}$$

Equation (1) provides a heuristic for determining the parameters N, R and Γ. Notice that a larger value of Γ results in a smaller number of *Strongly-Conflicted* sensors but in a larger overall time delay for the frame size. Further, the value of R is important, because if it is set too low, the network will be partitioned, and if it is set too high, there will be too many *Strongly-Conflicted* neighbors. We can use ad hoc network scaling law results to determine reasonable values of R. For instance, the work in [6] showed that with random placement of reliable nodes, an ad hoc network will remain connected as long as

$$R \sim \left(\sqrt{\frac{\log N}{N}} \right)$$

The above result assumes that the value of D is normalized to 1. Since w.l.g. we are assuming that the sensing range is equal to R, we use this result as a guide for determining R and N. Substituting this result into Equation (1) yields (ignoring lower order terms)

$$E[S_n] = \left(1 - \frac{3}{N}\right) \frac{\pi^2 (\log N)^2}{2\Gamma} \tag{2}$$

Equation (2) can be used as a guideline to find the balance between performance, coverage, and network partitioning.

5.2 Performance Evaluation

We have evaluated our protocols through replicated simulation experiments, using a large number of different configurations in terms of sensors, sinks, MOLS sizes, and traffic loads. By simulating packet transmissions, we take into account collisions both during the initialization phase and after the schedules have been set up, to account for both *Weakly-Conflicting* sensors and hidden terminal problems. We collected statistics for the size of the coverage region, the number of partitioned sensors and sinks, and the average number of collisions for different assignment schemes and traffic loads.

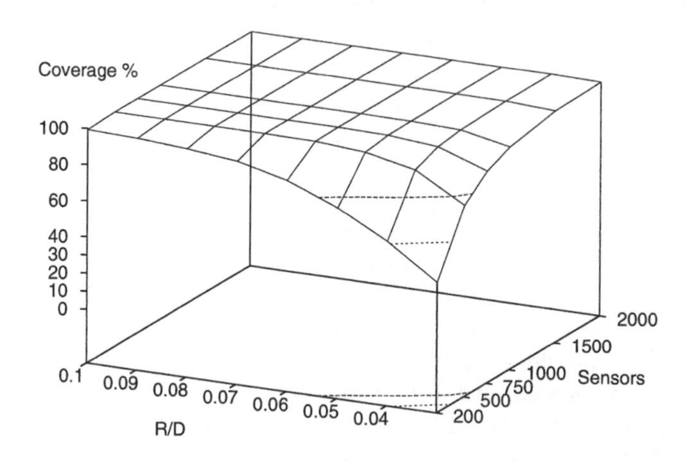

Fig. 5. Coverage Results.

We again assume without loss of generality that the sensing range is equal to the radio range R. For each experiment, we can express R as a function of the grid dimension D. With Equation (2) in mind, we use R/D ratios of 0.02, 0.04, 0.06, 0.08, and 0.1. Note, for example, that a run with an R of 20 and a D of 100, and a run with an R of 200 and a D of 1000 yield an equally dense network, evident by the close correlation in the resulting coverage and collision percentages for fixed numbers of sensors, assignment types, and loads. The values used in the graphs are averages of these closely correlated values, for sizes of D between 100 and 1000, in increments of 100.

Figure 5 provides an upper bound on the coverage area for given R/D ratios and a fixed numbers of sensors. The figure shows that it is possible to cover all of the area with reasonable R/D ratios. Figure 6 shows the actual coverage area, using the **STATIC** approach, when the effects of random sink placement and unreliable spanning tree formation are taken into account. This figure also

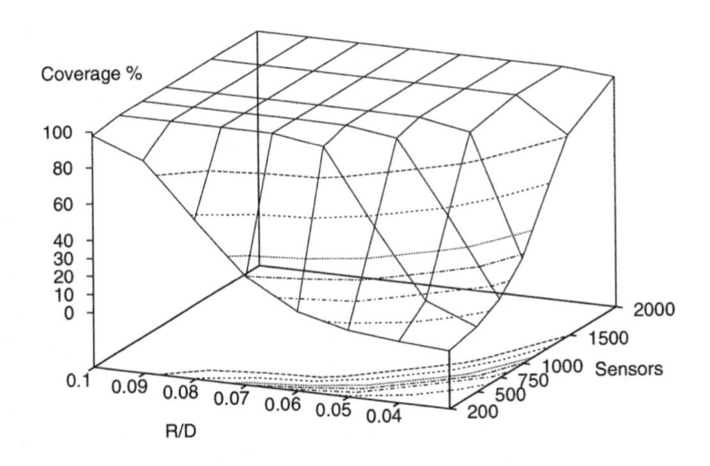

Fig. 6. Sink spanning tree coverage results.

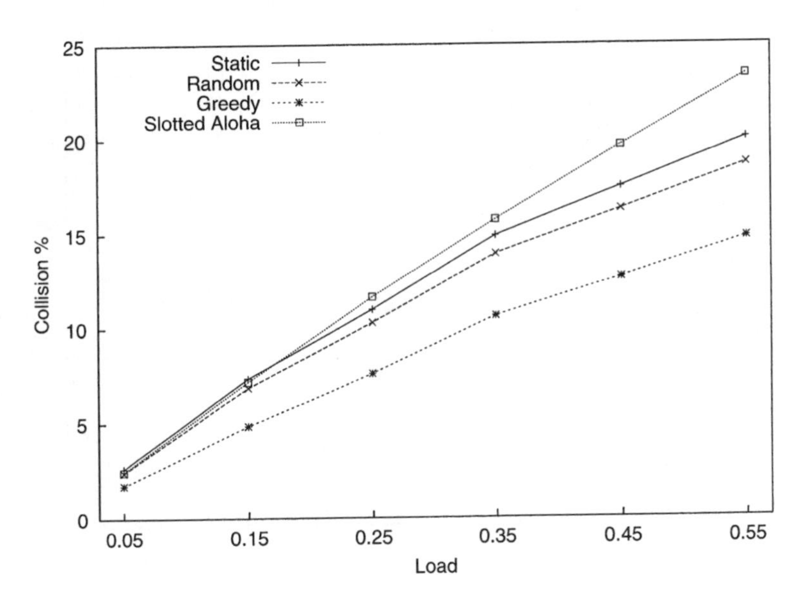

Fig. 7. Collisions for 500 sensors, 0.06 R/D.

aggregates the results of multiple experiments; for each experiment, the number of sinks is a fixed percentage of the number of sensors, sink locations are chosen randomly, and spanning trees are formed by the sinks based on the routing procedure shown in Section 4.2, incorporating the effects of both network partitions and packet collisions. The coverage of the spanning trees is then calculated. For

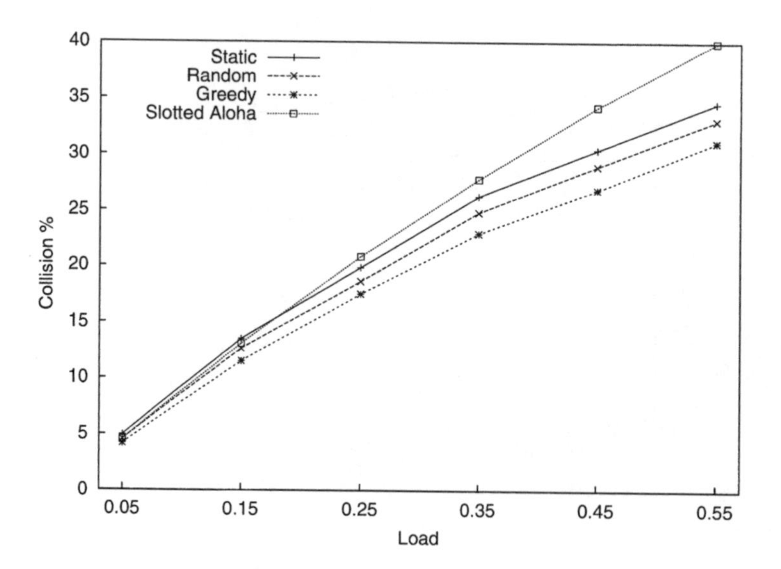

Fig. 8. Collisions for 1000 sensors, 0.06 R/D.

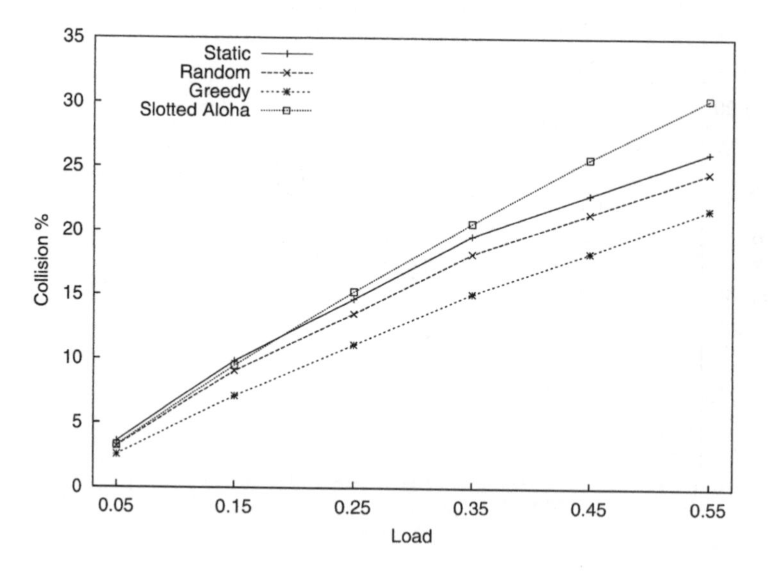

Fig. 9. Collisions for 1500 sensors, 0.04 R/D.

lower R/D ratios, the coverage from the point of view of an individual sink can be below 0.5. However, we expect that sinks will be able to share data, and adequate R/D ratios are selected to cover the environment. Furthermore, most of the low coverage results stem from already partitioned networks.

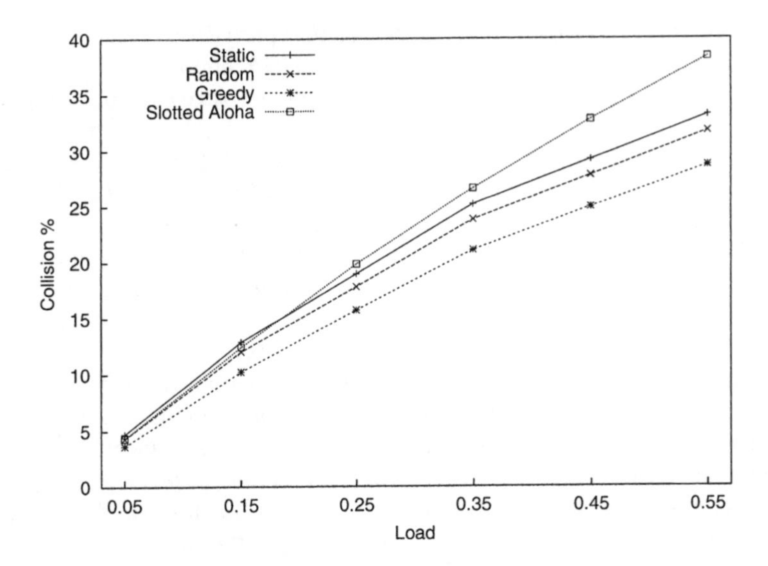

Fig. 10. Collisions for 2000 sensors, 0.04 R/D.

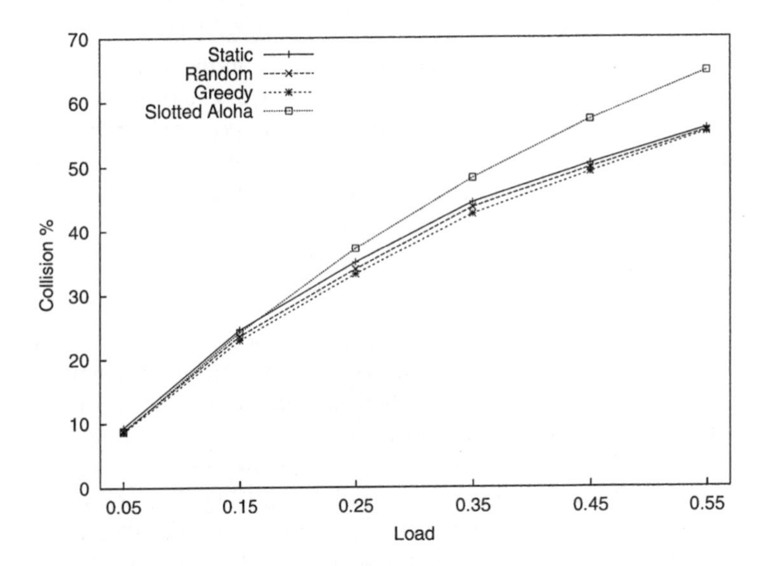

Fig. 11. Collisions for 2000 sensors, 0.06 R/D.

The sensors and sinks have **STATIC** symbol assignments in the coverage experiments. It is worth noting that although the absence of *DARK* sensors may indicate better coverage in the **STATIC** case, the number of collisions during spanning tree formation is higher. As a result, the size of the spanning trees would be smaller, yielding reduced coverage. Our experiments show that the coverage portrayed in Figure 6 of the **STATIC** approach is in fact worse than the coverage of both the **RANDOM** and **GREEDY** approaches, and therefore serves as a lower bound on the coverage area for our three symbol assignment schemes.

The next set of results shows traffic collision data for a wide variety of traffic loads, presented to demonstrate the effect of *Weakly-Conflicted* and two hop collisions. We ran experiments for all R/D ratios, grid sizes, and numbers of sensors between 200 and 2000 in increments of 200. Due to space limitations we only show a subset of these results. Traffic is generated according to a Poisson process at each node in the system. We count as a collision all hidden terminal (two hop) problems.

Results are shown for 500 sensors and 1000 sensors at an R/D of 0.06 (Figures 7, 8), and 1000, 1500, and 2000 sensors at an R/D of 0.04 (Figures 9, 10). These R/D ratios are presented because our experiments (Figure 6) show high coverage values for these ratios at these numbers of sensors. For purposes of comparison, we also simulated a contention based time slotted scheme, similar to slotted Aloha, whereby a node transmits at the beginning of the time slot if it has data. We note that for all runs, the **GREEDY** approach substantially outperformed both **RANDOM** and **STATIC**, and all three outperformed the comparison to slotted Aloha. Also, note that we expect the traffic load for most sensor systems to be less then 0.25, although we evaluated loads up to 0.55.

All of the other R/D runs showed similar trends, in particular the success of the **GREEDY** strategy. This can be seen by looking at the runs for 2000 sensors at an R/D of 0.06 (Figure 11), which leads to an extremely high amount of congestion and collisions, due to the large number of neighbors. The **GREEDY** approach outperforms the other schemes even under this extreme scenario.

6 Conclusion

This paper presented three different protocols for topology transparent link management in sensor networks. The basic idea is to assign sensors symbol-square combinations from a set of mutually orthogonal latin squares in order to produce conflict reduced transmission schedules between sensor nodes. The major advantage of our approach is the reduction or elimination of control message overhead for link configuration and management, while producing time scheduled sensor link access. We can therefore produce energy savings for sensor networks.

Our protocols are probabilistic, in the sense that they cannot guarantee complete coverage or complete collision freedom. However, we showed that by using simple dimensioning heuristics it is possible, with a high degree of certainty, to obtain complete coverage and to substantially reduce the number of collisions. Our technique is therefore well suited for large scale self-configuring sensor environments.

References

1. I.F. Akyildiz et al., "A Survey on Sensor Networks," IEEE Communications Magazine, August 2002, pp. 102–114.
2. L. Bao and J.J. Garcia-Luna-Aceves, "A New Approach to Channel Access Scheduling for Ad Hoc Networks," ACM/IEEE MobiCom, July 2001, pp. 210–221.
3. I. Chlamtac and A. Farago, "Making transmission schedules immune to topology changes in multi-hop packet radio networks," IEEE/ACM Transactions on Networking, Vol. 2, February 1994, pp. 23–29.
4. A. Woo and D. Culler, "Transmission Control Scheme for Media Access in Sensor Networks," ACM/IEEE MobiCom, July 2001, pp. 221–235.
5. J. Denes and A. D. Keedwell, Latin Squares and Their Applications. New York, Academic Press, 1974.
6. P. Gupta and P. R. Kumar, "Critical power for asymptotic connectivity in wireless networks," in Stochastic Analysis, Control, Optimisation and Applications, Edited by W.M. McEneany, G. Yin, and Q. Zhang. Boston, Birkhauser, 1998, pp. 547–566.
7. J. Elson and D. Estrin, "Time Synchronization for Wireless Sensor Networks," IPDPS Workshop and Parallel and Distributed Computing Issues in Wireless Networks and Mobile Computing, April 2001, pp. 1965–1970.
8. J.-H. Ju and V. O. K. Li, "An optimal topology-transparent scheduling method in multihop packet radio networks," IEEE/ACM Transactions on Networking, Vol. 6, June 1998, pp. 298–306.
9. J.-H. Ju and V. O. K. Li, "TDMA scheduling design of multihop packet radio networks based on latin squares," IEEE Journal on Selected Areas in Communication, Vol. 17, No. 8, August 1999, pp. 1345–1352.
10. LAN MAN Standards Committee of the IEEE Computer Society, *Wireless LAN medium access control (MAC) and physical layer specification*, IEEE N.Y., N.Y., USA, IEEE Std 802011-1997, 1997.
11. K. Sohrabi, J. Gao, V. Ailawadhi and G. Pottie, "Protocols for Self-Organization of a Wireless Sensor Network," IEEE Personal Communications Magazine, October 2000, pp. 16–27.
12. S. Tilak, N. Abu-Ghazaleh and W. Heinzelman, "Infrastructure Tradeoffs for Sensor Networks," ACM 1st International Workshop on Sensor Networks and Applications (WSNA '02), September 2002, pp. 28–36.
13. W. Ye, J. Heidemann and D. Estrin, "An Energy-Efficient MAC Protocol for Wireless Sensor Networks," INFOCOM, June 2002, pp. 1567–1576.

Structured Communication in Single Hop Sensor Networks*

Amol Bakshi and Viktor K. Prasanna

Department of Electrical Engineering - Systems,
University of Southern California,
Los Angeles, CA 90089 USA
{amol, prasanna}@usc.edu

Abstract. We propose a model for communication in single-hop wireless sensor networks and define and evaluate the performance of a robust, energy balanced protocol for a powerful and general routing primitive - (N, p, k_1, k_2) routing. This routing primitive represents the transfer of N packets among p nodes, where each node transmits at most k_1 packets and receives at most k_2 packets. Permutation routing is an instance of this primitive and has been recently studied by other researchers in the context of single-hop wireless radio networks. Our proposed protocol is the first to exploit the availability of a low-power control channel in addition to the "regular" data channel to achieve robustness in terms of handling node and link failures - both permanent and transient. We use a dynamic, distributed TDMA scheme for coordination over the control channel, which leads to collision-free transmissions over the data channel. The objective is to minimize overall latency, reduce per-node and overall energy consumption, and maintain energy balance. Our protocol is robust because the coordination is decentralized, there is no single point of failure, and one node failure affects only the packets destined for (and transmitted from) that particular node. Simulation results for different scenarios are presented.

1 Introduction

Wireless networks of smart sensors have the potential to revolutionize data collection and analysis for a host of applications. Each smart sensor or 'sensor node' is composed of one or more environment sensors, a processing unit, storage, limited power supply, and a transceiver. Most wireless sensor networks (WSNs) are designed and optimized for a specific application such as target tracking, habitat monitoring, etc. In other words, the nature of tasks and interactions between tasks is already known at design time. The challenge facing the designers is to customize the in-network computation and communication for maximum energy efficiency, which prolongs the lifetime of the network as a whole. Our primary

* This work is supported by the DARPA Power Aware Computing and Communication Program under contract no. F33615-C-00-1633 monitored by Wright Patterson Air Force Base.

H. Karl, A. Willig, A. Wolisz (Eds.): EWSN 2004, LNCS 2920, pp. 138–153, 2004.

concern is with the design and optimization of application-specific *communication* (or routing) protocols in WSNs.

A number of application-specific routing protocols have been proposed over the past few years. Most of these have focused on specific application-level communication patterns, and customize the network protocol stack for efficiency. We believe that for large-scale sensor networks, it is unreasonable to expect the end user to adopt such a bottom-up approach and manually customize the routing, medium access, and/or physical layer protocols for each application. As WSNs evolve from simple, passive data collection networks to more sophisticated sensing and/or real-time actuation, the design complexity will correspondingly increase. The right balance has to be achieved between system efficiency and ease of design. One of the approaches to achieving such a balance is to design energy-efficient and robust protocols for a set of *structured communication primitives*, and export these primitives to the end user[1]. "Structured communication" refers to a routing problem where the communication pattern is known in advance. Example patterns include one-to-all (broadcast), all-to-one (data gather), many-to-many, all-to-all, permutation, etc. This layer of abstraction provided by the library of communication primitives will allow the user to focus on high level design issues, instead of low-level hardware and networking details. Unlike the Internet, where point-to-point communication was the basic primitive, the collaborative in-network processing in sensor network applications (and its knowledge at design time) enables such a modular, composable approach. Feature extraction [2] and other complex image processing kernels are examples of sensor network applications that can benefit from using simple high-level system models and computation/communication primitives to hide the less relevant system details. The UW-API [3] is also a representative of the community's interest in defining building blocks for collaborative structured communication and using them for applications such as target tracking.

The primary contribution of this paper is an energy-efficient, robust protocol for (N, p, k_1, k_2) *routing in single-hop network topologies.* N denotes the number of packets to be transferred, p is the number of nodes, k_1 is the maximum number of packets that can be transferred by a node, and k_2 is the maximum number of packets that a node can expect to receive from others. We adopt a pragmatic approach towards measuring the efficiency of our protocol; based on the fact that typical *single-hop* WSNs will likely consist of a few tens or at most a few hundreds of nodes. Larger networks of thousands or tens of thousands of nodes will almost never have single-hop connectivity for various reasons. Permutation routing protocols such as those proposed in [4] are asymptotically near-optimal but have large overheads for small-size networks (see Sec. 2). Our protocols are not asymptotically optimal, but are efficient for single-hop network sizes that we expect will represent the overwhelming majority of real-world deployments.

(N, p, k_1, k_2) routing is a very general and powerful primitive whose special cases include almost all of the commonly encountered communication patterns

[1] The Message Passing Interface [1] standard for communication in parallel and distributed systems has a similar motivation.

- scatter, gather, all-to-all, etc. Our protocol tolerates permanent and/or transient node and link failures, dynamically adjusts the transmission schedule for maximum energy and latency savings, and is robust because all packets are transferred directly between the source and destination, without intermediaries. Therefore, node failures do not affect unrelated transmissions. Finally, our protocol is energy-balanced. For the scenario where $k_1 = k_2 = \frac{N}{p}$, each node spends almost the same amount of communication energy (see Sec. 5.3). For unequal k_1 and k_2, the energy consumed is proportional to the number of packets transmitted and received by the node.

Our protocol is also the first, to our knowledge, that explicitly uses a low-power control channel for dynamic scheduling and fault tolerance. Such dual channel architectures are being explored in the sensor networking community [5, 6]. The driving force behind the exploration of multiple radio architectures is the concern with saving energy. A separate, low-power radio facilitates the offloading of many control and coordination tasks that would otherwise need to be performed on the more heavy-duty data radio. The design of energy-efficient algorithms that acknowledge and exploit the existence of a low-power control channel promises to yield greater energy savings, in addition to optimizations over the "regular" data channel.

Hierarchical, clustered topologies have been advocated because of the quadratic relation between transmission range and transmission energy. By using short-range (and hence low power) radios and organizing the sensor network into clusters – with single-hop routing within clusters and multi-hop routing between clusters – significant energy savings can be realized. [7]. A hierarchical approach to data routing in such topologies is to first route packets to the suitable cluster that contains the destination node(s) and then redistribute packets within each cluster using efficient single-hop (N, p, k_1, k_2) routing.

The remainder of this paper is organized as follows. In Section 2, we analyze the recently proposed energy-efficient [4] and fault-tolerant [8] protocols for permutation routing in radio networks, and differentiate our work. Section 3 defines our system model, and Section 4 presents our basic protocol using a low-power control channel and one data channel. This protocol is further enhanced in Section 5 to handle permanent and transient node and link failures. Simulation results are used to estimate the energy and latency savings achievable through our protocol in the face of faults. We conclude in Section 6.

2 Permutation Routing in a Single-Hop Topology: State-of-the-Art

Energy-efficient protocols for permutation routing in single-hop radio networks were presented in [4] using Demand Assignment Multiple Access (DAMA). In [4], time is divided into fixed length slots, and the objective of the protocol is to avoid collisions by alternating reservation phases with the actual data transfers. There is no separate control channel, and reservation as well as data transfer is

Table 1. Simple Protocol in [4]

Nodes p	Packets n	%age Latency overhead	%age Energy overhead
50	1500	160	160
	2000	120	120
	2500	96	96
100	6000	163	163
	8000	122	122
	10000	98	98
200	20000	198	198
	30000	132	132
	40000	99	99

done using in-channel signaling. As a consequence, the same time and energy is required to transmit/receive a control packet as is required for data packets.

The "simple protocol" in [4] for $n \geq p^2$ requires $n + p^2 - 2p$ time slots and every station is awake for at most $2\frac{n}{p} + 2p - 4$ time slots. The more general, recursive protocol in [4] performs the task of permutation routing involving n items on a single-channel network in fewer than $2dn - 2(p-1)$ time slots with no station awake for more than $\frac{4dn}{p}$ time slots, where $d = \lceil \frac{\log p}{\log \frac{n}{p}} \rceil$. Tables 1 and 2 show the latency and energy overheads of these two protocols for typical single-hop network sizes. Energy and time overheads are evaluated using the following rationale. The actual transfer of n packets of data requires n time slots. Under a uniform energy cost model – i.e., transmitting or receiving a packet requires one unit of energy – the actual transfer of n packets requires $2n$ units of energy. If T is the total number of time slots required for the routing, we define *latency overhead as* $\frac{T-n}{n}$. If node i in the network is awake for $e(i)$ slots, the total energy E required for the routing is $\sum_{i=1}^{p} e(i)$. We define *energy overhead as* $\frac{E-2n}{2n}$. As seen in the table, if the protocols are implemented in (fault-free) single-hop WSNs with a few hundred nodes, the energy and latency overheads are unacceptably high.

It is instructive to note that in state-of-the-art (contention-based) MAC protocols for wireless networks (IEEE 802.11) and wireless sensor networks [9], the source broadcasts an RTS (Request To Send) control message and the destination responds with a CTS (Clear To Send) control message. Other eavesdropping nodes adjust their schedules accordingly. The result is that actual data is never transferred unless (i) both the sender and receiver are guaranteed (with high probability) exclusive access to the channel, and (ii) other nodes in the vicinity are silenced for the duration of the transmission through the virtual carrier sense mechanism. The control messages are smaller (hence cheaper in time and energy) than the regular data packets. For instance, the S-MAC protocol [9] for sensor networks uses special 8-byte packets for RTS/CTS on the Berkeley Motes, where regular data packets can be upto 250 bytes [10]. Of course, contention based pro-

Table 2. General Recursive Protocol in [4]

Nodes p	Packets n	%age Latency overhead	%age Energy overhead
50	1500	293	300
	2000	295	300
	2500	96	100
100	6000	296	300
	8000	297	300
	10000	98	100
200	20000	298	300
	30000	298	300
	40000	99	100

tocols cannot be compared with DAMA (TDMA) schemes. Even then, *the basic idea of a cheap negotiation phase preceding each data transfer is attractive* and can be exploited in TDMA-like synchronization mechanisms, as we show in later sections. If this negotiation is performed over a low-power radio (as against in-channel), the benefits in terms of energy saving are even greater.

The protocol in [4] is not designed to handle any type of communication failures among the participating nodes. A fault-tolerant protocol for the same problem is outlined in [8], which assumes the following:

- nodes can fail either due to damage sustained during deployment, or by the eventual depletion of energy reserves, and
- after the first phase of the routing protocol has identified a node as 'fault-free', it remains fault-free till the protocol completes.

Theorem 3.2. in [8] states that permutation routing of n packets in a single-hop network with p nodes and k channels can be solved in $\frac{2n}{k} + (\frac{p}{k})^2 + \frac{p}{k} + \frac{3p}{2} + 2k^2 - k$ slots. Also, the upper bound on energy consumption for all p nodes is $4nf_i + \frac{2n}{+}\frac{3p^2}{k} + pk^2 + \frac{p^2}{2k} + \frac{p^2}{2} + 4kp$ slots, where f_i is the number of faulty stations in a group of $\frac{p}{k}$ stations. The total number of faulty stations is therefore $f_i k$. Table 3 shows the latency and energy overhead of this protocol for typical single-hop network sizes.

In addition to the high overheads (which could be greatly reduced by running the same protocol on a dual channel communication architecture), the fault model is too optimistic for many sensor network deployments. Communication links between nodes can be prone to intermittent failures (transient, high error rates) due to environment dynamics. Therefore, a node that is identified as faulty at one stage of the protocol cannot be completely discounted, especially if the non-responsiveness was transient. Also, nodes could fail at any time, not just at the beginning of a routing phase. Our protocol is based on a general fault model that anticipates these possibilities.

Table 3. Fault-tolerant Permutation Routing Protocol in [8]

Nodes p	Faults $k.f_i$	Packets n	%age Latency overhead	%age Energy overhead
50	6	1500	48	807
		2000	36	755
		2500	29	724
100	10	6000	45	1197
		8000	33	1148
		10000	27	1118
200	16	20000	52	1831
		30000	34	1754
		40000	26	1715

We use RTS/CTS-like signaling on a time synchronized control channel, i.e., there are no collisions while sending the RTS/CTS messages. Compared to contention-based RTS/CTS signaling, this mechanism is more energy efficient[2]. Requiring a (contention-free) reservation phase before every single transmission does incur an overhead compared to protocols that schedule all packet transmissions initially and require no coordination during the data transfer phase (e.g., [4]). However, the messages exchanged in the reservation phase allow the protocol to be highly robust (see Sec. 5). Also, if these messages are exchanged over a low-power radio, the cost is further reduced.

In the following sections, we define a model for communication in single-hop WSNs that uses separate data and control channels. We then define and evaluate our efficient and robust protocol for permutation routing.

3 Our System Model

1. The network consists of p nodes labeled $S(1), S(2), \ldots, S(p)$. Each node has a unique identifier (ID) in the range $[1, p]$. The network is *homogeneous*, i.e. all nodes in the network are identical in terms of computing and communication capability.
2. All nodes are initially time-synchronized with each other and remain time-synchronized for the duration of the routing.
3. There are two communication channels: a data channel for transfer of data packets and a low power control channel for coordination purposes. Every node is correspondingly equipped with a data radio and a control radio. Note that *channel* is sometimes used to denote different frequencies on the same transceiver unit, and not separate transceivers. In our system architecture,

[2] As an example, the multiaccess protocol in the WINS system [11] is motivated by the advantages of synchronicity over random access in improving energy performance of a wireless sensor network.

the control channel is always a separate transceiver unit. There can be multiple frequencies available to the data radio, and these will constitute different *data channels*. In this paper, we consider a single data channel. Therefore, radio and channel can be used interchangeably.

4. A data or control radio can be in one of three states: Transmit, Receive, or ShutDown. We assume that energy consumption per packet is same for the Transmit and Receive states; and zero (negligible) for ShutDown.
5. Instantaneous signaling is possible between the two radios on the same node. For example, the data radio can signal the control radio on successful receipt of a data packet. The control radio can likewise signal the data radio to shift the data radio from one mode to another.
6. The network has single-hop connectivity. Any data (control) radio can directly communicate with any other data (control) radio in the network.
7. If a data (control) radio in Receive state receives simultaneous transmissions from two or more data (control) radios in the Transmit state, a collision occurs and neither transmission succeeds.

Costs: Transmitting one packet over the data channel takes t_{dp} time and e_{dp} energy. Similarly, a packet transmission over the control channel requires t_{rs} time and e_{rs} energy. We assume that control packets sent over the control channel are smaller than data packets sent over the data channel. $t_{dp}/t_{rs} = r_1$ and $e_{dp}/e_{rs} = r_2$, where r_1 and r_2 are positive integers. Based on state-of-the-art hardware and projections for the near future, it is reasonable to assume r_1 to be in the range of 5-10, and r_2 to be atleast a few tens.

We also consider two types of faults in the network:

- **Transient faults:** Communication links both into and out of k $(0 \leq k \leq p)$ nodes are subjected to transient outages. Let TF be the set of these k nodes. p_{tf} is the probability that a node $S(i) \in TF$ will be unavailable at a given time slot. p_{tf} is calculated independently for each of the nodes in TF, and also for each time slot.
- **Permanent faults:** Any of the p nodes can fail with a certain probability p_{pf} in any time slot and remain unresponsive to all future requests. Failure probability is uniformly distributed across all nodes in the network.

Values of k, p_{pf}, and p_{tf} will depend on the particular sensor network deployment. The choice of permanent or transient fault model for algorithm design is also up to the end user. Algorithms designed for transient faults will work correctly for permanent faults, although with a larger overhead (see Sec. 5).

Since coordination takes place on the control channel, the above fault models really apply to the (non-) availability of the control radios. We assume that failure of the control channel between a pair of nodes implies a failure of their data channel. Also, any fault that occurs is assumed to be *non-malicious*, i.e., the node simply stops communicating with other nodes for the duration of the fault.

For sake of simplifying the analysis, we assume that if a sends a request to b and does not receive a response (i.e., regards b as failed), *the packet addressed to b is removed from a's buffer*. Otherwise, the running time of the protocol could be non-deterministic and unbounded. We analyze (and simulate) the protocol for N reservation phases, each possibly followed by a data phase, where N is the total number of packets to be routed. In real sensor networks, data samples will be relevant only for a short duration because they represent a physical phenomenon to be detected in real-time. Enough spatio-temporal redundancy can be expected in a robust sensor network deployment for occasional packet loss to be tolerated at the application level. Even if this may not always be the case, we believe it is a good policy to drop a packet if the destination is not responding than to keep transmitting to a possibly dead sensor node and hold up useful data packets meant for other live destinations.

4 An Energy-Efficient Protocol Using a Low-Power Control Channel

Collisions over the data channel lead to significant energy overhead due to retransmission. Since the control channel is low-power, collisions of control packets might not cause much of an energy overhead, but are still undesirable due to the nondeterministic latency of conflict resolution. Instead of contention-based mechanisms, we use *TDMA-like coordination over the control channel* and asynchronous transfers over the data channel. For energy efficiency, it is also important that a node's radios are active only when they are the sender or recipient of an ongoing transmission, and asleep at all other times.

Our basic protocol ($p_{pf} = p_{tf} = 0$) has the following properties:

- there are never any collisions between control packets over the control channel
- there are never any collisions between data packets over the data channel
- data radios are ShutDown at all times except when a 'useful' transmission or reception is in progress
- control messages are exchanged in aTDMA fashion over the control channel
- data packets are transmitted asynchronously over the data channel
- all data packets are transmitted directly from the source to the destination, without using intermediate nodes
- all coordination is decentralized (peer-to-peer)

Description: Time is divided into frames each consisting of exactly one reservation phase (RP) and at most one data transfer phase (DP). Let the RPs be numbered in increasing order, with the first reservation phase being $RP[1]$. Similarly, let the DPs be labeled such that a DP that immediately succeeds $RP[i]$ is labeled $DP[i]$. $RP[i]$ is said to be *owned* by $S(i)$ if no other node is supposed to send an RTS in $RP[i]$. $RP[i]$ is said to be *successful* if $S(i)$ sends an RTS addressed to some $S(j)$ and $S(j)$ sends a CTS in acknowledgment. $DP[i]$ is said to be *owned* by nodes $S(i)$ and $S(j)$ if $RP[i]$ is successful. Data is transferred in $DP[i]$ iff $RP[i]$ is successful. Node $S(i)$ owns $RP[\frac{N(i-1)}{p}+1]$ through $RP[\frac{N.i}{p}]$.

Every node $S(j)$ maintains a counter c_j initialized to 0. The algorithm followed by each node is given in Fig. 1. One $S(i)$ owns the RP and broadcasts RTSes. All non-owner nodes which have not yet received their N/p packets ($c_j < N/p$) monitor the control channel in the Receive mode during this RP. $S(i)$ switches its data radio to Transmit mode for $DP[x]$ and the destination $S(j)$ switches its data radio to the Receive mode for $DP[x]$. $S(i)$ transmits the packet to $S(j)$ in $DP[x]$. Data radios of all $S(k)$, where $k \neq i$ and $k \neq j$, stay in ShutDown mode in $DP[x]$.

```
If I am owner of current RP
    remove next packet from transmit queue
    send RTS to the destination
    wait for CTS from destination
    transmit packet in the following DP
else
    if I have received less than N/p packets
        listen to RTS sent my owner of RP
        if I am the destination of the RTS
            reply with a CTS
            receive packet in the following DP
            increment count of received packets
        else
            sleep in the following DP
    else
        do nothing (ShutDown)
```

Fig. 1. The Basic Protocol

Pre-processing: A pre-processing phase is carried out only over the control channel, if k_1 is different for different nodes. This phase consists of exactly p slots, where each slot is equivalent to an RP in terms of time and energy cost. In the i_{th} RP, $S(i)$ broadcasts the value of k_1^i. All p nodes remain awake for the p RPs. At the end of this phase, each node knows the owner of each of the N RPs and DPs to follow. It might also be desirable for each node to know how many packets are destined for itself, i.e., for each $S(i)$ to know the value of k_2^i. When this value is known, $S(i)$ can shut off completely once it has received k_2^i packets (except in the RPs it owns). Now, a node knows the destinations of the packets it has to transmit, but does not know how many packets are destined for itself, and no means of computing this number except by explicit communication with every other node. For cases where k_2 is not the same for all nodes, this information can be exchanged over the control channel in a series of p rounds, each round consisting of p slots[3]. In the jth slot of the ith round, $S(j)$ broadcasts the number of packets in its buffer destined for $S(i)$. This phase requires p^2 slots, with each node remaining awake for $2p-2$ slots. If $n \gg p$, and r_1 and r_2 are of the order of tens or a hundred, the overhead for the pre-processing phase is a small fraction of the total energy requirements for the actual routing process.

[3] This approach is similar in some respects to the reservation protocol in [4].

The pre-processing phase of p^2 slots, where each node computes the value of its k_2 might not be useful in real-world scenarios when either the transient or permanent fault model is applicable and values of p_{pf}, p_{tf}, r_1 and r_2 are non-negligible. The primary purpose of each node knowing its k_2 is to allow it to shut down after it has received k_2 packets, knowing that there will be no more packets destined for itself. However, consider a scenario where some $S(i)$ fails with N/p packets in its buffer, and those packets are addressed to the other $p-1$ nodes. These $p-1$ nodes will never receive their k_2 packets and will remain awake for the entire N slots, thereby negating any usefulness of knowing their k_2. The utility of the advance collaborative computation of k_2^is for each $S(i)$ is only in ideal, fault-free scenarios which are unrealistic in typical sensor networks. In the protocol description and analysis in this paper, we assume that each node knows its k_2 and the value of k_2 is the same for all nodes.

Heterogeneity: We assume that clocks on all nodes are perfectly synchronized. Control radios of all nodes participating in $RP[x+1]$ go to sleep for t_{dp} time from the beginning of $DP[x]$. Depending on the error rate of the channel, the actual transmission time could be more or less than t_{dp}. Now, the control radio cannot sense if the data channel is busy or not. Therefore, some signaling mechanism is required to coordinate between the control radios and the data radios. *We describe a simple extension to the basic protocol to handle heterogeneity.* The concept is essentially the same as the backoff mechanism in MAC protocols, except that all nodes backoff for a predetermined time period as against the random backoff used to resolve contention in most MAC protocols.

Suppose a data transfer is in progress in $DP[x]$, $S(i)$ and $S(j)$ being the source and destination respectively. Exactly two data radios will be active; $S(i)$'s in Transmit and $S(j)$'s in Receive. We require each of these radios to inform the control radio (through an interrupt mechanism) about the successful completion of the packet transfer – which could occur in more or less than t_{dp}.

In the enhanced protocol, the duration of RP is extended by a small time t_{wait} – added to the beginning of the RP. Let $S(k)$ be the owner of $RP[x+1]$. At the beginning of $RP[x+1]$, $S(i)$'s control radio will know if the packet transfer has completed in t_{dp} time or not. If it has completed, there will be no transmission in the initial t_{wait} time. After t_{wait}, $S(k)$ will broadcast an RTS and the protocol will proceed as usual. If the transfer has not completed, $S(i)$'s control radio will immediately broadcast a $WAIT$ packet over the channel without waiting for t_{wait}. All control radios that are monitoring the channel will go to sleep for a predetermined time t_{ht} to allow $DP[x]$ to complete. The same procedure is followed when the control radios once again switch from ShutDown to Receive mode after t_{ht}. We assume that t_{dp} and t_{ht} are determined *a priori* (maybe through empirical means) and are provided to the protocol. Depending on the accuracy of estimates for t_{dp} and t_{ht}, the number of wait periods will be different.

Robustness: The basic protocol protocol is robust because all packets are transferred directly between the source and destination and there is no single point of failure. The fact that packets are not routed through intermediate nodes means that a single node failure affects only the packet transmissions destined for that

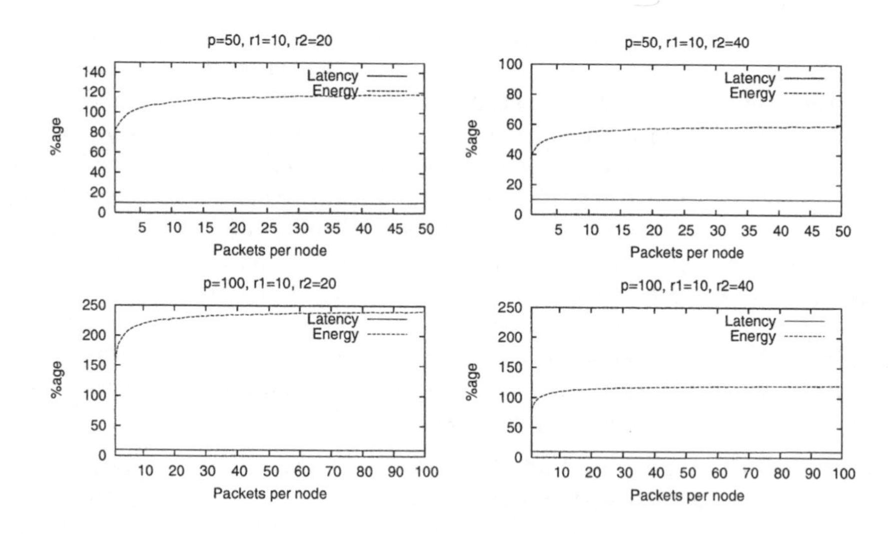

Fig. 2. Latency and Energy Overheads: Ideal Case

node and the packets that were supposed to be transmitted by that node. *Failures are therefore isolated and their effect on unrelated packet transmissions is minimized.*

Performance: Total latency of our routing is $N.(t_{dp} + t_{rs})$. Since $r_1 = t_{dp}/t_{rs}$, the latency overhead of the reservation phase is $\frac{1}{1+r_1}$. For $r_1 = 20$, the reservation phase requires approximately 5% of the total latency.

Every node stays awake for exactly $2.N/p$ DPs - consuming $2.N.e_{dp}/p$ energy. This is the minimum energy required for data transmission in the permutation routing. *The protocol is therefore optimal in terms of* e_{dp} *(and* t_{dp}*)*. Every node participates in the N/p RPs that it owns. Also, a node stops participating in RPs (not owned by itself) as soon as it has received N/p packets. Because a node does not know in which RP it will receive an RTS, every node stays awake for an average of $N(p+1)/2p$ RPs. The *average energy* spent by each node in the RP is therefore $e_{rs}.N(p+1)/2p$.

As defined earlier in Sec. 2, if routing one packet takes one unit of time, and T is the total time taken to route n packets, we define latency overhead as $\frac{T-n}{n}$. Similarly, if E is the total energy required to route n packets and transmitting (or receiving) one packet takes one unit of energy, we define energy overhead as $\frac{E-2n}{2n}$.

We simulated our protocol for a 100-node network ($p = 100$) and varied the packets per node (N/p) from 1 through 100 – i.e., the total number of packets in the network vary from 100 through 10,000. Values of r_1 and r_2 will depend on the radio electronics - typical values of r_1 will be a few tens (see Sec. 2), and the value of r_2 could be upto a hundred (see transmit/receive energies of different radios in [12]).

The top two graphs in Fig. 2 show latency and energy overheads of our protocol for a 50 node network for two different values of r_2. The bottom two graphs show the same for a 100 node network. As expected, *latency overhead is constant, regardless of the size of the network or the number of packets.* A greater r_2 results in a lesser energy overhead, because nodes spend less energy in the RPs. Note that a 0% overhead represents the ideal, prescient algorithm whose performance is an absolute upper bound on the performance of any (N, p, k_1, k_2) protocol. The 0% overhead is unachievable in practice (in our model) and used here only as an absolute benchmark. For a 50-node network with $r_2 = 40$, our energy overhead is an almost 10-fold improvement over those in [8] (see Table 3) and for a 100-node network, the improvement is about 4-fold for $r_2 = 20$ and 11-fold for $r_2 = 40$.

Energy overhead does not appear to vary significantly beyond a certain (low) value of N/p because the factor of r_2 difference between e_{dp} and e_{rs} makes the incremental energy overhead less significant as the total energy consumption increases. We discuss the energy balance of our protocol in Sec. 5.3.

5 Permutation Routing in a Faulty Network

5.1 Handling Permanent Faults

If the nodes can fail permanently with a non-zero probability p_{pf}, one of the following three scenarios can occur in some $RP[x]$.

- $S(i)$ *sends* RTS, $S(j)$ *sends* CTS: This means the sender and receiver are both alive, and data is transferred in $DP[x]$.
- $S(i)$ *sends* RTS, $S(j)$ *does not respond*: This means that $S(j)$ has failed. Since the RTS contains the ID of $S(j)$, all listeners (including $S(i)$) know that $S(j)$ has failed. As the fault is assumed to be permanent, all nodes can assume that remaining transactions to $S(j)$ can be safely canceled. However, only the node that owns a packet knows the identity of the destination. Hence, the RPs where some $S(k)$ transmits an RTS to $S(j)$ (knowing that $S(j)$ will not respond) cannot be proactively eliminated because other nodes do not know exactly when such RPs will occur. The real savings in this case will occur if $S(j)$ fails in some $RP[x]$, where $x < \frac{N.j}{p} + 1$. In that case, all nodes have information about the exact N/p RPs that are owned by $S(j)$, and all nodes know that $S(j)$ has failed. As soon as $DP[N.j/p]$ terminates $S(j+1)$ can start its transmissions. This policy saves $(t_{dp} + t_{rs})N/p$ latency and at most $2.N.(e_{dp} + e_{rs})/p$ energy per permanent fault.
- $S(i)$ *does not send* RTS: Since the RPs are assigned to specific nodes based on their IDs, all listeners in $RP[x]$ know the ID of the owner $S(i)$ of $RP[x]$. A lack of RTS allows all listeners to collaboratively and implicitly cancel $DP[x]$, and also all remaining RPs and DPs owned by $S(i)$, proceeding directly to $RP[\frac{N(i+1)}{p} + 1]$. The best case is when $S(i)$ fails in the very first RP owned by $S(i)$, leading to a cancelation of the remaining $\frac{N}{p} - 1$ RPs and

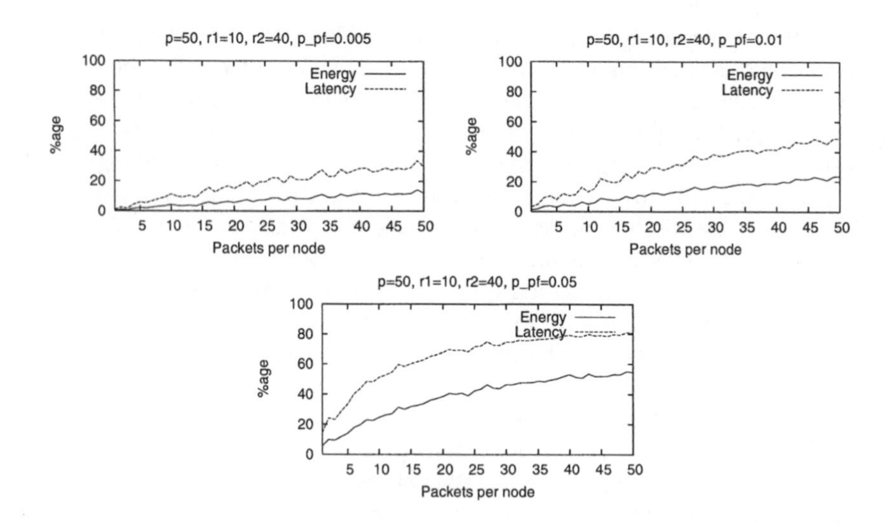

Fig. 3. Latency and Energy Savings: Permanent Faults

N/p DPs. The latency and energy savings are comparable to those for the previous scenario.

We simulated this protocol for a 50-node network for three different values of p_{pf} shown in Fig. 3. For each case, *we measure the energy and latency savings obtained by our dynamic rescheduling*, compared to an approach that does not dynamically compress the schedule in response to the knowledge of permanent failure of one or more nodes. As seen from the graphs, energy and latency savings become very significant as the failure probabilities increase. Even for a very low p_{pf} of 0.5%, upto 15% energy and upto 30% latency can be saved through dynamic rescheduling and elimination of wasteful transmissions and receptions.

5.2 Handling Transient Faults

Transient faults preclude most of the collaborative dynamic rescheduling that is possible in the case of permanent faults, because non-responsiveness of a node in some RP/DP cannot be used to infer permanent node failure. Consider the same three possibilities during some $RP[x]$ as in the previous protocol.

- *$S(i)$ sends RTS, $S(j)$ sends CTS*: This means the sender and receiver are both alive, and data is transferred in $DP[x]$.
- *$S(i)$ sends RTS, $S(j)$ does not respond*: Since the fault could be transient, the only optimization possible in this scenario is for all listeners to collaboratively eliminate $DP[x]$, i.e., $RP[x+1]$ immediately succeeds $RP[x]$. The RPs and DPs owned by the failed $S(j)$ cannot be eliminated (as was possible for a permanent fault) because $S(j)$ can recover at any time slot. A latency saving of t_{dp} and energy saving of $2.e_{dp}$ can be achieved per transient fault.

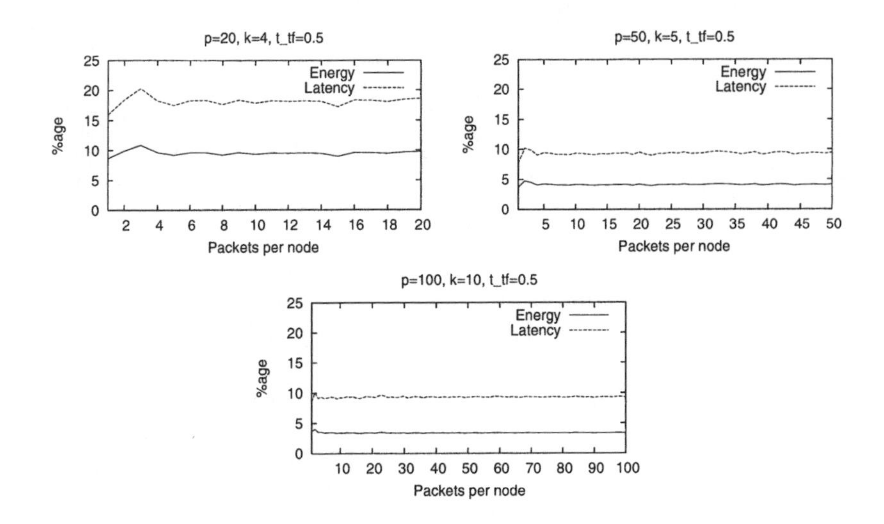

Fig. 4. Latency and Energy Savings: Transient Faults

- $S(i)$ *does not send* RTS: The policy to be followed in this case is the same as that for the previous scenario. $DP[x]$ is eliminated and $RP[x+1]$ follows $RP[x]$. Latency and energy savings are same as for the previous scenario.

Fig. 4 shows simulation results for the energy and latency savings that are achievable in face of transient failures. Three different network sizes were chosen – the values of p, k, and p_{tf} are provided with the corresponding graphs.

In the ideal case (no node failures), a node can stop participating in future RPs and DPs if it has received N/p packets and transmitted N/p packets. In case of failures, every failed RP create one more 'receiver' (in the worst case) whose count of received packets will never reach N/p, and who will therefore participate for all N RPs, waiting for a packet that will not be delivered. This is the primary reason for the energy overhead of the reservation phase for both the transient and permanent fault models.

5.3 Energy Balance

In addition to total latency and total energy, energy balance is a critical measure of algorithm performance in wireless sensor networks. If the distribution of workload (especially communication) among nodes is unequal, there will be a rapid depletion of the energy reserves of the overused nodes and possibly a network partition. Since the time to network partition is a measure of the lifetime of a sensor network, equal distribution of workload is an important measure of the efficiency of algorithms designed for WSNs.

Permutation routing protocols in [4,8] are not designed for energy balance. As shown in the first graph of Fig. 5, our protocol is *energy balanced*, i.e., all

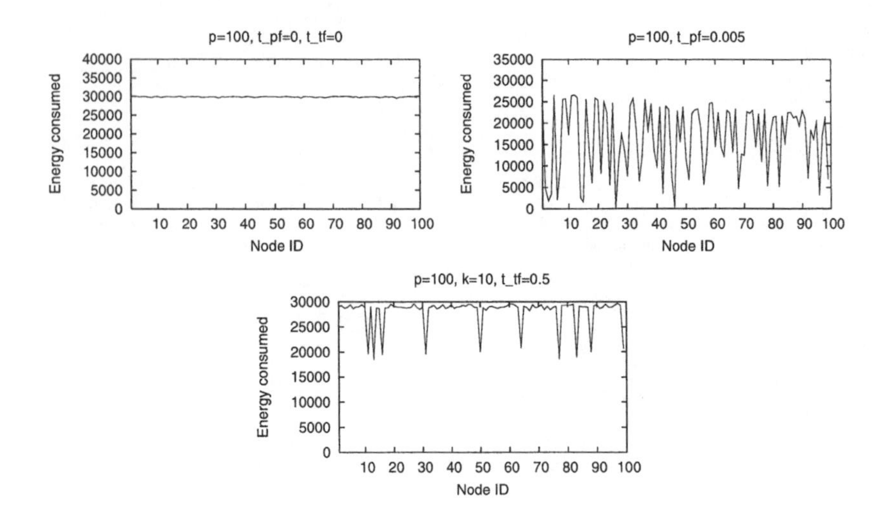

Fig. 5. Energy Balance

nodes spent approximately the same amount of energy. The second and third graph in Fig. 5 show the distribution of energy consumption in the presence of permanent and transient faults respectively. For the permanent fault model, the energy distribution reflects the energy savings that result from a detection of the faults and elimination of future transmissions to and from the faulty nodes. In the transient fault model (third graph in Fig. 5), the 10 points of low energy consumption correspond to the k (=10) faulty nodes in the network. The energy balance of the protocol can still be observed for the non-faulty nodes.

6 Concluding Remarks

The protocol described in this paper is designed specifically for one control channel and one data channel. We are investigating extensions to the protocol which will allow efficient and robust (N, p, k_1, k_2) routing with one control channel but multiple data channels.

By defining 'clean', high-level models for (classes of) sensor networks, a large body of research on parallel and distributed systems can be leveraged - including structured communication on regular and irregular topologies [13,14]. Although a majority of work in embedded networked sensing is from the traditional networking perspective, there is a growing body of research that is approaching sensor networking from a parallel and distributed systems point of view [15,16, 17]. Our work in this paper is a contribution towards a top-down approach to sensor network application design based on successive layered abstractions.

References

1. MPI Forum, MPI: A message-passing interface standard. International Journal of Supercomputer Applications and High performance Computing **8** (1994)
2. Krishnamachari, B., Iyengar, S.S.: Efficient and fault-tolerant feature extraction in sensor networks. In: 2nd Workshop on Information Processing in Sensor Networks. (2003)
3. Ramanathan, P., Wang, K.C., Saluja, K.K., Clouqueur, T.: Communication support for location-centric collaborative signal processing in sensor networks. In: DIMACS Workshop on Pervasive Networks. (2001)
4. Nakano, K., Olariu, S., Zomaya, A.: Energy-efficient permutation routing protocols in radio networks. IEEE Transactions on Parallel and Distributed Systems **12** (2001) 544–557
5. Guo, C., Zhong, L.C., Rabaey, J.M.: Low power distributed MAC for ad hoc sensor radio networks. In: Proceedings of IEEE GlobeCom. (2001)
6. Power Aware Sensing, Tracking and Analysis (PASTA), http://pasta.east.isi.edu/
7. Bandopadhyay, S., Coyle, E.: An energy efficient hierarchical clustering algorithm for wireless sensor networks. In: Proceedings of IEEE InfoCom. (2003)
8. Datta, A.: Fault-tolerant and energy-efficient permutation routing protocol for wireless networks. In: Proc. 17th International Parallel and Distributed Processing Symposium. (2003)
9. Ye, W., Heidemann, J., Estrin, D.: An energy-efficient MAC protocol for wireless sensor networks. Technical Report ISI-TR-543, USC/ISI (2001)
10. Ye, W., Heidemann, J., Estrin, D.: A flexible and reliable radio communication stack on motes. Technical Report ISI-TR-565, Information Sciences Institute, University of Southern California (2002)
11. Pottie, G., Clare, L.: Wireless integrated network sensors: Towards low-cost and robust self-organizing security networks. In: Proceedings of SPIE, Vol. 3577. (1999) 86–95
12. Estrin, D., Sayeed, A., Srivastava, M.: Wireless sensor networks (tutorial). In: Eighth ACM Intl. Conf. on Mobile Computing and Networking (MOBICOM). (2002)
13. Leighton, T.: Methods for message routing in parallel machines. In: Proceedings of the 24th annual ACM symposium on Theory of computing, ACM Press (1992) 77–96
14. Barth, D., Fraigniaud, P.: Approximation algorithms for structured communication problems. In: Proc. of the 9th Annual ACM Symposium on Parallel Algorithms and Architectures. (1997)
15. Bhuvaneswaran, R.S., Bordim, J.L., Cui, J., Nakano, K.: Fundamental protocols for wireless sensor networks. In: International Parallel and Distributed Processing Symposium (IPDPS) Workshop on Advances in Parallel and Distributed Computational Models. (2001)
16. Singh, M., Prasanna, V.K., Rolim, J., Raghavendra, C.S.: Collaborative and distributed computation in mesh-like wireless sensor arrays. In: Personal Wireless Communications. (2003)
17. M. C. Pinotti and C. S. Raghavendra (Co-Chairs), 3rd International Workshop on Wireless, Mobile and Ad Hoc Networks (WMAN), IPDPS 2003.

ACE: An Emergent Algorithm for Highly Uniform Cluster Formation*

Haowen Chan and Adrian Perrig

Carnegie Mellon University
Pittsburgh PA 15213, U.S.A
{haowenchan,perrig}@cmu.edu

Abstract. The efficient subdivision of a sensor network into uniform, mostly non-overlapping *clusters* of physically close nodes is an important building block in the design of efficient upper layer network functions such as routing, broadcast, data aggregation, and query processing.
We present ACE, an algorithm that results in highly uniform cluster formation that can achieve a packing efficiency close to hexagonal close-packing. By using the self-organizing properties of three rounds of feedback between nodes, the algorithm induces the emergent formation of clusters that are an efficient cover of the network, with significantly less overlap than the clusters formed by existing algorithms. The algorithm is scale-independent — it completes in time proportional to the deployment density of the nodes regardless of the overall number of nodes in the network. ACE requires no knowledge of geographic location and requires only a small constant amount of communications overhead.

1 Introduction

Large-scale *distributed sensor networks* are becoming increasingly useful in a variety of applications such as emergency response, real-time traffic monitoring, critical infrastructure surveillance, pollution monitoring, building safety monitoring, and battlefield operations. Such networks typically consist of hundreds to tens of thousands of low cost *sensor nodes*, deployed via individual installation or random scattering. The nodes are usually highly power-constrained and have limited computation and memory resources. They typically utilize intermittent wireless communication. The sensor network is usually organized around one or more *base stations* which connect the sensor network to control and processing workstations or to an external communications network.

* This research was supported in part by the Center for Computer and Communications Security at Carnegie Mellon under grant DAAD19-02-1-0389 from the Army Research Office, and by gifts from Bosch and Intel Corporation. The views and conclusions contained here are those of the authors and should not be interpreted as necessarily representing the official policies or endorsements, either express or implied, of ARO, Bosch, Carnegie Mellon University, Intel, or the U.S. Government or any of its agencies.

H. Karl, A. Willig, A. Wolisz (Eds.): EWSN 2004, LNCS 2920, pp. 154–171, 2004.

Clustering is a fundamental mechanism to design scalable sensor network protocols. A clustering algorithm splits the network into disjoint sets of nodes each centering around a chosen cluster-head. A good clustering imposes a regular, high-level structure on the network. It is easier to design efficient protocols on this high-level structure than at the level of the individual nodes. Many efficient protocols rely on having a network partitioned into clusters of uniform size. Some examples of these protocols include routing protocols [14,23], protocols for reliable broadcast [19,20], data aggregation [10,26], and query processing [6]. We further discuss clustering in Section 3.

Conventional algorithms that use centralized control and global properties of the sensor network have inherent difficulties in the properties of scalability and robustness, which are two important design goals for protocols in large-scale sensor networks. Centralized, top-down algorithms often need to operate with knowledge of the conditions and variables at every point of the network. In a very large network, the network traffic and time delay induced by the collection of this large amount of data may be undesirable. Finally, since some specific nodes, commands or data are usually of higher importance in a centralized protocol, an error in transmission or a failure of a critical node could potentially cause a serious protocol failure.

As an alternative to centralized algorithms, *localized algorithms* reduce the amount of central coordination necessary and only require each node to interact with its local neighbors [6]. While sometimes harder to design, these algorithms do not have the limitations of centralized algorithms and are often highly scalable, fast and efficient.

A class of localized algorithms that are particularly promising are *emergent algorithms*. Emergent algorithms have the additional characteristic that the individual agents (i.e., the sensor nodes in the case of distributed sensor networks) only encode simple local behaviors and do not explicitly coordinate on a global scale. Through repeated interaction and feedback at the individual level, global properties *emerge* in the system as a whole. Emergent behaviors are being studied extensively in biological, physical and social systems — such systems are often collectively termed *complex adaptive systems*. Examples include ant colonies, ecosystems, and stock markets. It is possible that emergent algorithms have the potential to be more flexible than non-emergent localized algorithms, which are constrained by the fact that a complex global property may be difficult to directly encode in a program that can act only upon local information.

In this paper, we provide an introduction to the definitions and motivations of localized and emergent algorithms. To demonstrate the potential of emergent algorithms in sensor networks, we present a new emergent protocol for node clustering called ACE (for Algorithm for Cluster Establishment). ACE has high cluster packing efficiency approaching that of hexagonal close-packing, and only incurs a small constant amount of communications overhead. ACE is scale-independent (it completes in constant time regardless of the size of the network) and operates without needing geographic knowledge of node positions or any kind of distance or direction estimation between nodes.

2 Localized Protocols and Emergent Protocols

In this section we define localized and emergent protocols, and discuss the particular benefits and trade offs of using these protocols in sensor networks.

2.1 Localized Protocols

Estrin et al. [6] offer a broad definition of a localized protocol:

Definition 1. *A* ***localized protocol*** *for a sensor network is a protocol in which each sensor node only communicates with a small set of other sensor nodes within close proximity in order to achieve a desired global objective.*

In this paper, we use a narrower definition of localized algorithms that better conveys the intuition of localized algorithms being free from centralized control:

Definition 2. *A* ***strictly localized protocol*** *for a sensor network is a localized protocol in which all information processed by a node is either: (a) local in nature (i.e. they are properties of the node's neighbors or itself); or (b) global in nature (i.e. they are properties of the sensor network as a whole), but obtainable immediately (in short constant time) by querying only the node's neighbors or itself.*

This narrower definition captures the notion that in a good localized protocol, each node should be capable of independent simultaneous operation in the protocol at any period. For example, consider a protocol that involves building a spanning tree in time proportional to the diameter of the network by doing a distributed breadth-first search involving only local communication (e.g. the Bannerjee and Khuller clustering algorithm [3]). Such a protocol would be a localized protocol by the first definition but not a *strictly* localized protocol by the second definition since a spanning tree is a global data structure and the entire network must be traversed before it can be computed.

In this paper, when we mention "localized protocols" or "localized algorithms", we will be referring to *strictly* localized protocols and algorithms.

Localized protocols have the following benefits:

- **Scalability.** Localized protocols can enable nodes to act independently and simultaneously in various parts of the network. Hence, localized protocols often exhibit better scalability in large networks than centrally controlled protocols, which may have to wait for information to propagate across the network.

- **Robustness.** When information use is purely local and no centralized control infrastructure is needed, the chances for protocol failure due to transmission errors and node failure are reduced. It is also more likely for performance to degrade gracefully under communication error rather than simply fail or end up in an erroneous state. This is because if all information is local, then the impact of any datum of information is most likely also locally limited. For example, if no critical control messages need to be routed across the entire network in a localized algorithm, then if a node fails then it will most likely induce a failure of the protocol at most only within its own vicinity.

2.2 Emergent Protocols

In this paper, we make use of the definition of an emergent algorithm as outlined by Fisher and Lipson [7]:

Definition 3. *An* ***emergent algorithm*** *is any computation that achieves formally or stochastically predictable global effects, by communicating directly with only a bounded number of immediate neighbors and without the use of central control or global visibility.*

Hence, an *emergent protocol* for a sensor network is a localized protocol in which the desired global property is neither explicitly encoded in the protocol nor organized by a central authority, but emerges as a result of repeated local interaction and feedback between the nodes.

One of the main distinguishing characteristics of emergent protocols over other localized protocols is the existence of *feedback* during protocol operation. Feedback occurs when some node A affects some node B, which then directly or indirectly affects node A again. Due to the reliance on repeated feedback, emergent protocols are commonly *iterative* in nature, requiring several rounds of communication between a node and its neighbors before the network as a whole converges on the desired global property.

The main advantages of emergent protocols are:

- **Sophisticated applications.** Emergent algorithms have the potential for more easily expressing complex global properties than localized algorithms. Iterated feedback allows the algorithm to sidestep the explicit coordination and calculation required for such tasks as efficient cluster formation and pattern formation.
- **Increased robustness against transient faults.** The iterated nature of emergent protocols further improves robustness against transient node failure, since a small number of missing or incorrect interactions are unlikely to have a large effect due to the fact that all interactions are repeated several times. This may allow the protocol to tolerate some error in consistency and synchronization between nodes.

Emergent protocols are often harder to design effectively than localized algorithms, since the repeated feedback can create complex interactions that are difficult to analyze. However, their increased expressive power and robustness make them an important class of algorithms, particularly in large-scale distributed sensor networks.

3 Overview of Sensor Node Clustering and Applications

Efficiently organizing sensor nodes into clusters is an important application in sensor networks. Many proposed protocols for both sensor networks and ad-hoc networks rely on the creation of clusters of nodes to establish a regular logical structure on top of which efficient functions can be performed. For example, clustering can be used to perform data aggregation to reduce communications energy overhead [10,26]; or to facilitate queries on the sensor network [6]; clusters can be used to form an infrastructure for scalable routing [14,23]; clustering also can be used for efficient network-wide broadcast [19,20]. Single-level clustering is sufficient for many applications; for others, multi-level hierarchical clustering can be performed (by creating clusters of clusters, and so on).

The clustering problem is defined as follows. At the end of the clustering algorithm, the nodes should be organized into disjoint sets (*clusters*). Each cluster consists of a *cluster-head* (cluster leader) and several cluster *followers*, all of which should be within one communication radius of the cluster-head, thus causing the overall shape of the cluster to be roughly a circle of one communication radius, centered on the cluster-head. Each node belongs to exactly one cluster (i.e., every node chooses only one leader, even if there may be several leaders within range). Given these constraints, our goal is to select the smallest set of cluster heads such that all nodes in the network belong to a cluster. The problem is similar to the *minimum dominating set* problem in graph theory. We note that if every node is in *exactly* one cluster, then maximizing the average cluster sizes while maintaining full coverage is exactly equivalent to minimizing the number of clusterheads while maintaining full coverage. The purpose of minimising the number of cluster heads is to provide an efficient cover of the network in order to minimize cluster overlap. This reduces the amount of channel contention between clusters, and also improves the efficiency of algorithms (such as routing and data aggregation) that execute at the level of the cluster-heads.

For brevity, we have defined the clustering problem as obtaining a single-level clustering. We note that, assuming that clusterheads can establish multiple-hop communications to neighboring clusterheads of the same hierarchy level, it is possible to generalize any single-level clustering protocol to multi-level hierarchical clustering by repeatedly executing the clustering protocol on the cluster-heads of each level to generate the cluster-heads of the next level, and so on.

We summarize in brief a few simple examples of efficient protocols that rely on the effective solution of the single-level clustering problem. A straightforward example is in data aggregation. In an unclustered network, if an aggregate query of sensors over a given sub-area is desired, the query needs to be forwarded to every sensor in the sub-area, each of which then needs to individually send its reply to the base station. In contrast, in a clustered network, a query of sensors over a given sub-area needs only be forwarded to the relevant cluster-head which will then query its followers and send a single aggregated reply.

As an example of the importance of highly uniform clustering with low overlap, consider the clustered broadcast protocol described by by Ni et al.[19]. In this protocol, the broadcast message is relayed from cluster-head to cluster-head, which then broadcast the message to their followers. In a clustering with few clusterheads and large cluster sizes, the clusters have minimal overlap and provide the best coverage of the network with the fewest clusters. Hence, the number of repeated broadcast transmissions over any area will be small, thus reducing the amount of transmission collisions and channel contention, allowing communications to become faster, more efficient and more reliable. On the other hand, a poor clustering with much cluster overlap and many cluster-heads loses much of the benefits of clustering as transmissions will be repeated in areas of overlap with significant channel contention.

4 ACE — Algorithm for Cluster Establishment

In this section, we present ACE (the Algorithm for Cluster Establishment), an emergent cluster formation algorithm. The algorithm consists of two logical parts — the first controls how clusters can spawn (by having a node elect itself to be leader) and the second controls how clusters *migrate* dynamically to reduce overlap. In general, clusters are only created when the overlap of the new cluster with existing clusters is small. After creation, clusters will move apart from each other to minimize the amount of mutual overlap, thus yielding a near-optimal packing in very few iterations.

4.1 Overview of the ACE Protocol

We first present a high level overview of the protocol. ACE has two logical parts: the *spawning* of new clusters and the *migration* of existing clusters. New clusters are *spawned* in a self-elective process — when a node decides to become a cluster head, it will broadcast a RECRUIT message to its neighbors, who will become followers of the new cluster. A node can be a follower of more than one cluster while the protocol is running (it picks a single cluster for membership only at the end of the protocol). *Migration* of an existing cluster is controlled by the cluster head. Each cluster head will periodically POLL all its followers (i.e., all its neighbors) to determine which is the best candidate to become the new leader of the cluster. The best candidate is the node which, if it were to become cluster head, would have the greatest number of nodes as followers while minimizing the amount of overlap with existing clusters. Once the best candidate is determined by the current cluster head, it will PROMOTE the best candidate as the new cluster head and ABDICATE its position as the old cluster head. Thus, the position of the cluster will appear to *migrate* in the direction of the new cluster head as some of the former followers of the old cluster-head are no longer part of the cluster, while some new nodes near the new cluster head become new followers of the cluster.

4.2 Detailed Description of the ACE Protocol

In ACE, time synchronization is not required — the nodes may in fact start the protocol at slightly different times due to network delay or clock discrepancies. During the protocol, nodes respond immediately to communications from other nodes, but will only *initiate* actions at random intervals to avoid collisions. Each time that an action can be initiated for a node is called a node's *iteration*. The iterations of different nodes do not need to be synchronized. The duration of the random time interval between iterations (the *iteration interval*) is uniformly random distributed.

We now describe the operation of ACE is described in detail. A node can have three possible states: it can be unclustered (not a follower of any cluster), clustered (a follower of one or more clusters) or it may be a cluster-head. In the beginning of the protocol, all nodes are unclustered. Each node waits for its next iteration (i.e., by waiting for a random iteration interval) before deciding on what action to take on that iteration, if any. When a node's iteration arrives, its available choice of actions depends on what state it is currently in.

If a node A is unclustered when its next iteration arrives, it assesses its surroundings and counts the number l of *loyal* followers it would receive if it declared itself a cluster-head of a new cluster. A *loyal* follower is a follower of only one cluster. Hence, in this case, this number is the same as the number of unclustered neighbors that A has. A knows how long it has been since it started the protocol; call this time t. It then computes the *spawning threshold* function $f_{min}(t)$ (the design of f_{min} will be described later). If $l \geq f_{min}(t)$ then A will spawn a new cluster. It does so by generating a random (unique with high probability) cluster ID and broadcasting a RECRUIT message. A's neighbors will receive this message and become followers of the new cluster.

If a node A is a cluster-head when its next iteration arrives, it prepares to migrate its cluster. It POLLs all of its neighbors to find the best candidate for the new cluster-head. The best candidate leader for a cluster is the node with the largest potential number of *loyal* followers in its neighbor set (recall that a loyal follower is a member of only one cluster). Hence, the best candidate for the new cluster-head is the node which has the largest number of nodes in its neighbor set which are either unclustered or have A's cluster as their only cluster. By counting only loyal followers and not counting nodes that lie on the overlap of two or more clusters, the best candidate node is generally in the direction of least overlap with other clusters. This generates a *repulsion* effect between clusters which leads to good packing efficiency. If the best candidate for cluster-head is A itself, then A does nothing. Otherwise, suppose the best candidate is some node B. A will now MIGRATE the cluster onto the new cluster-head B. It does so by issuing a PROMOTE message to B. On receiving the PROMOTE message, B will issue a RECRUIT message with A's cluster ID. This is similar to spawning a new cluster except that an existing cluster ID is used instead of generating a new one. The effect of this is that the neighbors of B that were not in the cluster will now be added to the cluster (with B as the cluster-head), while the existing members of the cluster that are B's neighbors will realize that B is being promoted and thus update B as their new cluster-head. Once A observes B's RECRUIT message, it will then issue an ABDICATE message to its neighbors. The effect of this will be that common neighbors of A and B will have seen B's RECRUIT message beforehand and thus ignore the message; neighbors of A who are not neighbors of B will leave the cluster. The net effect of this sequence of actions is that leadership passes from A to B and the cluster as a whole *migrates* from being centered around A to being centered around B.

If a node is clustered (i.e., it is a follower in one or more clusters), then it does nothing during its iteration. It merely waits a random iteration interval for its next iteration to arrive.

Each node needs to be able to efficiently find out the number of loyal followers it may gain. This state can be efficiently maintained by having all nodes keep track of the list of clusters that each neighbor is in. Hence, whenever a node becomes a follower in a new cluster or leaves an existing cluster, it broadcasts an update locally to its neighbors. The overhead of this updating is low because clusters generally do not make drastic shifts in position during migration, hence the cluster sets of most nodes change only slowly with time. By keeping track of these periodic updates, each node can immediately compute how many loyal followers it can gain without needing to query its neighbors.

Each node runs the protocol for at least a time cI where c is the desired average number of iterations per node and I is the expected length of the iteration interval. After a node has completed its iteration, if it still has not passed time cI counting from when it started running the protocol, then it will wait another random iteration interval until its next iteration.

After a node has passed time cI since it started running the protocol, it is ready to terminate the protocol. If the node is a cluster-head, it terminates immediately and informs its neighbors that it is done. If the node is a clustered node, it waits until all its cluster-heads have terminated before choosing one at random to become its final cluster-head (it does not need to notify its neighbors that it has terminated). After termination, the node will respond with "N/A" to leadership polls from clusters that have migrated into its range to indicate its unwillingness to return to the protocol.

Parameter selection. In the protocol, an unclustered node will spawn a new cluster by declaring itself a cluster head whenever it finds that it can gain at least f_{min} loyal followers if it were to become a cluster head. The function f_{min} is called the *spawning threshold function* and is dependent on the time that has passed since the protocol was initiated for that node. In general, f_{min} should decrease as the algorithm proceeds. This causes fewer clusters to form near the beginning of the algorithm. Fewer clusters in the beginning means that the clusters have more room to maneuver themselves apart from each other, in order to form the basis for an efficient clustering. As time advances, the algorithm then causes the gaps between the clusters to be filled in by spawning new clusters more and more aggressively. We observe that the unclustered gaps between clusters decrease roughly exponentially in size when cluster migration is taking place. Hence, in our implementation, we used an exponentially decreasing function for f_{min}:

$$f_{min} = (e^{-k_1 \frac{t}{cI}} - k_2)d$$

In this formula, t is the time passed since the protocol began and cI is the duration of the protocol as described earlier. d is the estimated average degree (number of neighbours) of a node in the network, and is pre-calculated prior to deployment. k_1 and k_2 are chosen constants that determine the shape of the exponential graph.

In practice, we have empirically found that $k_1 = 2.3$ and $k_2 = 0.1$ have produced good results. In this case, f_{min} starts at $0.9d$ at the beginning of the protocol and reduces to 0 by the final iteration. This ensures that any node left unclustered at the end of the protocol will declare itself a cluster head. A node A may (rarely) find itself unclustered at the end of the protocol if its cluster-head migrates away from A after A has completed its last iteration. To cover this case, an additional "clean-up" iteration should be run after the algorithm has completed for every node. During this final clean-up iteration, cluster migration is disabled, and any node that is still unclustered should declare itself as a cluster-head. This will ensure that evey node in the network is covered by a cluster.

An alternative parameter setting is $k_1 = 2.3$ as before, but setting $k_2 = 0$. In this case the function starts near d when the protocol commences and reduces to $0.1d$ at the end of the protocol. Since $0.1d > 1$ if $d > 10$, it is possible that there will be a small number of nodes that will not be within one hop radius of any cluster-head at the end of the algorithm. This means that this algorithm would not strictly satisfy the problem statement described in Section 3. However, this setting still has practical relevance because the number of unclustered nodes at the end of the algorithm is small. We observed in simulation that the number of nodes not within one-hop radius of a cluster-head is, on average, less than 4% of the total number of nodes in low node deployment densities, and around 2% for moderate to high node deployment densities (20 or more neighbors per node). These nodes that are not within one hop radius of any cluster-head can simply pick a clustered neighbor to act as their bridge to the cluster-head, thus becoming *two-hop followers* (because they take 2 hops to communicate with the cluster-head, instead of the usual 1 hop).

It remains to determine c, the number of iterations the algorithm should execute. Figure 1 reflects how the performance of ACE changes as it is given a longer number of iterations to operate. ACE was simulated in a 2D area with a uniform random distribution with an average deployment density d of 50 nodes per circle of one communication radius. Results for the simulation with $k_1 = 2.3$ and $k_2 = 0.1$ are shown (results for $k_2 = 0$ have similar characteristics). We note that increasing the number of iterations above 3 yielded only very slight improvements in average cluster size. In our simulations, the total number of iterations did not significantly affect the standard deviation in cluster sizes,

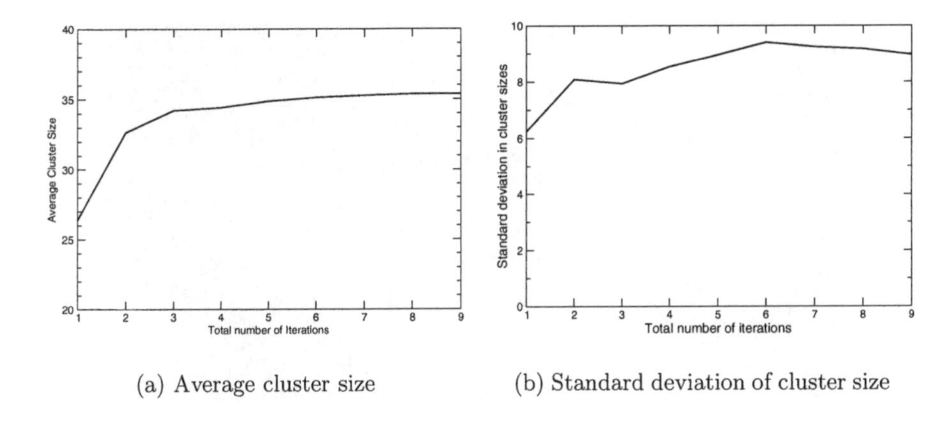

(a) Average cluster size

(b) Standard deviation of cluster size

Fig. 1. Performance of ACE at various maximum iterations, $d = 50, k_1 = 2.3, k_2 = 0.1$

which was between 6 and 10 for all iterations > 1. Based on these results, we choose $c = 3$ as a number of iterations for ACE that provides a good tradeoff between communication overhead and cluster size.

Figure 2 illustrates the ACE algorithm operating in simulation (with $k_1 = 2.3$ and $k_2 = 0$). The little circles represent nodes. Cluster-heads are highlighted in black, and their range is indicated with a large black circle (nodes within the circle are in that cluster). The clusters migrate away from each other in successive iterations to produce a highly efficient cover of the area. Clusters tend to center over areas where nodes are dense. The clusters overlap minimally, and when they do overlap, they tend to overlap in areas where nodes are sparse. Figure 2d provides a qualitative visual comparison of the Node ID algorithm with ACE . It can be observed that ACE provides a packing with significantly less cluster overlap than Node ID.

5 Performance Evaluation of ACE

To assess ACE's performance, ACE was simulated and its performance was compared with a well-known 2D packing (hexagonal close packing) as well as two other clustering algorithms, the Node ID algorithm and the Node Degree algorithm. In our simulations we simulated both ACE with full coverage ($k_1 = 2.3, k_2 = 0.1$), which we called ACE-1 and also ACE with parameters that leaves a small fraction of nodes uncovered ($k_1 = 2.3, k_2 = 0$), which we call ACE-2.

Hexagonal close-packing (HCP) is the well-known honeycomb packing that minimizes overlap between uniform circular clusters while ensuring full coverage. In general this packing is difficult to achieve unless nodes have very specific information about their geographic locations, e.g. as assumed Zhang and Arora [27], and even then a centralized algorithm needs to be used to coordinate the hon-

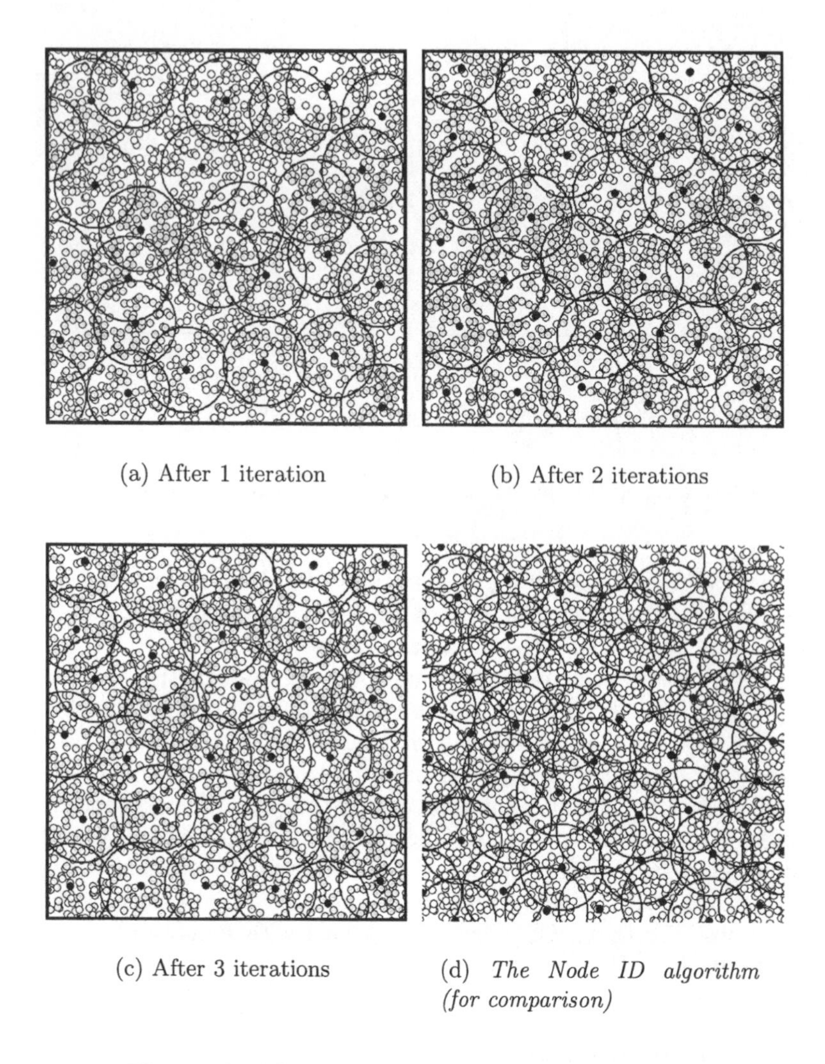

(a) After 1 iteration

(b) After 2 iterations

(c) After 3 iterations

(d) *The Node ID algorithm (for comparison)*

Fig. 2. The ACE algorithm (with $k_1 = 2.3$ and $k_2 = 0$)

eycomb structure, which leads to lower scalability. The Node ID algorithm is a generic name for the class of algorithms related to LCA (Linked Cluster Architecture) [2,5]. In this algorithm, the node with the highest ID elects itself as a cluster-head, followed by the node with the next highest ID that is not already a follower, and so on until all the nodes have been covered. The basic concept has been revisited in various architectures such as those described by Lin et al. and Gerla et al. [9,17]. Amis et al. [1] improved the algorithm for multi-hop clusters by adding a second pass in which clusters with low node IDs are expanded to better balance cluster size; however this improvement has no effect on 1-hop clusters hence we do not simulate it here. Nagpal and Coore propose a variation

Algorithm 5.1 ACE

```
procedure SCALE_ONE_ITERATION()
  if myTime > 3× EXPECTED ITERATION LENGTH then
    if myState = CLUSTER-HEAD then
      return DONE
    else if myState = CLUSTERED then
      wait for my cluster-heads to terminate, then pick one as my cluster-head
      return DONE
    else if myState = UNCLUSTERED then
      pick a random clustered node to act as my proxy after it terminates
      wait for it to terminate, then return DONE
    end if
  else if myState = UNCLUSTERED
  and numLoyalFollowers() ≥ f_min(myTime) then
    myClusterID ← generate_New_Random_ID()
    locally_broadcast(RECRUIT, myID, myClusterID)
  else if myState = CLUSTER-HEAD then
    bestLeader ← myID
    bestFollowerCount ← numLoyalFollowers
    for all n where n is a potential new cluster-head do
      followerCount = Poll_For_Num_Loyal_Followers(n, myClusterID)
      if followerCount > bestFollowerCount then
        bestLeader ← n
        bestFollowerCount ← followerCount
      end if
    end for
    if bestLeader is not myID then
      send(bestLeader, PROMOTE, myClusterID)
      wait for bestLeader to broadcast it's RECRUIT message
      locally_broadcast(ABDICATE, myID, myClusterID)
    end if
  end if
end procedure
```

of the Node ID algorithm where the nodes generate a random number and start counting down from it; when the counter reaches zero and the node is not already a follower in some cluster then the node elects itself as cluster-head [18]. This algorithm is similar to the Node ID algorithm with the benefit that it can be repeated for re-clustering on successive epochs. Using the degree of connectivity instead of node ID as a metric for which nodes to elect as cluster-heads has been proposed Basagni [4] and Gerla et al. [9]. This causes nodes in denser areas to become cluster-heads first. We model this algorithm as the Node Degree algorithm.

The various algorithms were simulated on various deployments of 2500 nodes in a square area where each node's coordinates were uniformly random. In our

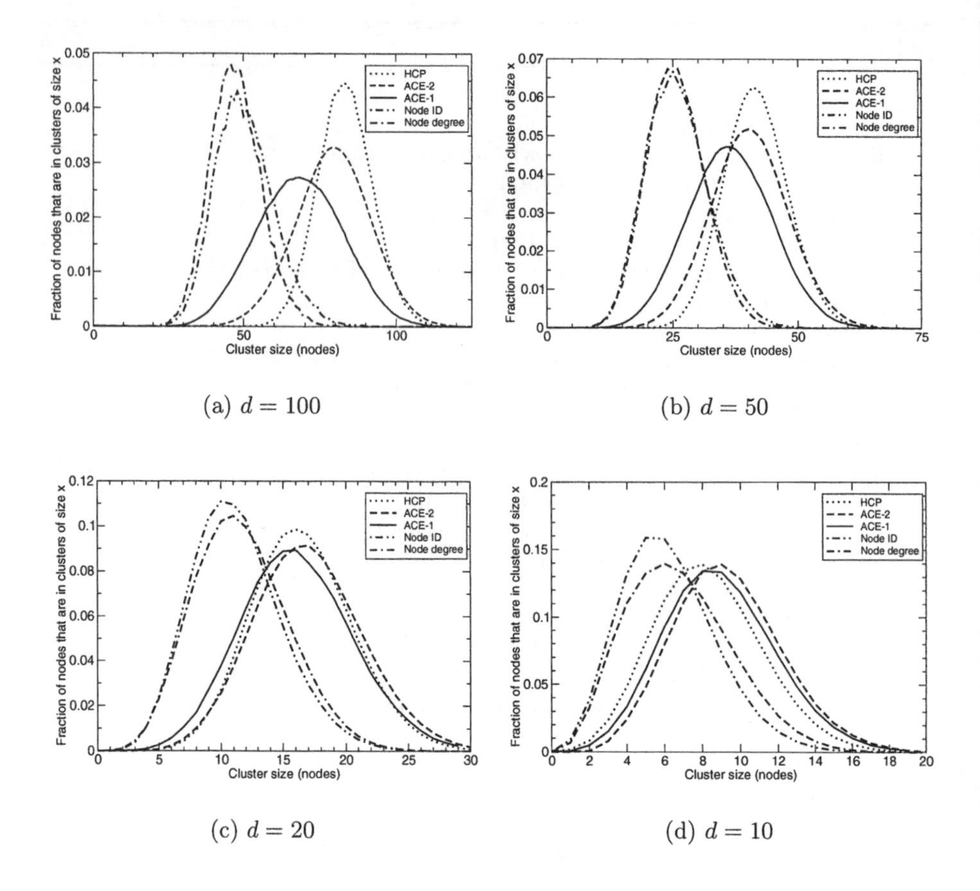

(a) $d = 100$

(b) $d = 50$

(c) $d = 20$

(d) $d = 10$

Fig. 3. Distribution of cluster sizes for various clusterings

simulation, we assume that the communication links were bi-directional and that the communication range of all the nodes is uniform. 500 simulations per algorithm were run for each of the node densities (expected number of neighbors in a circle of one communication radius) of $d = 10, 20, 50, 100$.

Figure 3 shows the relative distributions of cluster sizes for the various algorithms under the various node densities simulated. Figure 4 compares the average cluster sizes of the various algorithms as d varies.

It is clear that ACE exhibits superior packing efficiency to either the Node ID or Node Degree algorithms. ACE-1 exhibits consistent performance of around $0.7d$ average cluster size for all node densities. ACE-2 exhibits performance around $0.8d$ average cluster sizes. For reference, the average cluster size for the ideal 2D packing of HCP is $0.83d$. ACE sometimes exceeds the ratio for HCP because it intelligently forms clusters around areas where nodes are most densely distributed, while choosing areas of overlap where the nodes are least densely dis-

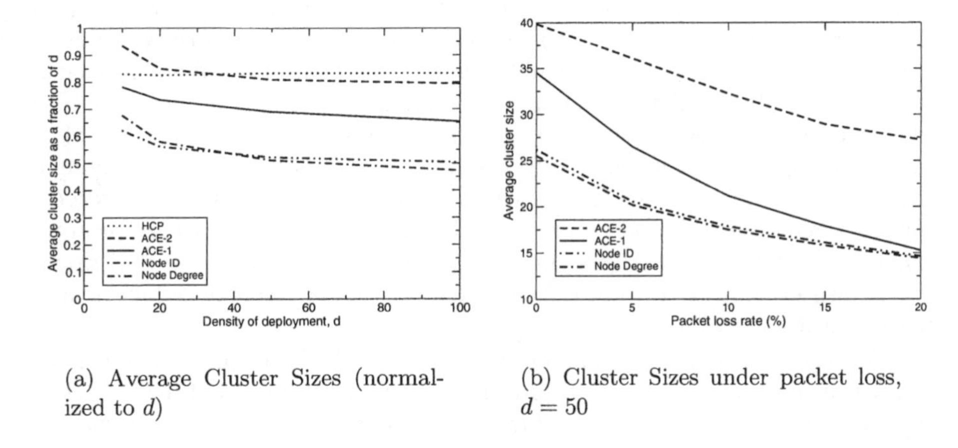

(a) Average Cluster Sizes (normalized to d)

(b) Cluster Sizes under packet loss, $d = 50$

Fig. 4. Average cluster sizes of various schemes

d	10	20	50	100
ACE-1	6.68	6.80	7.07	7.32
ACE-2	5.41	4.96	4.47	4.56
Node ID	1.17	1.09	1.04	1.02
Node Deg.	1.17	1.09	1.04	1.02

Fig. 5. Average communications overhead (per node per epoch)

tributed. In comparison, both Node ID and Node Degree converge towards only $0.5d$ for large d and never perform better than $0.7d$ even at low node densities.

Figure 3 shows that the variance in cluster sizes for ACE is small and only slightly larger than the baseline variance of the number of nodes in a given area (this is reflected in the variance of cluster sizes for HCP). The low variance and high average cluster sizes reflect that the ACE algorithm produces good packing efficiency.

We investigated the performance of the various algorithms under conditions of packet loss. Various deployments of $d = 50$ were simulated with packet loss rates ranging from 0 to 20%. Packet loss was simulated by having the simulated nodes ignore an incoming message with a probability corresponding to the packet loss rate. Figure 4b reflects the results. The performance of each protocol degrades gracefully under packet loss. ACE-2 maintains its large advantage over Node ID and Node Degree even under conditions of heavy packet loss. ACE-1 degrades at a higher rate and approaches the performance of Node ID and Node Degree under conditions of high packet loss, but never actually performs worse than either algorithm under the simulated conditions. We note further that since the ACE algorithm is localized and require no central direction, it is highly resistant to transmission errors and random node failure. For example, a centralized algorithm utilizing a BFS tree (e.g. Bannerjee and Khuller's algorithm [3]) could suffer the loss of an entire subtree if one of the nodes high in the tree suffers a failure, thus leading to failure of the protocol. In our protocols, in the worst case, the loss of a cluster-head node would leave at most one cluster of nodes unclustered. If a cluster-head node fails while executing the protocol, and the failure is detected by the followers, they can reset their own states to "unclustered", thus allowing neighboring clusters to migrate into the new vacant space or allowing a new clusterhead to spawn a cluster within the vacant space. Hence, the protocol has an innate self-stabilization property. These additional adaptations for unreliable communications were not simulated; if they were implemented they would likely further improve the protocol's resilience towards random node failures and communication errors.

We also measured the communications overhead of the various algorithms. Each transmission was considered one unit of communication, and the final cluster handshake where all nodes confirm membership with their cluster-heads was also considered an additional unit of communication. The results are tabulated in Figure 5. Because of the low number of iterations needed by ACE (only 3 iterations), the communications overhead is small, only averaging around 4 to 8 communications per node per epoch. Each communication is brief (at most a message identifier, a node and cluster identifier, and a number). Hence the overall communications overhead is small compared with the normal communications load for the sensor network.

ACE exhibits scale independence (perfect scalability). The protocol takes a fixed amount of time, $O(d)$, to complete regardless of the total number of the nodes in the network. This is because it is a *strictly localized* algorithm (see definition in Section 2), where each node is capable of operating immediately and independently on local information without needing any global information to be computed by the network. As a result, both running time and per-node communications overhead of ACE are independent of the total size of the network.

6 Related Work

In this section, we review related work in localized and emergent algorithms in sensor networks, as well as clustering algorithms in general.

Currently, few practical emergent algorithms have been developed for use in sensor networks. Henderson suggests using Turing's reaction-diffusion equations [24] for forming patterns in sensor networks [11]. These approaches are promising and indicative of the future potential of emergent algorithms.

We now discuss related work in clustering. Many clustering protocols currently known are *self-elective* protocols, where a node creates a new cluster by declaring itself as a cluster-head. They differ in the heuristic used to select the nodes which will declare themselves. The node ID and node degree heuristics have been discussed in Section 5. Examples of node ID based clustering protocols include [1,2,5,9,17].Basagni proposes a node degree based clustering protocols [4]. Some researchers propose using a random number as a heuristic for cluster-head selection [8,12,18,25]. Estin et al. propose using the remaining energy level of a node as another heuristic for cluster-head selection [6].

Ramanathan and Steenstrup [21], and Krishnan, Ramanathan, and Steenstrup [15] propose a clustering algorithm that controls the size of each cluster and the number of hierarchical levels. Their clustering approach follows the node ID approach. In general, these self-elective protocols all suffer from the same problem of being unable to prevent two nodes which are just over one cluster radius apart from simultaneously electing themselves as cluster-heads, thus leading to a large overlap in their clusters. Such overlap occurs sufficiently frequently to make the resultant cluster packing inefficient, as can be seen in our evaluation in Section 5.

The *minimum dominating set* (MDS) problem in graph theory has been addressed by several algorithms. The clustering problem is a special case of the MDS problem applied to random geometric graphs. While these algorithms have provable theoretical *asymptotic* bounds on performance on arbitrary graphs, their *actual* average performance on a random geometric graph is undetermined. We implemented in simulation two algorithms described by Jia et al. [13] and Kuhn et al. [16], however neither of them had comparable performance to even the simple Node-degree or Node-ID algorithms under our particular simulation conditions. We speculate that the relatively poor performance of these algorithms in simulation may be due to the fact that they are designed for arbitrary graphs while dedicated clustering algorithms are optimized for random geometric graphs. Hence, we did not run full simulation comparisons against these algorithms.

Many centralized (non-localized) clustering algorithms are known, which deal with the topology of the entire network as a whole. This class of algorithms often uses graph-theoretic properties for clustering. In general, such algorithms are not as robust or scalable as localized algorithms, eventually requiring significant communications or computation overhead for very large networks. For example, Krishna et al. [14] proposes a technique where each cluster forms a clique, however their approach has $O(d^3)$ overhead. Some researchers proposed tree-based constructions for network partitioning. Thaler and Ravishankar propose to construct a top-down hierarchy, based on an initial root node [22]. Banerjee and Khuller also propose a tree-based clustering algorithm [3]. A drawback for using their algorithm in sensor networks is that only one node needs to initiate the clustering, and that the protocol still requires $O(n)$ time in linear networks. Zhang and Arora present a centralized scheme to produce an approximate hexagonal close packing [27]. However, they assume that each node knows its precise location, which may be difficult to achieve in sensor networks. In general, besides scalability issues, most these nonlocalized algorithms also suffer from increased vulnerability of the protocol to node failure in certain key parts of the network (usually near the root of the tree, or near the base station).

7 Conclusion

We present ACE , the Algorithm for Cluster Establishment. ACE is an emergent algorithm that uses just three rounds of feedback to induce the formation of a highly efficient cover of uniform clusters over the network. This efficiency of coverage approaches that of hexagonal close-packing. ACE is fast, robust against packet loss and node failure, and efficient in terms of communications. It completes in constant time regardless of the size of the network and uses only local communications between nodes. The algorithm does not require geographic location information or any kind of distance or directional estimation between nodes. Besides its practical usefulness, ACE is a good demonstration of the power and flexibility of *emergent* algorithms in large-scale distributed systems.

References

1. Alan D. Amis, Ravi Prakash, Thai H.P. Vuong, and Dung T. Huynh. Max-Min D-Cluster Formation in Wireless Ad Hoc Networks. In *Proceedings of IEEE INFOCOM 2000*, pages 32–41, 2000.
2. D.J. Baker, A. Ephremides, and J.A. Flynn. The Design and Simulation of a Mobile Radio Network with Distributed Control. *IEEE Journal on Selected Areas in Communication*, 2(1):226–237, January 1984.
3. Suman Banerjee and Samir Khuller. A Clustering Scheme for Hierarchical Control in Wireless Networks. In *Proceedings of IEEE INFOCOM 2001*, April 2001.
4. Stefano Basagni. Distributed Clustering for Ad Hoc Networks. In *Proceedings of the IEEE International Symposium on Parallel Architectures, Algorithms, and Networks (I-SPAN)*, pages 310–315, June 1999.
5. A. Ephremides, J.E. Wieselthier, and D.J. Baker. A Design Concept for Reliable Mobile Radio Networks with Frequency Hopping Signaling. *Proceedings of IEEE*, 75(1):56–73, 1987.
6. Deborah Estrin, Ramesh Govindan, John Heidemann, and Satish Kumar. Next Century Challenges: Scalable Coordination in Sensor Networks. In *Proceedings of the Fifth Annual ACM/IEEE International Conference on Mobile Computing and Networking (MobiCom '99)*, pages 263–270, August 1999.
7. David A. Fisher and Howard F. Lipson. Emergent Algorithms: A New Method for Enhancing Survivability in Unbounded Systems. In *Proceedings of the Hawaii International Conference On System Sciences*, January 1999.
8. M. Gerla, T.J. Kwon, and G. Pei. On Demand Routing in Large Ad Hoc Wireless Networks with Passive Clustering. In *Proceedings of IEEE Wireless Communications and Networking Conference (WCNC 2000)*, September 2000.
9. M. Gerla and J.T. Tsai. Multicluster, Mobile, Multimedia Radio Network. *ACM/Kluwer Journal of Wireless Networks*, 1(3):255–265, 1995.
10. W. Heinzelman, A. Chandrakasan, and H. Balakrishnan. Energy-Efficient Communication Protocol for Wireless Microsensor Networks. In *Proceedings of the 33rd Hawaii International Conference on System Sciences (HICSS '00)*, January 2000.
11. T. C. Henderson, M. Dekhil, S. Morris, Y. Chen, and W. B. Thompson. Smart Sensor Snow. In *Proceedings of IEEE Conference on Intelligent Robots and Intelligent Systems (IROS)*, October 1998.

12. X. Hong, M. Gerla, Y. Yi, K. Xu, and T. Kwon. Scalable Ad Hoc Routing in Large, Dense Wireless Networks Using Clustering and Landmarks. In *Proceedings of IEEE International Conference on Communications (ICC 2002)*, April 2002.
13. Lujun Jia, Rajmohan Rajaraman, and Torsten Suel. An Efficient Distributed Algorithm for Constructing Small Dominating Sets. In *Proceedings of the 20th Annual ACM Symposium on Principles of Distributed Computing*, pages 33–42, 2001.
14. P. Krishna, N. H. Vaidya, M. Chatterjee, and D. Pradhan. A cluster-based approach for routing in dynamic networks. *ACM SIGCOMM Computer Communication Review*, 27(2):49–65, April 1997.
15. Rajesh Krishnan, Ram Ramanathan, and Martha Steenstrup. Optimization Algorithms for Large Self-Structuring Networks. In *IEEE INFOCOM '99*, 1999.
16. Fabian Kuhn and Roger Wattenhofer. Constant-Time Distributed Dominating Set Approximation. In *Proceedings of the 22nd Annual ACM Symposium on Principles of Distributed Computing*, pages 25–32, 2003.
17. Chunhung Richard Lin and Mario Gerla. Adaptive Clustering for Mobile Wireless Networks. *IEEE Journal of Selected Areas in Communications*, 15(7):1265–1275, 1997.
18. Radhika Nagpal and Daniel Coore. An Algorithm for Group Formation in an Amorphous Computer. In *Proceedings of the 10th International Conference on Parallel and Distributed Computing Systems (PDCS'98)*, October 1998.
19. Sze-Yao Ni, Yu-Chee Tseng, Yuh-Shyan Chen, and Jang-Ping Sheu. The Broadcast Storm Problem in a Mobile Ad Hoc Network. In *Proceedings of the Annual ACM/IEEE International Conference on Mobile Computing and Networking (MobiCom '99)*, pages 151–162, August 1999.
20. Elena Pagani and Gian Paolo Rossi. Reliable broadcast in mobile multihop packet networks. In *Proceedings of the Annual ACM/IEEE International Conference on Mobile Computing and Networking (MobiCom '97)*, pages 34–42, 1997.
21. R. Ramanathan and M. Steenstrup. Hierarchically-Organized, Multihop Mobile Wireless Networks for Quality-of-Service Support. *ACM/Baltzer Mobile Networks and Applications*, 3(1):101–119, June 1998.
22. David G. Thaler and Chinya V. Ravishankar. Distributed Top-Down Hierarchy Construction. In *Proceedings of IEEE INFOCOM*, pages 693–701, 1998.
23. P. F. Tsuchiya. The landmark hierarchy: a new hierarchy for routing in very large networks. In *Symposium Proceedings on Communications Architectures and Protocols (SIGCOMM '88)*, pages 35–42, 1988.
24. Alan Turing. The chemical basis of morphogenesis. *Philosophical Transactions of the Royal society of London B237*, pages 37–72, 1952.
25. Kaixin Xu and Mario Gerla. A Heterogeneous Routing Protocol Based on a New Stable Clustering Scheme. In *IEEE MILCOM 2002*, October 2002.
26. Ya Xu, Solomon Bien, Yutaka Mori, John Heidemann, and Deborah Estrin. Topology Control Protocols to Conserve Energy in Wireless Ad Hoc Networks. Technical Report 6, University of California, Los Angeles, Center for Embedded Networked Computing, January 2003.
27. H. Zhang and A. Arora. GS^3: Scalable Self-configuration and Self-healing in Wireless Networks. In *21st ACM Symposium on Principles of Distributed Computing (PODC 2002)*, 2002.

Improving the Energy Efficiency of Directed Diffusion Using Passive Clustering*

Vlado Handziski, Andreas Köpke, Holger Karl,
Christian Frank, and Witold Drytkiewicz

Telecommunication Networks Group, Technische Universität Berlin
Sekr. FT 5, Einsteinufer 25
10587 Berlin, Germany
{handzisk,koepke,karl,chfrank,drytkiew}@ft.ee.tu-berlin.de
http://www.tkn.tu-berlin.de

Abstract. Directed diffusion is a prominent example of data-centric routing based on application layer data and purely local interactions. In its functioning it relies heavily on network-wide flooding which is an expensive operation, specifically with respect to the scarce energy resources of nodes in wireless sensor networks (WSNs).
One well-researched way to curb the flooding overhead is by clustering. Passive clustering is a recent proposal for on-demand creation and maintenance of the clustered structure, making it very attractive for WSNs and directed diffusion in particular.
The contribution of this paper is the investigation of this combination: Is it feasible to execute directed diffusion on top of a sensor network where the topology is implicitly constructed by passive clustering?
A simulation-based comparison between plain directed diffusion and one based on passive clustering shows that, depending on the scenario, passive clustering can significantly reduce the required energy while maintaining and even improving the delay and the delivery rate. This study also provides insights into the behavior of directed diffusion with respect to its long-term periodic behavior, contributing to a better understanding of this novel class of communication protocols.

1 Introduction

Wireless ad hoc networks and wireless sensor networks (WSNs) are similar in many respects: the network has to be self-configurable, scalable, and energy-efficient. The most important difference is the communication model: ad hoc networks typically are built to support standard communication interactions between nodes, where WSNs are centered around data that can be observed from the environment. Owing to the high redundancy of nodes in WSNs, any

* This work has been partially sponsored by the European Commission under the contract IST-2001-34734 – Energy-efficient sensor networks (EYES).

H. Karl, A. Willig, A. Wolisz (Eds.): EWSN 2004, LNCS 2920, pp. 172–187, 2004.

individual node is not particularly important—it might not even have a unique address—rather, the data is. One appropriate programming model for such a network is publish/subscribe, with data-centric routing as the supporting communication layer.

Directed diffusion [1] is a prominent instance of data-centric routing where the selection of routes is based on application level data. By using purely local interactions and in-network processing it is well tailored to the specific needs of WSNs.

In its operation, directed diffusion relies heavily on performing network-wide broadcasts. Consequently, the overall performance of the protocol can be strongly influenced by the efficiency of this elementary operation.

One of the most straightforward ways to realize a network-wide broadcast is to use the simple flooding algorithm where every node in the network forwards each new message to all of its neighbors. But this approach is rather inefficient in wireless networks where the significant overlap between the neighborhoods of two nodes in immediate vicinity can lead to a large number of unnecessary rebroadcasts. This additionally results in increased channel contention and waste of bandwidth that take further toll on the scarce energy resources of the nodes.

The problem of identifying a set of re-forwarding nodes that still guarantees the distribution of the messages to all of the nodes in the network is very well researched. The proposed techniques vary from probability-based approaches to area- and neighborhood-knowledge methods [2].

The neighborhood-knowledge based group mainly comprises the distributed algorithms for efficient approximation of the minimum connected dominating set of nodes. They are based on the partitioning of the network into logical substructures called *clusters* that confine and mask local interactions from the rest of the network. In the process of defining these substructures (*clustering*) a single node (*clusterhead)* can be selected from every cluster to exert local control and coordination. The interfacing between the different clusters is then realized only by dedicated *gateway* nodes. The selection of the clusterheads and the gateways is such as to guarantee the connectivity of the resulting, reduced topology.

Clustering reduces the flooding overhead by limiting the re-forwarding of the messages to this reduced topology. Yet, for the approach to be overall effective, the cost of building and maintaining the clustered structure has to be lower then the energy savings from reducing the number of sent and received messages.

This requirement is not easily met if clustering is performed for the sole reason of limiting the flooding overhead, especially if one has in mind the pro-active and resource intensive nature of the majority of clustering algorithms. To overcome this problem, Kwon and Gerla [3] proposed the Passive Clustering algorithm (PC) for on-demand creation and maintenance of the clustered substrate.

In this paper, we study if and how directed diffusion can benefit from a passive-clustering-based network topology to reduce its flooding overhead. To do so, we first summarize in Sect. 2 the protocol mechanisms of passive clustering and directed diffusion and discuss our combination of these two mechanisms.

Section 3 then describes the investigation methodology we used to compare original directed diffusion with directed diffusion over passive clustering: the choice of metrics, simulators, and scenarios as well as some implementation details. Section 4 presents the simulation results, and Sect. 5 puts our approach and these results into context with related work. Section 6 concludes the paper.

2 Protocol Description

2.1 Passive Clustering

Passive clustering has several unique properties that increase its viability as a flooding overhead control mechanism for on-demand wireless networking protocols. In the following, we briefly summarize these characteristics.

Unlike in "classical" clustering algorithms, the formation of clusters here is dynamic and is initiated by the first data message to be flooded. In this way, the potentially long initial set-up period is avoided, and the benefits of the reduction of the forwarding set can be felt after a very small number of data message rounds.

Because the main function of the clusters is to optimize the exchange of flooded messages, there is no point in wasting valuable resources to pro-actively maintain such an elaborate structure between floods, when there is no traffic that can make use of it. Consequently, passive clustering refrains from using explicit control messages to support its functionality and all protocol-specific information is piggybacked on the exchanged data messages. This approach joins the frequency of exchanged messages with the quality of the clustered substrate resulting in a graceful trade-off between the "freshness" of the clustered substrate and the introduced overhead for its maintenance.

An additional trade-off is also made over the issue of selection of a clusterhead between several potential candidates. Passive clustering does not have the benefit of exchanging specific control messages to optimally resolve this conflict, as it is usually done in other clustering schemes. It introduces a simple novel rule, called *first declaration wins*. Under this rule, the first aspirer to become a clusterhead is immediately and without further neighborhood checks declared as such and allowed to dominate the radio coverage area. This may at first sound suboptimal, but is more then compensated for by the decreased latency, the prevention of "chain re-clusterings" and the natural correlation that emerges between the traffic flow patterns and the resulting clustered topology.

The effectives of any clustering algorithm as a tool for reducing the number of redundant flood messages directly depends on its ability to select the minimal number of gateways while still maintaining the connected property of the topology. In order to avoid the large overhead of collecting full two-hop neighborhood information that is required for optimal solution of this clustering subproblem, passive clustering resorts to a counter-based gateway and distributed gateway selection heuristic. Under this heuristic, the probability of a node that belongs to two or more clusters to become a gateway is directly proportional to the number of clusterheads in its radio range and inversely proportional to the number of gateways already present in the same area. By controlling the coefficients of proportionality, one can make trade-offs between the flooding overhead reduction and the level of connectivness in the reduced topology.

As demonstrated by the authors in [4], the combination of these design decisions results in a lightweight and flexible algorithm that can significantly lower the overhead associated with flooding in wireless networks.

2.2 Directed Diffusion

Directed diffusion is in great detail described in [1]. Here, only an overview is given to enable the reader to understand the basic mechanisms as used in our performance evaluation, where passive clustering changes it, and the influence of some protocol parameters.

Directed diffusion is a data-centric routing protocol. In a Wireless Sensor Network (WSN) the data that the network can provide is interesting, not specific nodes. In order to receive data, an interested node (a "sink") floods the network with an interest message. This message contains a detailed description of the event it is interested in (e.g. cars in a certain geographical region). When this interest is received by a node, it sets up a gradient to the neighbor from which it heard the interest. If it hears the same interest form several neighbors, gradients will be set up for each one of them. This focus on neighbors is a specific feature of directed diffusion, which allows the protocol to scale with the number of nodes in the network.

When a gradient has been set up and the node can potentially contribute to this interest (e.g. it can sense data and is in the requested region), the node starts to sense its environment for matching events. When it observes a matching event, it becomes a source and sends the data to all neighbors to which it has gradients. This data is flagged as "exploratory". Exploratory data is forwarded along gradients towards the sink and eventually reaches it; intermediate nodes follow essentially the same rules regarding gradients as sources and sinks. The sink then sends a reinforcement message to only one of its neighbors, namely the one where it heard the exploratory data first. When the source receives a reinforcement, it starts to send data more often, for instance every 2 s. At the same time, it continues to send exploratory data periodically (e.g. every 60 s) to all neighbors to which it has gradients. The gradients are stored only for a limited amount of time (e.g. 125 s), unless a new interest is received. The interests are also sent periodically (e.g. every 20 s).

Directed diffusion maintains a data cache in order to limit flooding. Messages are repeated only once. This (together with some additional protocol messages not described here) ensures that directed diffusion often chooses a single, a "reinforced" path with empirically low delay between sources and sinks.

The periodic flooding of interests as well as the sending of exploratory data (which can easily amount to flooding, as directed diffusion keeps more than one neighbor per received interest) is important to maintain or repair the paths between sources and sinks.

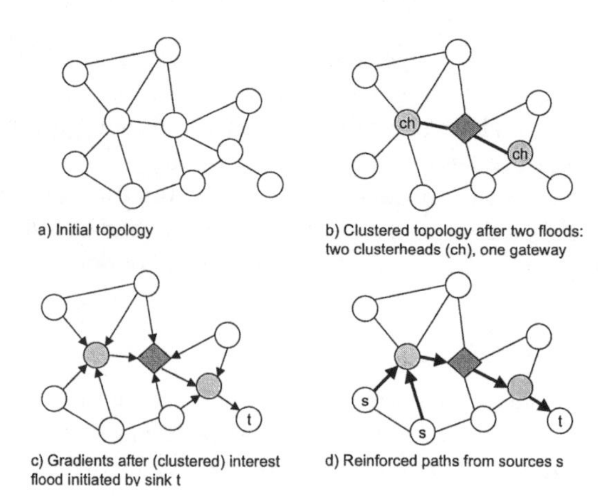

Fig. 1. Establishing the clustered diffusion structure (gray circles indicate clusterheads, diamonds indicate gateways)

2.3 Directed Diffusion and Passive Clustering

From the description in Sect. 2.2 it is evident that the propagation of interests and exploratory data are one of the most resource-demanding operations in directed diffusion. Depending on the actual network topology, their implementation using simple flooding rules can lead to significant and unnecessary overhead that reduces the operational lifetime of the sensor network.

We believe that passive clustering is an appropriate mechanism to increase the efficiency of these crucial diffusion phases and by that the efficiency of the whole protocol. The protocols share some common properties:

1. They are on-demand mechanisms that maintain the operation only while there is application traffic in need of their services
2. They rely on purely local information for performing their functions

As a result, their combination is perfectly adapted to the operating conditions that are common in WSNs.

For the initial exploration, we decided to limit the interaction between them just to a single point in the operation of the original directed diffusion:

Ordinary (non-clusterhead, non-gateway) nodes do not forward the interest and exploratory data messages that they receive.

To explain the effects of this interaction, we demonstrate in Fig. 1 the establishment of the directed diffusion routing structures when used in combination with passive clustering.

Let the operation start with the topology depicted in Fig. 1(a). In passive clustering, nodes that are first able to piggyback information onto an outgoing message (see Sec. 2.4) will declare themselves as clusterheads (assuming that they have not heard of other clusterheads before). Nodes that have heard of two clusterheads but of no other gateway assigned to these same clusterheads will become a gateway, resulting in the clustered topology shown in Fig. 1(b). Assuming this constellation, the next diffusion *interest* flood will establish a sparse gradient structure shown in Fig. 1(c). The reinforced data paths are then shown in Fig. 1(d).

Based on this description, we can make a couple of observations:

1. The number of redundant transmissions during the broadcasting tasks is significantly reduced
2. The size of all directed diffusion messages that are broadcasted is increased by the size of the passive clustering header
3. Passive clustering reduces the level of connectivity available to directed diffusion

This means that the overall effectiveness of our approach depends on the answers to the following crucial questions:

- Does the combination of the reduced number of messages with the increased message size result in net gain in energy efficiency?
- Does the reduced level of connectivity result in degradation of the data distribution capability?
- Does the reduced level of connectivity result in increase of the latency as sometimes less optimal paths have to be followed?

We shall answer these questions in Sect. 4. But first, a few remarks on the implementation of directed diffusion with passive clustering are in order.

2.4 Implementation

The interfacing between our modules and the diffusion core was realized using the provided Filter API [5]. The node's passive clustering component is declaring two filters to the diffusion core, which bracket the diffusion gradient filter: One filter with high priority to intercept messages before gradient processing (*pre-filter*) and a second filter with lower priority for messages from gradient to the network (*post-filter*). The interaction of passive clustering with the diffusion routing core is shown in Fig. 2.

Due to the reactive nature of the protocol, all state changes happen upon interception of a diffusion message. Passive clustering attaches a header to outbound messages and inspects it in incoming messages.

Here, the notion of inbound and outbound messages in diffusion is somewhat different from the usual network stack view: Messages arrive at the diffusion routing core (point (1) in Fig. 2), both from a local application or from the network. After passing the filters (8), the message destination may again be either the network or a local application. All messages pass the filter in the same

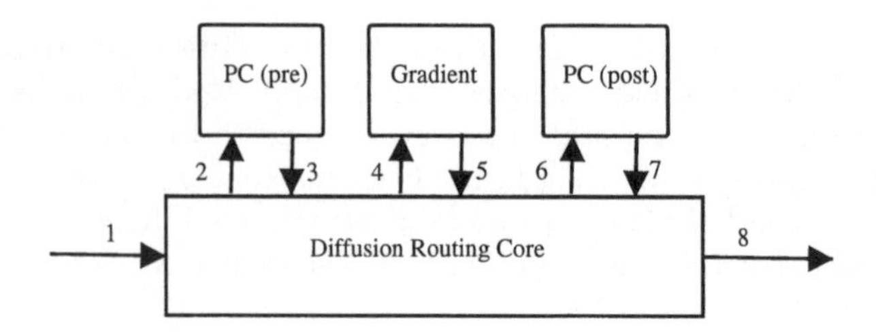

Fig. 2. Integration of Passive Clustering and Diffusion

order (according to the given numbering in Fig. 2), independent of their source. In a network stack view, passive clustering would be located below the gradient layer.

If a message has been received from the network, the *pre-filter* processes the message header and changes the node's internal state (Tab.1) according to the newly available information about a neighbor. This includes entering certain "candidate" states, which denote that the new information is relevant for the node to change its external state on the next outbound message. Pre-filter does not process any messages coming from the local application. After inspecting a message for clustering-relevant information, the message is then forwarded to the diffusion gradient filter.

The final state-change to an externally visible state like clusterhead *(CLUSTER_HEAD)* or gateway *(GW_NODE, DIST_GW)* is performed on the next outbound message addressed to the network that is intercepted by the *post-filter.*

Table 1. Passive clustering FSM states

Internal States	External States
CH_READY	INITIAL_NODE
GW_READY	CLUSTER_HEAD
	GW_NODE
	DIST_GW
	ORDINARY_NODE

A passive clustering header (Fig. 3) is piggybacked onto this message *only* if the external state of the node has changed or if it has not announced its current state for more than a certain parametrized number of seconds. This keeps passive clustering overhead to a minimum, as it does not grow with the number of messages but is bounded by O(#nodes).

The last two fields (CH1_ID and CH2_ID) are present only in the messages issued by gateway nodes and represent the IDs of the cluster heads they connect.

Fig. 3. Structure of the piggybacked passive clustering header

Including the 4 byte overhead caused by attaching the header as an attribute to the diffusion message, the fields sum up to a total of 17 bytes in the worst case.

The essential task of discarding the unnecessarily forwarded messages is also performed by the *Post-Filter*: both *Interest* and *Exploratory Data* messages are discarded if the node is in (or has changed to) *Ordinary Node* state. This happens unless, of course, messages originate from a local application that is resident on this node or are addressed to it.

3 Methodology

Designing performance evaluation experiments for wireless sensor networks is faced with a number of practical and conceptual difficulties. This section summarizes our main choices for the simulation setup.

3.1 Choice of Protocol Versions

We simulated the original directed diffusion and our passive clustering extension using the ns-2 simulator [6], version 2.26. Directed diffusion is implemented for this simulator in two versions. The first version is a simplified implementation of directed diffusion that was used by Intanagonwiwat et al. [1] for the evaluation of the protocol. Additionally, directed diffusion 3 is available; it is a complete protocol implementation, not a simplified version. The implementation is described in [1] and allows a more realistic evaluation of the protocol.

The protocol versions have some major differences. First of all, the simplified version uses somewhat arbitrary lengths for different message types. The interest messages, for example, are half the size (32 bytes) of data messages (64 bytes). For the protocol implementation, the difference is not as large, since the interest contains a fairly complete description of the event the sink is interested in. In our case, the interest message is 88 bytes long.

As this newer version better reflects reality, we decided to base our simulations on it, although this makes comparison with the published results difficult. We used the "ping" application which adds a sequence number and a time-stamp to the payload of a data message, resulting in messages that are 116 bytes long.

3.2 Load Model

A second major difference are the timer settings of these protocols. The simplified version floods an interest message every 5 s and an exploratory message every 50 s. On a reinforced gradient, the sources send data every 0.5 s. The standard settings of these parameters for the protocol implementation are different. Interests are sent every 25 s, exploratory data every 60 s and the sources send data every 5 s. Here, only five data messages per interest flood are received by a sink. In this case, the protocol overhead for directed diffusion is very large, making it questionable whether it pays off at all compared to plain flooding.

In our simulation we changed the ping application to send data every 2 s and interest messages every 20 s, which results in 10 data messages per interest flood, the same as in the simplified version. As in [1] we regard every such message as a "new distinct event" and we do not use duplicate suppression. Duplicate suppression, or aggregation as it is called in [1], allows nodes in the network to drop messages if it is an event that they have already forwarded. However, duplicate suppression is difficult in simulations, as there is no commonly used model that produces the "right" amount of duplicates.

3.3 Energy Model

In a WSN, energy is a scarce resource and should be saved. Hence, energy consumption per data packet (or per distinct event) is our main metric. The problem is how to measure this energy. Clearly, every sending of a message consumes power and the number of sent messages should be reduced as far as possible. In addition, every message that was received by a node when it is not the final destination is overhead and the power used for receiving messages should be reduced, too.

A further problem is the idle power, the power a node consumes when it is listening for transmissions. This is tightly connected with the used MAC; in the case considered here, the IEEE 802.11 implementation of ns-2 as used in the original directed diffusion investigations. For such a CSMA MAC the node has to listen to the channel permanently. Hence, for a CSMA MAC, the idle power is comparable to the receive power. Although this would be technically correct, the idle power could mask the effects at the network layer. With long application related inactivity periods it would dominate the energy budget making the comparison between the two protocols difficult. Since we wanted to concentrate on the relative performance of the protocols, we used a zero idle power in our simulations.

For comparability we used the same energy model as in [1] that is the PCMCIA WLAN card model in ns-2. This card consumes 0.660 W when sending and 0.395 W when receiving.

3.4 Metrics

The characteristics of the ping application result in two performance metrics. One metric is the delivery ratio, which compares the number of messages arriving at the sinks with the number of messages sent by the sources. Another metric is the delay—how much time does a message need to travel from the source to the sink.

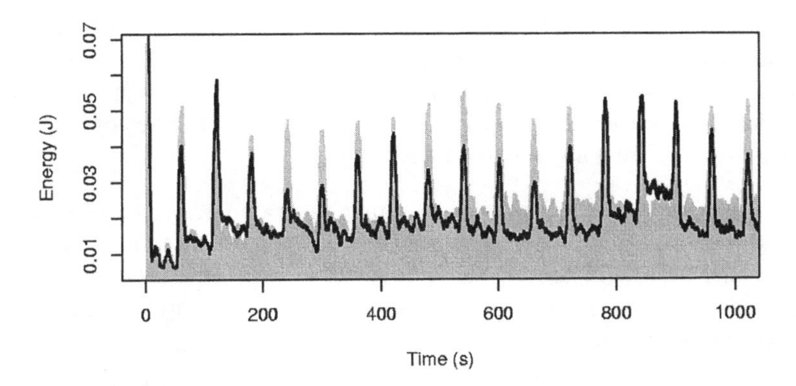

Fig. 4. Time series of the energy per distinct event and sink, showing the initial transient. Plain directed diffusion – gray in the background, diffusion with passive clustering – black in the foreground.

Every time a new distinct event is seen, the energy consumed since the last distinct event is computed for every node in the network. From these values the average is computed and reported—this constitutes our third performance metric, average dissipated energy (per new distinct event). While the energy consumption is reported only once per distinct event, the delay and delivery ratio are reported by each sink separately, as this is more natural for these metrics.

Figure 4 shows an example for the energy consumption per distinct event and sink as a function of time (to display this figure properly, the actual data has been slightly smoothed by a moving average with a small window size). A close inspection of the time series of the energy consumption revealed a long initial transient phase, which lasts for about 300 s in the example shown in Fig. 4. This transient phase is in part due to the staggered subscriptions and publications performed by all nodes. To correctly handle the initial transient, we discarded the observation values produced during the first 600 s from every simulation run; to compute the average energy consumption per new distinct event, we took into account only the values from the remaining 1000 s of simulated time. The passive clustering structure is refreshed every 45 s, so discarding of the first 600 s does not skew the results in favor of the passive clustering approach.

3.5 Scenarios

Every node has a fixed transmission power resulting in a 40 m transmission range. The sources and sinks were spread uniformly over the entire area; the size of the area varies between simulations.

We used two different types of source/sink setups.

Five sources and five sinks. With five sources and sinks each distributed as described above, either the area or the average density of the network are kept constant: an increasing number of nodes is placed either in an area of fixed size (160 m by 160 m) or on a growing area.

Variable number of sources/sinks. Using a 160 m by 160 m area with 100 nodes, we used either a single sink with a variable number of sources or a single source with a variable number of sinks.

For each parameter setting, we randomly generated 10 scenarios with different placements of nodes. For each scenario, we computed (after removing the initial transient) the means for all three metrics. From these ten (stochastically independent) means we computed the means over all scenarios and their 95% confidence intervals for all metrics.

4 Results and Discussion

4.1 Effects of the Network Topology

We start the presentation of our results with the five sources/five sinks case in networks of either fixed area or fixed density. Figure 5 depicts the average dissipated energy per unique event for the original directed diffusion protocol (DD) and for the directed diffusion with passive clustering (PCDD) under three different network sizes. For each different network size, the simulation area is scaled so that the average node degree in the network remains constant at around 10. This enables us to examine the scaling properties of the protocols as a function of the number of the nodes in the network in isolation. In all of the topology experiments the load of the network is generated by five sources sending unique events every two seconds.

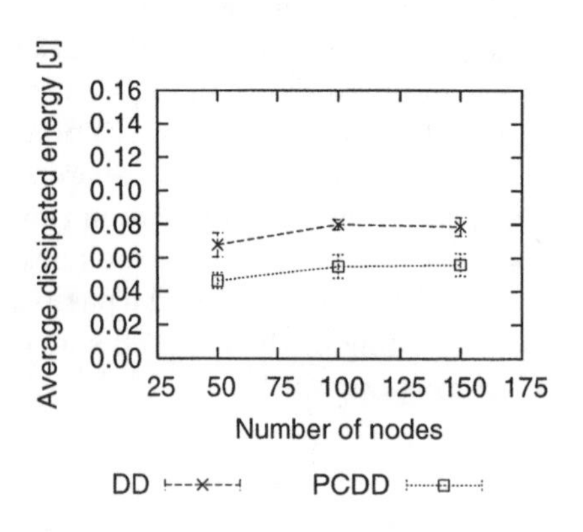

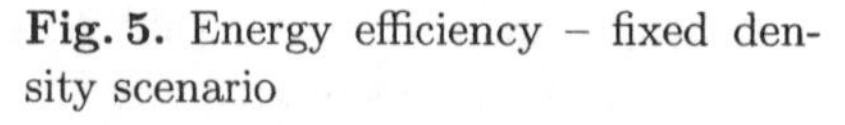

Fig. 5. Energy efficiency – fixed density scenario

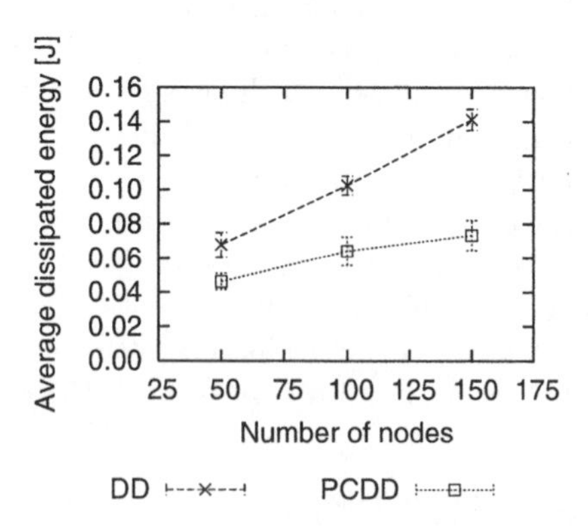

Fig. 6. Energy efficiency – fixed area scenario

The results clearly show the effectiveness of passive clustering in controlling the unnecessary re-forwarding of messages in directed diffusion. While both protocols show slow degradation of their performance with the increase in number of nodes, the relative savings of the passive-clustering augmented directed diffusion over the original protocol is evident even for the 50 nodes scenario and is maintained as the size of the network increases. This is a confirmation to our assumption that the reduced number of flooded messages results in a net gain in the energy metric, despite the increased size of the remaining messages.

Importantly, this gain was not achieved at the cost of either the delivery ratio (remaining above 0.95 for both protocols) or the delay (remaining below 0.6 s for both protocols). The corresponding figures are omitted due to space constraints but it is obvious that the resulting connectivity after the passive filtering allowed these metrics to remain close to their optimal values for the whole range of simulated network sizes.

As described in Sect. 2, both directed diffusion and passive clustering are based on pure local interaction. Many of the elementary operations of which they are comprised operate over the set of neighboring nodes that are in radio range. As a result, their complexity, and by that, their performance is strongly determined by the size of that set, i.e., on the average node degree in the network. Again, we have examined our performance metrics for three different network sizes as previously, but now keeping the simulation area constant to the initial 160 m by 160 m. This gives us an insight into the performance of the evaluated protocols under an average number of around 10, 20, and 30 neighbors.

In Fig. 6 we can see the average dissipated energy per unique event for an increasing number of nodes in this fixed area. The original directed diffusion shows rather linear increase in the required energy with the increase of the number of neighbors. This was to be expected, as the increased number of neighbors results in higher number of flooded interest messages and an increase in the number of unicasted exploratory data messages. Passive-clustering-supported directed diffusion, on the other hand, behaves much better in dense networks. Like in the fixed degree case, the increased overhead due to the PC headers is more then compensated by the reduction of the costly send and receive operations. The denser the network, the larger the advantage over pure directed diffusion as more redundant paths can be suppressed by passive clustering.

Moreover, these advantages in energy efficiency are not paid for in a reduced delivery rate. As Fig. 7 shows, the contrary is true: the delivery rate actually is considerably better with passive clustering. Compared with the fixed degree scenario, the delivery ratio of plane directed diffusion experiences significant degradation as the size of the neighborhood increases. By reducing the superfluous message exchanges, passive clustering leads to a decrease in the collision rate that results in a very good delivery ratio even for the highest density case.

Similarly, Fig. 8 shows that also the delay is much better with passive clustering, despite the potential problem of the longer paths. The reduced load in the network more than compensates for this potentially negative aspect. Without this reduction in the amount of traffic in the network, plain directed diffusion faces extensive delays in the higher degree scenarios.

Overall, we can conclude that passive clustering augmented directed diffusion scales much better with respect to the neighborhood size, maintaining a satisfactory performance in the progressively harsher network environment.

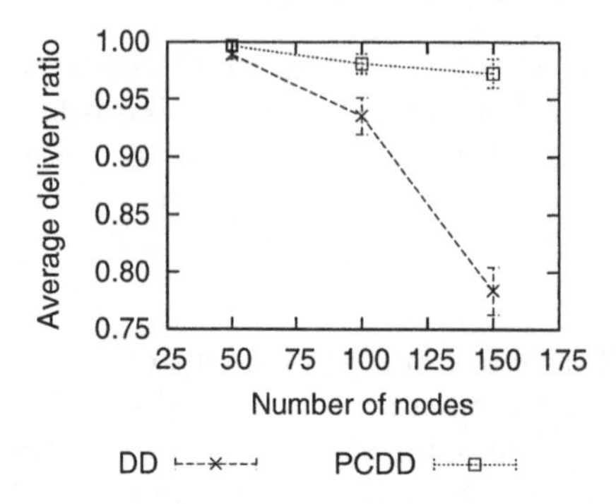

Fig. 7. Delivery ratio – fixed area scenario

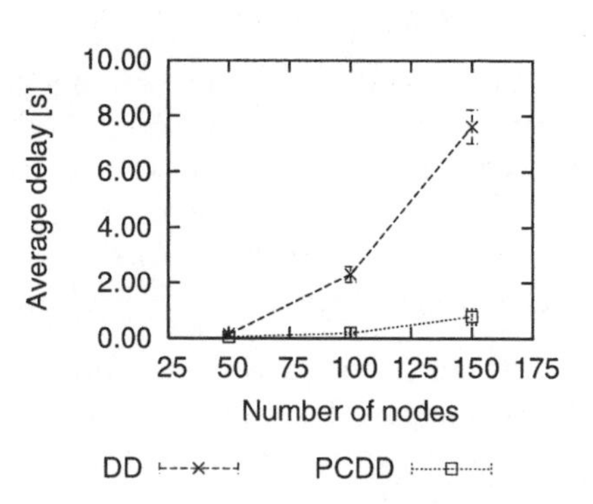

Fig. 8. Delay – fixed area scenario

4.2 Effects of the Traffic Pattern

What happens when the number of sinks or sources is varied, as described in the second type of scenarios in Sect. 3.5?

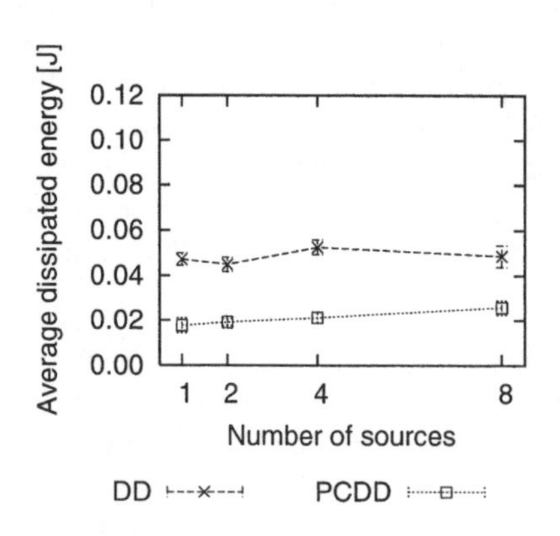

Fig. 9. Energy efficiency – single sink scenario

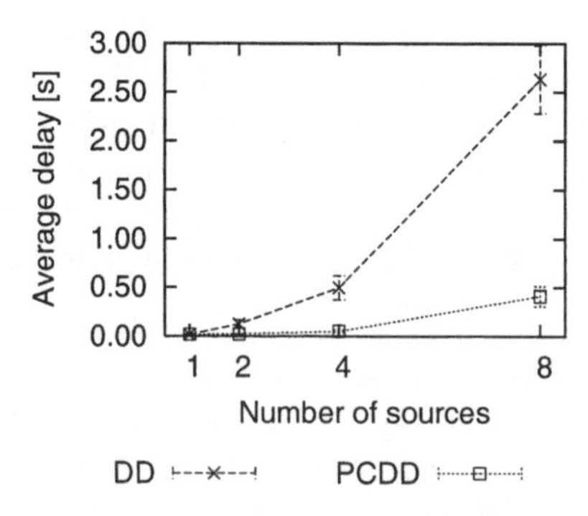

Fig. 10. Delay – single sink scenario

As one could expect, the relative gain of the passive clustering enhancements compared to plain directed diffusion increases with the number of sources: More sources issue more exploratory data floods, which in turn benefit from passive clustering. Our energy metric shown on Fig. 9 somewhat masks these effects because the average dissipated energy *per event* tends to drop as the number of generated events due to the multiple sources increases. The effects of the increased load and the performance gains from the passive clustering are much more visible on the delay plot shown on Fig. 10.

A similar pattern becomes apparent when the number of sinks is varied while having only one source. Here the amount of "sensed" data is not increased, but each data message needs to be reported to more locations in the network. This, combined with the increase in the "interests" traffic, leads to a rise in the energy consumption as shown in Fig. 11. The delay (Fig. 12) also ncreases with the number of sinks, but not as dramatically as with the number of sources.

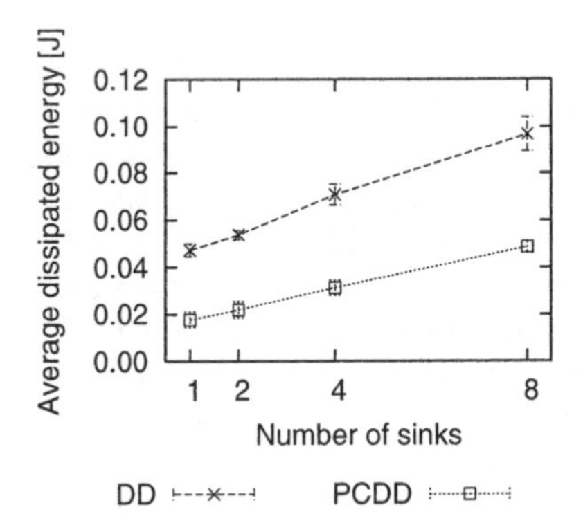

Fig. 11. Energy efficiency – single source scenario

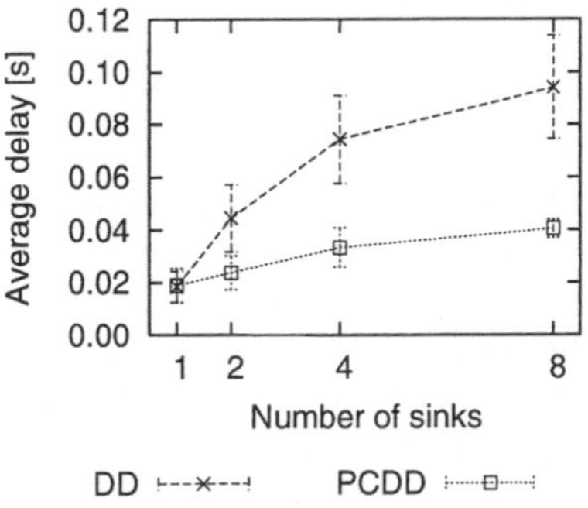

Fig. 12. Delay – single source scenario

5 Related Work

The optimization of the data distribution process in WSNs has been the focus of interest in a couple of recently published papers; the most relevant ones are discussed here.

In [7], the authors concentrate on the problem of efficient distribution of queries that do not excite large amount of data in response. For these *one-shot* queries it is usually more efficient to directly send data along an even non-optimal path, foregoing optimization overhead. They present the *rumor routing* algorithm where the query is sent on a random walk until it intersects an already established path to the event. In contrast to this active networking approach, the passive clustering and directed diffusion combination is better suited for the more long-lived data gathering applications in WSNs. Under these conditions, it is worthwhile to explore the network topology in order to find the optimal paths between the sinks and the sources, as they are going to be used by a (likely) large amount of data during the lifetime of the application.

When the nodes have information about their own and their neighbors' geographic location, geo-routing and geographical overlays become another potential tool for limiting the overhead. Ye et al. [8] present the *Two-tier data dissemination* approach for efficient distribution in large sensor networks with multiple sources and multiple *mobile* sinks. They create grid distribution overlays over which the sources pro-actively announce the availability of new data. The flooding of the queries, initiated by the moving sinks, is then confined to a single cell of this overlay and does not create a network wide restructuring of the routing structure. Our approach does not assume the availability of location information in the network. Instead, we are indirectly sensing the density of the nodes in our neighborhood and try to dynamically adjust the forwarding topology such that the redundant rebroadcasts are eliminated as much as possible.

The authors of [9] describe the *Cluster-based Energy Conservation* protocol for topology control in WSNs, which is closest to our work. Their clustering algorithm also results in the formation of 2-hop clusters. Instead of the *first declaration wins* rule that is used in passive clustering, CEC selects the clusterheads based on the estimated lifetime of the nodes. The gateway selection heuristics is similar as it is influenced by the number of clusterheads and already promoted gateways in the vicinity. But major differences also exist. The CEC is positioned as a general-purpose topology and scheduling mechanism for WSNs. Our approach, on the other hand, is specifically focused on limiting redundant retransmissions during the flooding phases of directed diffusion. One of the major benefits of the passive clustering is the fact that it does not require an active maintenance of the clustering structure, something that is crucial for the operation of CEC. We believe that its properties are a better match to the problem at hand then any other "active" clustering approach.

A number of other cluster-based routing protocols have also been proposed, e.g., LEACH [10]. These approaches share the disadvantages of active clustering or are otherwise not directly applicable to the support of directed diffusion.

6 Conclusion

Several distinct conclusions can be drawn from the results in this paper. First of all, we showed that flooding consumes a large part of the energy in directed diffusion. When flooding of interests and exploratory messages is limited using passive clustering, this will result in large energy savings. But not only the energy savings are important. The decreased load also leads to a better delivery ratio and lower delay, as fewer messages are lost due to collisions.

While the passive clustering in the form presented here is very beneficial, it can introduce hot spots in the network, namely the gateways between two clusterheads and the clusterheads themselves. Their energy expenditure is much higher than those of ordinary nodes. How to deal with this asymmetry between nodes on top of an intentionally passive clustering approach is an open problem. We intend to compare our approach with active clustering approaches under this perspective.

Further extensions could make use of the clusterheads as natural points to aggregate the data. This aggregated data is more important than not aggregated data, which could make it beneficial to transmit it redundantly using multiple gateways instead of using a reliable MAC.

In this paper, we have consciously decided to focus on network-level improvements of directed diffusion and ignored any possible gains from a modification of the MAC layer. For example, the decision to unicast exploratory data to all neighbors, although logical from the routing protocol point of view, harms the performance. We have performed some preliminary experiments which used MAC broadcasts instead of MAC unicasts for these packets. These experiments show a substantial improvement over the unicast case.

This simple modification, however, does not remove all remaining problems, as some of the difficulties stem from the traffic patterns. For the highly correlated and periodic traffic produced by directed diffusion, the IEEE 802.11 MAC is unsuitable. Thus, the interaction and the integration of directed diffusion with the underlying MAC merits further investigation.

References

1. Intanagonwiwat, C., Govindan, R., Estrin, D., Heidemann, J., Silva, F.: Directed diffusion for wireless sensor networking. IEEE/ACM Transactions on Networking (TON) **11** (2003) 2–16
2. Williams, B., Camp, T.: Comparison of broadcasting techniques for mobile ad hoc networks. In: Proceedings of the third ACM international symposium on Mobile ad hoc networking & computing, ACM Press (2002) 194–205
3. Kwon, T.J., Gerla, M.: Efficient flooding with passive clustering (PC) in ad hoc networks. ACM SIGCOMM Computer Communication Review **32** (2002) 44–56
4. Yi, Y., Gerla, M.: Scalable AODV with efficient flooding based on on-demand clustering. ACM SIGMOBILE Mobile Computing and Communications Review **6** (2002) 98–99
5. Silva, F., Heidemann, J., Govindan, R.: Network routing application programmer's interface (API) and walk through 9.0.1. Technical report, USC/ISI (2002)
6. UCB/LBNL/VINT: Network simulator - ns-2. (http://www.isi.edu/nsnam/ns/)
7. Braginsky, D., Estrin, D.: Rumor routing algorithm for sensor networks. In: Proceedings of the first ACM international workshop on Wireless sensor networks and applications, ACM Press (2002) 22–31
8. Ye, F., Luo, H., Cheng, J., Lu, S., Zhang, L.: A two-tier data dissemination model for large-scale wireless sensor networks. In: Proceedings of the eighth annual international conference on Mobile computing and networking, ACM Press (2002) 148–159
9. Xu, Y., Bien, S., Mori, Y., Heidemann, J., Estrin, D.: Topology control protocols to conserve energy in wireless ad hoc networks. Technical Report 6, University of California, Los Angeles, Center for Embedded Networked Computing (2003) submitted for publication.
10. Heinzelman, W.R., Chandrakasan, A., Balakrishnan, H.: Energy-efficient communication protocol for wireless microsensor networks. In: Proc. of the 33rd Hawaii Intl. Conf. on System Sciences. (2000)

The XCast Approach for Content-Based Flooding Control in Distributed Virtual Shared Information Spaces—Design and Evaluation

Jochen Koberstein, Florian Reuter, and Norbert Luttenberger

Communication Systems Research Group
Computer Science Dept., Christian-Albrechts-University in Kiel
{jko,flr,nl}@informatik.uni-kiel.de

Abstract. In this paper, we assume that sensor nodes in a wireless sensor network cooperate by posting their information to a distributed virtual shared information space that is built on the basis of advanced XML technology. Using a flooding protocol to maintain the shared information space is an obvious solution, but flooding must be tightly controlled, because sensor nodes suffer from severe resource constraints. For this purpose we propose a content-based flooding control approach ("XCast") whose performance is analyzed in the paper both analytically and by ns2 simulations. Results show that already a generic XCast instance effectively controls the number of messages generated in a sensor network. It is argued that application-specific XCast instances may expose an even better behavior.

1 Introduction

The ongoing increases in computing power, memory size and communication capabilities of small and even very small electronic devices, and the (considerably slower) increase in battery lifetime together enable new kinds of applications. In this article, we are interested in applications where a number of such devices act together as a "swarm": They follow a common operation goal, are—in some scenarios—mobile, are connected by a wireless network in ad-hoc mode, and take care of their self-organization. Examples for such applications are sensor networks deployed in ad-hoc mode (e.g. for disaster relief), applications in traffic telematics (e.g. for spreading warning messages like "traffic jam ahead"), or applications for collaboration support (e.g. for participants in ad-hoc meetings).

In such environments, the classical client/server co-operation paradigm does not longer seem to be adequate for a number of reasons:

- The client/server paradigm is built upon the assumption that services offered by some nodes or even a single node remain accessible in an at least partially static network topology—something which cannot be guaranteed in mobile ad-hoc networks.

H. Karl, A. Willig, A. Wolisz (Eds.): EWSN 2004, LNCS 2920, pp. 188–203, 2004.

- For mobile networks, service discovery and complex routing protocols have been proposed. Unfortunately, the related mode of operation normally causes a significant increase in bandwidth and power consumption—an approach hardly acceptable for resource-constraint sensor networks.
- The typical request/response protocols for client/server co-operation can be characterized as point-to-point protocols. The "natural" mode for communication in wireless networks though is "broadcast": radio waves propagate in all directions in space. A co-operation paradigm suitable for swarms should exploit this fact.

For all these reasons, we propose a different co-operation paradigm for nodes in a swarm, which we call "information sharing". Swarm nodes provide locally acquired information to each other such that every swarm node can enhance its local view to a global view of the (physical or virtual) area in which the swarm is acting. Swarm nodes remain autonomous, but when initiating actions, they can base their decisions on a more profound knowledge of their environment. To share information, a swarm of nodes establishes a *distributed virtual shared information space* (dvSIS). Every swarm node holds a local instance of the dvSIS, which may be incomplete, partially obsolete, or inconsistent with the local instances of other nodes. The dvSIS is looked upon as a virtual entity, as it exists as an abstraction only. To contribute to the dvSIS, every swarm node broadcasts recently acquired information (e.g. from reading sensors) to the swarm nodes in its vicinity. Swarms nodes may help each other by relaying data, i.e. by re-broadcasting data received from others. From the point-of-view of routing schemes, information in a dvSIS comes close to information flooding.

This immediately raises severe concerns: Can a co-operation paradigm that is based on flooding become effective in terms of bandwidth and power consumption? How can the number of messages usually associated with flooding be tightly controlled? This article tries to answer these questions in the following steps:

In the next section, we deepen the notion of a "distributed virtual shared information space" by modelling it as an XML Schema instance. We then introduce a general model for communication in a XML-based distributed virtual shared information spaces that we call XCast. In the following section we evaluate XCast's flooding control performance by giving analytical bounds and simulation results for a simple XCast instance. And finally, we point to some related work and present our on-going research activities.

2 Information Sharing in a Distributed Virtual Shared Information Space

2.1 Comparison to Publisher/Subscriber

At first glance, the information sharing paradigm for co-operation of nodes in a swarm has similarities with the well-known publisher/subscriber co-operation paradigm, because in publisher/subscriber systems, publishers share their information with all subscribers that declare their interest in some type of information that is going to be provided by the publishers. But two important features are different from information sharing in a dvSIS:

- In most publisher/subscriber systems, we find a centralized "message hub" that accepts subscriptions from subscribers and emits notifications to these whenever a publisher announces the availability of new information. As outlined above, central elements must by any means be avoided in swarms. We therefore consciously introduced the dvSIS as a truly distributed virtual shared information space.
- Protocols for publisher/subscriber systems can be designed such that except for short amounts of time, all subscribers share the same view on their environment. We do not designate any means for this purpose in a swarm. Swarm nodes must take precautions themselves to cope with missing or out-of-date information, and they cannot rely on a consistent view on their common environment.

The information sharing co-operation paradigm is best supported by network-wide flooding of information. Though often criticised for its excessive communication cost when not tightly controlled, flooding has a number of pros on its side:

- Flooding inherently makes communication more reliable because of message replication.
- Reliability is increased without the cost of elaborate ACK schemes.
- No route acquisition overhead is required. (This overhead normally pays off in static networks in a short time frame, but remember that we include mobile sensor networks into our discussion, too.)
- No TTL expiration errors are experienced by messages sent in flooding networks.

In the next section, we propose an XML-based coding for the dvSIS, and we will show how communication cost can be reduced by a novel flooding control scheme that is based on the XML-coding for the dvSIS.

2.2 XML-Coding for Distributed Virtual Shared Information Spaces

In our approach, the dvSIS is represented by an XML-coded, semi-structured document for two reasons: First, the semi-structuredness of XML documents lets the designer appropriately reflect variations in the configuration of a swarm, and second, XML enables the exchange of self-describing information elements between swarm nodes. (We do not cover the increased overhead coming with XML encoding; a number of projects is working on this subject.)

The dvSIS is formally specified by an XML Schema. The specification is such that three different kinds of dvSIS instance documents are covered by the schema:

1. a kind of instance document that can hold the sum of all information components being shared among the swarm nodes ("global dvSIS instance"),
2. several kinds of instance documents that describe the information being held by a single swarm node ("local dvSIS instances"), and finally
3. several kinds of instance documents that can hold one or more document elements to be encapsulated into messages that can be broadcast by swarm nodes to their neighbors ("message instances").

The local dvSIS instances may differ from each other. Differences in the structure of local instances are "legal" if they are compatible with the given dvSIS schema, whereas differences in document element content or element attribute values obviously result from delays in information spreading or from temporary unreachability of swarm nodes.

Merging a received document element into a local dvSIS instance may be confined to updating some document element content or some attribute value, or it may require a structural enhancement of the local dvSIS instance. For merging, a "contains" relation $\sqsubseteq$ is defined; it describes a half-order on a set of documents. The merge operation $\sqcup$ aggregates two document instances into one document instance. We may e.g. write $D_1 \sqsubseteq D_1 \sqcup D_2$.

To formally define the "contains" half order we assume that the coding of an instance document can be described by a simplified XML instance model (C, E, A, r) as follows:

$$
\begin{aligned}
&C \sqsubseteq \mathit{charset} * \text{(content)}\\
&A \sqsubseteq \{(\mathit{name}, \mathit{value}) : \mathit{name}, \mathit{value} \sqsubseteq \mathit{charset}*\}\text{(attributes)}\\
&E \sqsubseteq \{< \mathit{name}, A_n, < e_1, \ldots, e_k >: \mathit{name} \sqsubseteq \mathit{charset}*, A_n \sqsubseteq A,\\
&\qquad e_i \sqsubseteq C \sqcup E\}\text{(elements)}\\
&r \in E\text{(root node)}.
\end{aligned}
$$

A general definition of the "contains" half-order reads as follows:

$$
\begin{aligned}
&c_1 \sqsubseteq c_2 : c_1 = c_2 \text{with} c_1, c_2 \in C\\
&a_1 \sqsubseteq a_2 : a_1 = a_2 \text{with} a_1, a_2 \in A\\
&\qquad < \mathit{name}_1, A_{n1}, e_{11}, \ldots, e_{1j} > \sqsubseteq < \mathit{name}_2, A_{n2}, e_{21}, \ldots, e_{2k} > \text{if}\\
&\qquad 1.\ \mathit{name}_1 = \mathit{name}_2\\
&\qquad 2.\ \forall a_1 \in A_{n1} : \exists a_2 \in A_{n2} : a_1 \sqsubseteq a_2\\
&\qquad 3.\ \exists f : \mathbb{N} \longrightarrow \mathbb{N} : \forall i 1 \leq i \leq j : e_{1i} \sqsubseteq e_{2f(i)} \wedge f \text{ injective}
\end{aligned}
$$

The merge algorithm can be described as follows:

1. Concatenate both document instances to a single document instance.
2. Now reduce the resulting document instance: delete every element E_1 for which an element E_2 exists ($E_1 \neq E_2$) if $E_1 \sqsubseteq E_2$.

To consider non-valid documents we define the merge operation such that it returns an error if merging is not possible, e.g. $I_1 \sqcup I_2 = \perp$.

2.3 XCast: An Abstract Model for Information Sharing

To distribute their information to others, swarm nodes broadcast message instances. Receiving swarm nodes validate a received message instance against the dvSIS schema, "merge" the received document element(s) into their local dvSIS instance if it is not yet contained therein, and possibly re-broadcast the received information. We now introduce an abstract model (called "XCast") for the description of flooding control, and later show a sample instance for this model.

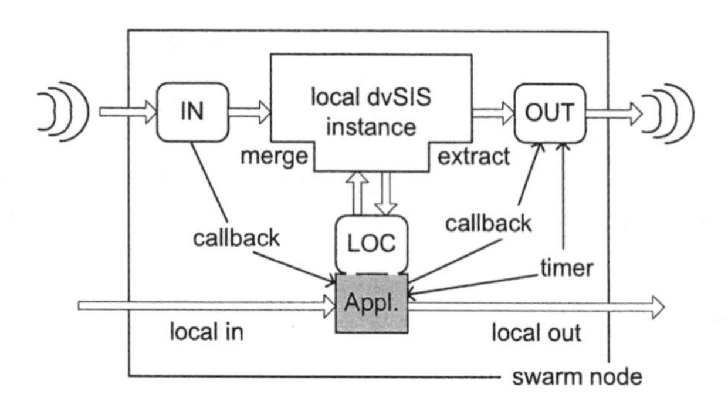

Fig. 1. XCast model with IN, OUT, and LOC filters

The XCast model (Fig. 1) foresees a collection of three filters IN, OUT, LOC (the latter for "local"). The filters are responsible for handling XML document structures, especially for parsing incoming messages, for merging these with the local dvSIS instance, and for extracting data from the dvSIS. A novel kind of finite state machines called "Cardinality Constraint Automata" (CCA) has been developed for this purpose. CCAs are very efficient and at the same time extend the semantics of the W3C XML Schema [1,2]. Preparing work was done by the authors and may be reviewed in [3]. We do not deepen this aspect, because in this paper we are more interested in how these filters contribute to flooding control in the swarm's ad-hoc network.

The IN filter filters message instances coming from the wireless network. IN has to ensure that merging local information with new information will result in a new valid dvSIS instance. Incoming information not complying with this requirement must be converted or discarded. At the same time IN should be somewhat tolerant to allow a big set of accepted informations, because any information may be important. This consideration is well known in conjunction with the internet architecture [4]. IN may activate an application object upon successful operation.

Information to be sent to other nodes is filtered by OUT, which thus determines the node's behavior as seen from other nodes. Due to mobility, a node cannot expect that messages sent are received by any other node. No mechanisms for acknowledgements are provided. Probabilistic approaches for resending locally stored information have been proven advantageous. OUT may be e.g. triggered by a probabilistic timer signal or by an application object, the latter, whenever the application has acquired "fresh" information over its local input.

The application uses the LOC filter to read from or write to the local dvSIS instance. It enables the application e.g. to update locally measured or generated values or some kind of meta-information like e.g. a time value that must be adjusted to a local clock.

Filters may contain additional content based constrains such that document elements may be treated based upon e.g. an attribute containing time or location information. These constraints are application specific and may be necessary to avoid ambiguities while merging document instances.

We now develop the XCast model more formally. Suppose n swarm nodes $\{1, ..., n\}$ cooperate in an ad-hoc network. Each of them may send a message at time t. For node j, the subset $inRange^t(j) \sqsubseteq \{1, ..., n\}$ gives all swarm nodes which have node j inside their radio coverage area at time t; i.e. the $inRange^t$ subset of swarm nodes can reach node j by a single broadcast. Thus, a message sent by node k arrives at node j if $k \in inRange^t(j)$. Further we denote a node's local dvSIS instance as I_j. Let E denote the set of document elements to be wrapped into message instances except the message root element.

Posting information from node j to node k at a time t is described by the *put* function:

$$\begin{aligned} put_j^1 &: E \to E \times E \\ put_j^1(I) &:= (OUT(I), LOC(I)) \end{aligned} \tag{1}$$

For incoming information the node behavior is described by the get function:

$$\begin{aligned} &get_{j,k}^t : E \times E \to E \\ &get_{j,k}^t(I_{extern}, I_{intern}) := \begin{cases} I_{intern} & j = k \\ IN(I_{extern}) & j \neq k \wedge k \in inRange^t(j) \\ \emptyset & j \neq k \wedge k \notin inRange^t(j) \end{cases} \end{aligned} \tag{2}$$

Combining 1 and 2 leads to a description of the entire communication of a node j at time t:

$$comm_j^t(I_1, ...I_n) := \bigcup_{k=1}^{n} get_{j,k}^t(put_k^t(I_k)) \tag{3}$$

Exchange of information over time is modeled by the *XCast* function. Assuming an initial state $XCast^{t=0}(I_1, ..., I_n) := (I_1^0, ..., I_n^0)$ the *XCast* function is recursively defined as:

$$\begin{aligned} XCast^t\ &(I_1, ..., I_n) := \\ &(comm_1^t(XCast^{t-1}(I_1, ..., In)), ..., comm_n^t(XCast^{t-1}(I_1, ..., In))) \end{aligned} \tag{4}$$

Assuming that there is only a finite set of information in the network and that the filter functions are designed adequately, a fixed point of the *XCast* function exists. A tuple $(I_1^*, ..., I_n^*)$ is a fixed point iff $XCast^t(I_1^*, ..., I_n^*) := (I_1^*, ..., I_n^*)$ with $t \geq 0$ holds, i.e. calling the XCast function does not change the memory of nodes anymore.

The application program developer has to design the XCast filters such that the *XCast* function fixed point can be reached. If in certain application contexts, a swarm node is interested in regional information only or if a swarm node has to cope with permanently changing information (e.g. when reading periodically from a temperature sensor) the *XCast* function fixed point may be a "regional" or "interval" fixed point. This is often an acceptable view: If e.g. a temperature is measured in a room, the measurement interval can be extended such that during an interval there is only a finite set of update messages to be exchanged between swarm nodes. If the network is not fully connected, the fixed point may not depict the situation in which every node holds all information. Thus the fixed point may be affected by mobility of nodes.

A framework called "swarmWorks" which is under development in our group is going to provide templates for the three mentioned filters, which can be adapted to application needs by the application programmer. The swarmWorks framework calls the application objects' methods as callback functions, whenever new information arrives. And viceversa the application may call swarmWorks methods whenever it has acquired new information from local input.

2.4 Design of Content-Based Flooding Control for XCast

The IN filter shown in Fig. 1 decides which incoming message are to be merged into the local dvSIS instance, and above that, it decides if any of the received messages or components thereof are to be re-broadcasted (flooding control). The OUT filter takes care of broadcasting messages that have been elected by the IN and LOC filters. A simple approach for controlling the OUT filter behaviour is to trigger it by a timer and make it broadcast messages in probabilistic intervals. This control scheme guarantees that all nodes remaining in the vicinity of a sending node for some time can share the information already collected by the sending node.

To introduce the XCast flooding control approach, we compare it with the simple flooding control mechanisms known from text books. In *simple flooding* it is the goal to forward a message not more than once. For this purpose, every message is tagged in its protocol header with a network-unique node ID and a message ID. The message ID is chosen such that it is unique within some time interval. Receiving nodes store the (node ID, message ID) tuples of received messages in an internal list. A node forwards a received message, unless its (node ID, message ID) tuple can be traced in the internal list. If the internal list is too short or if it has to be overwritten too early, messages are sent more then once. The simple flooding procedure is obviously not well suited for resource-constraint systems, because memory and processing resources have to be devoted to flooding control exclusively.

Things are different for *content-based flooding control.* Roughly spoken: messages are forwarded only if their content has been locally unknown before message arrival. By application of the $\sqsubseteq$ half-order, the IN filter checks if the content of a received message is contained in the local dvSIS, and passes only "fresh" information to the OUT filter for forwarding to other nodes. Through selecting messages for forwarding by their content and not by some additional protocol header fields, content-based flooding control needs no extra resources, as the message content of all messages received before is contained in the local dvSIS instance anyway.

IN filters may apply the above given general definition of the $\sqsubseteq$ half-order, but we can also design IN filters that exploit application-specific semantics for the $\sqsubseteq$ half-order. (Thus we talk of XCast as an abstract control model that can have different instances.) Take, for example, a sensor network whose nodes continuously measure their environments' humidity and exchange the resulting meter readings. If sensors were not interested in recording a "humidity history file" the $\sqsubseteq$ half-order could be defined as follows:

```
<Humid time="t1">val-1</Humid>⊑<Humid time="t2">val-2</Humid>
  :⇔ t1 ≤ t2
```

And finally it is worth noting that, to reduce the packet overhead incurred when sending a message, a node may piggyback its own "news" onto news coming from others, i.e. a message instance can be composed of dvSIS elements from different sources and wrapped into a single over-the-air message.

2.5 An XCast Instance for Content-Based Simple Flooding

To illustrate our ideas further we show a very simple algorithm for flooding control called Content-based Simple Flooding (CBSF). Here, the XCast filters IN_{CBSF} and OUT_{CBSF} use a common send_buffer structure as output queue (Fig. 2).

A related code fragment for the IN filter may look as follows:

```
if ¬(I ⊑ I_j) then send_buffer := send_buffer ⊔ I;
```

The `send_buffer` content is merged with the initial information I_j^0 of the node with probability p_s. This probability may be used to tune the CBSF algorithm's behaviour. Code for OUT_{CBSF} may look as follows:

```
local_buffer := send_buffer;
send_buffer := ∅;
local_buffer := local_buffer ⊔ I_j^0 (merge with probability ps);
send local_buffer;
```

Since every node tries to gather all information it receives CBSF leads to a fixed point of the *XCast* function. It should nevertheless be emphasised that the proposed algorithm does not claim to be optimal and may be insufficient for some applications needing additional application specific adopted filter rules. But it is possible to do some analytic work because of its simplicity. Related analysis and simulations are presented in the following chapter.

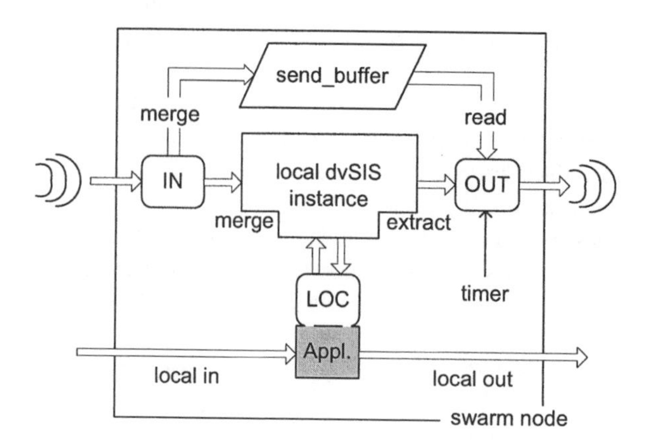

Fig. 2. CBSF, a simple XCast instance

3 Evaluation

For evaluation purposes, we use a simple scenario which may described as follows: The participants of a conference are equipped with wireless-enabled address books. The electronic address books communicate over an 802.11b wireless network in ad-hoc mode to distribute all address entries over all participants. Initially, each device contains an address entry for its owner consisting of a 30 byte name field and a 30 byte e-mail address field. To exclude possible benefits of mobility (see [5,6]) we assume static positions of nodes. Further details may be reviewed in Tab.1.

Table 1. Parameters for CBSF example scenario

number of nodes	n = 50
area	$500\,m \times 500\,m$
distribution of nodes over area	random
node mobility	none
initial information I_j^0	2×30 byte (one address entry)
probability for sending	$p_s = 0.03$
random send interval	$t_{min} = 8ms$ to $t_{max} = 12ms$
observation time	$T = 10s$

3.1 Analytical Evaluation

With the following calculation, we calculate an upper bound for the number of generated messages. We assume the worst case that no aggregation of document elements takes place.

To derive an upper bound for the number of messages generated in the network we calculate both the number of messages generated per send interval by OUT filters and the number of messages generated per send interval by IN filters (which contain newly received document elements). The sum of both gives us the maximum number of generated messages:

$$\sum_{t=0}^{T}(N_{OUT}(t) + N_{IN}(t)) \tag{5}$$

The number of messages generated by OUT is

$$N_{OUT}(t) \leq n \cdot p_s(t) = n \cdot p_s \tag{6}$$

because p_s is invariant in time. We define $p_r(t)$ as probability to receive a new document element, i.e. a document element which is not contained in the local dvSIS instance. The upper bound on the number of messages generated by IN filters can be written as follows:

$$N_{IN}(t) \leq n \cdot p_r(t) \tag{7}$$

For this XCast instance, $p_r(t)$ converges towards zero for increasing t, because nodes collect information from a finite amount of information. In this XCast instance every node re-broadcasts each document element only once, and thus when reaching the fixed point of the XCast function the sum of messages generated by IN is at most n^2.

This result raises the concern that at least in the presented example, there seems to be no performance difference between Simple Flooding and CBSF, since both initially require n^2 transmissions to distribute all available information to all nodes in the network. But now consider the case, when the OUT filter is triggered by a probabilistic timer signal to re-send locally stored information, for instance, to accommodate late-comers in the sensor network area: In Simple Flooding, a total number of n^2 messages would result from such an action since the flooding system generates a fresh (node-ID, message-ID) tuple per message. In CBSF, on the other side, we can afford such message re-sending, since only a node accidentally not finding the message content in its local dvSIS instance re-broadcasts it itself. All other nodes discover that they hold the information already in their local dvSIS instances. Re-sending of information is neither covered by our analytical calculations, nor by simulation.

In general we have to use a more sophisticated approach which accounts the content based character of the algorithm. Hence we transform the time axis to a document axis: Let $docs(t)$ denote the number of document elements received by a filter function IN of one node at time t. The possibility can be converted to

$$p_r(t) = p_{new\_doc}(docs(t)) \tag{8}$$

$p_{new\_doc(d)}$ is the probability that the dth information received is a new one. Thus it is possible to transform the axis and write

$$\sum_{t=0}^{T} p_r(t) = \sum_{t=0}^{T} p_{new\_doc}(docs(t)) = \sum_{d=1}^{docs(T)} p_{new\_doc}(d) \tag{9}$$

For further considerations in full generality it would be necessary to define a probability relation $\chi(I, d)$ which gives us the probability to receive a certain information I as d^{th} document. Further on χ may depend on more parameters and this is getting very complex. We assume for simplicity that the probability is uniformly distributed over the set of informations (which holds for this example scenario).

The precondition that a new document element is comprised in a received message instance is that this document element is not in the prefix of sequence of received document elements. Considering this last element we have to make the choice between n different document elements each with a possibility of $\frac{1}{n}$. For the prefix we may choose out of $n-1$ document elements each with a possibility of $\frac{1}{n}$. With a prefix length of $d-1$ we may calculate the sum of $p_{new\_doc(d)}$ as follows:

$$\begin{aligned} \sum_{d=1}^{docs(T)} p_{new\_doc}(d) &= \sum_{d=1}^{docs(T)} \underbrace{\left(\frac{1}{n}\right)^{d-1} \cdot (n-1)^{d-1}}_{prefix} \cdot \underbrace{\frac{1}{n} \cdot n}_{postfix} \\ &= \frac{n}{n-1} \sum_{d=1}^{docs(T)} \left(\frac{n-1}{n}\right)^{d} \\ &= \frac{n}{n-1} \sum_{d=1}^{docs(T)} \frac{1}{k^d} \text{ with } \frac{1}{k} = \frac{n-1}{n} < 1 \\ \text{for } T \to \infty &: \sum_{d=1}^{docs(T)} p_{new\_doc}(d) = n = D \end{aligned} \tag{10}$$

Hence the sum converges and it is possible to show that it converges against n (see Fig. 3). This equals to the result above. We call D the document-boundary which depends on the structure of transmitted documents.

Using equations 6 and 10 in 5 leads to

$$\text{number of transmitted messages} \leq T \cdot n \cdot p_s + n \cdot D \tag{11}$$

This result points out that we may estimate the number of transmitted messages without considerations based upon time. Hence there is a direct interdependency between representation of information and the number of packets. This enables us to optimize the flooding control algorithm by using an efficient information encoding.

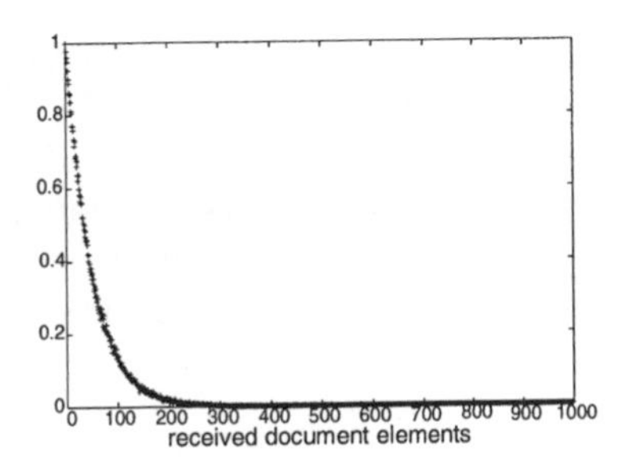

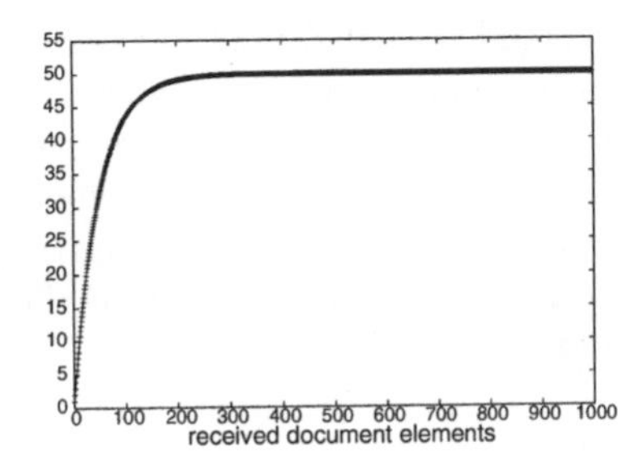

Fig. 3. Probability to receive a new document element (left) and the convergence of the sum against the document-boundary (right)

3.2 Simulation Results

Using the network simulator ns2 we have simulated the scenario described above. Figure 4 shows that all address entries are fully distributed after approx. three seconds. It depicts the number of messages scheduled for retransmission by the IN filter. According to our analytical results we expect about $n^2 = 2500$ messages but experience only about 2400 in the simulation. This difference can be explained by packet losses at the MAC layer because of limited buffer space. Figure 5 shows the effect of information aggregation, i.e. a node merging its own

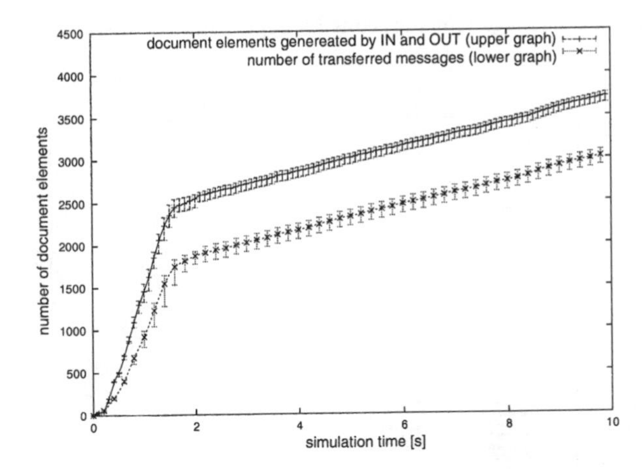

Fig. 4. Average number of document elements scheduled for retransmission by filter IN with corresponding minima and maxima over the number of simulations

"news" with news coming from others. The efficiency of information aggregation (i.e. the gap between the number of address entries in the upper graph and the number transmitted messages in the lower graph) strongly depends on the duration of the send intervall: the longer the send intervall, the more address entries are merged into the `send_buffer`. As result the percentage of saved messages varies in simulations with different send intervalls between 10% and 50%. The steep graph at the beginning of each simulation nearly equals to the results of simple flooding since most document elements are new to the receivers and are rebroadcasted. The reduced gradient of the displayed graph after approximatly 2 seconds reflects the savings of the CBSF algorithm. The resulting messages sizes are depicted in figure 7.

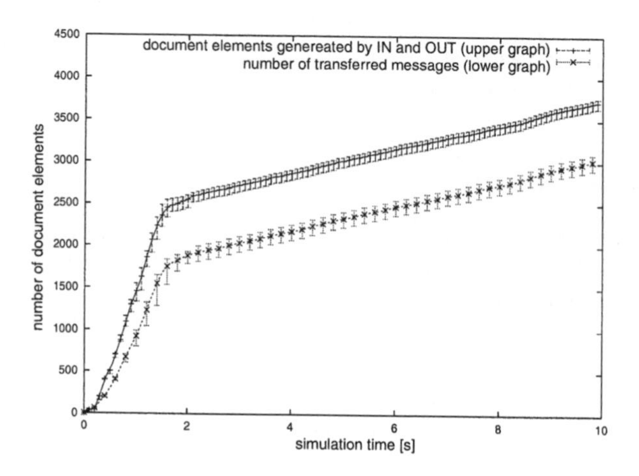

Fig. 5. Saving message transmissions by aggregation of document elements

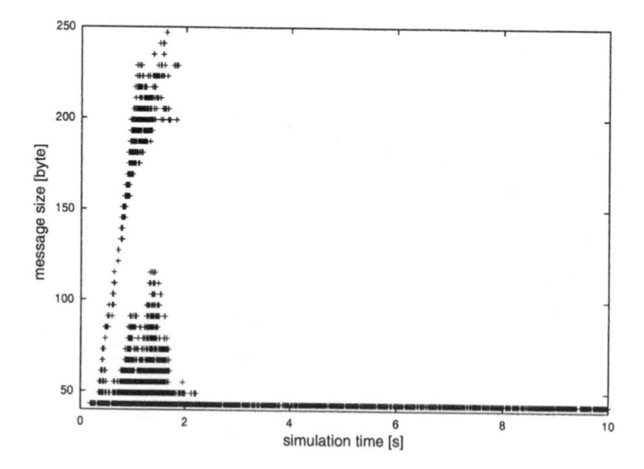

Fig. 6. Message size

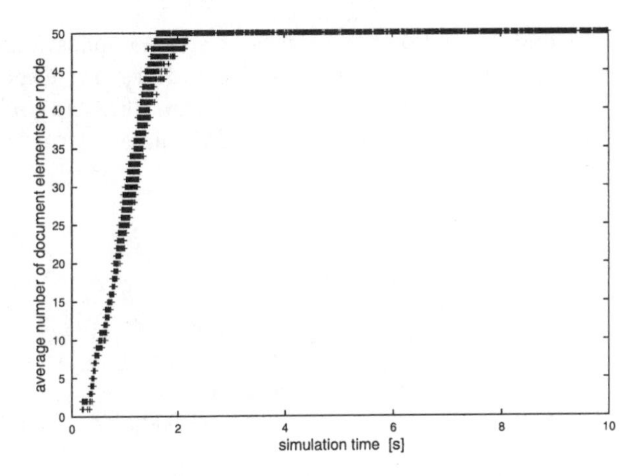

Fig. 7. Average number of different document elements per node

4 Related Work

A well known paradigm for distributed measurement, communication and calculation applications in the area of sensor networks is "Directed Diffusion" [7]. It is comparable to our approach, because (name,value)-tuples are transmitted in the network. Many research projects are based on Directed Diffusion but as far as we know none examines the problem of information coding and its influence on performance.

An overview of non-content based algorithms for flooding control is given in [8]. Content-based protocols which try to reduce the communication overhead are presented in [9]. Using meta-information for filtering is the most important idea in this approach.

A well known technique in the area of databases is "Rumor Spreading" [10]. In analogy to rumors and epidemics each node tries to "infect" some selected neighbors with its information to synchronize distributed databases. In contrast to XCast it is not based on broadcasts.

Applications in the range of XML-based middleware like UPnP [11] and Web-Services [12] or XML-based tuple spaces [13,14,15] are mostly using client/server paradigms and are not suitable for sensor networks without adaptations. Crashing a node running any kind of server may crash the whole application running in the network.

5 Outlook

Based on the results shown, new and more efficient algorithms for sensor networks based on the XCast model can be developed. As shown above paramters of filter functions like p_s or the random send interval of the CBSF algorithm affect the efficiency and depend on the scenario. Certain applications require application specific filter rules to avoid ambiguities and preserve scalability. Thus further research is going to concentrate on the design of filters and the understanding of the influence of filter parameters and document structure.

References

1. Reuter, F., Luttenberger, N.: Cardinality Constraint Automata: A core technology for Efficient XML Schema-aware Parsers. http://www.swarms.de/publications/cca.pdf (2002)
2. Thompson, H.S., Beech, D., Maloney, M., Mendelsohn, N.: XML Schema Part 1: Structures. W3C. (2001) http://www.w3.org/TR/xmlschema-1/.
3. SWARMS: Software Architecture for Radio-based Mobile Systems, http://www.swarms.de. (2002)
4. Carpenter, B.: RFC 1958: Architectural principles of the internet. (1996) Status: INFORMATIONAL.
5. Beaufour, A., Leopold, M., Bonnet, P.: Smart-tag based data dissemination. First ACM International Workshop on Wireless Sensor Networks and Applications (WSNA) (2002) 68–77
6. Vahdat, A., Becker, D.: Epidemic routing for partially-connected ad hoc networks. Technical report, Duke University, Durham, North Carolina (2000) ftp://ftp.cs.duke.edu/dist/techreport/2000/2000-06.ps.
7. Intanagonwiwat, C., Govindan, R., Estrin, D.: Directed diffusion: A scalable and robust communication paradigm for sensor networks. In: Proceedings of the sixth annual international conference on Mobile computing and networking, ACM Press (2000) 56–67
8. Williams, B., Camp, T.: Comparison of broadcasting techniques for mobile ad hoc networks. In: The Third ACM International Symposium on Mobile Ad Hoc Networking and Computing (MobiHoc), Lausanne, Switzerland, ACM (2002)
9. Heinzelmann, W.R., Kulik, J., Balakrishnan, H.: Adaptive protocols for information dissemination in wireless sensor networks. In: Proceedings of the Fifth Annual ACM/IEEE International Conference on Mobile Computing and Networking (MobiCom), Seattle, Washington, USA, ACM/IEEE (1999)
10. Demers, A., Green, D., Hauser, C., Irish, W., Larson, J., Shenker, S., Sturings, H., Swineheart, D., Terry, D.: Epidemic algorithms for replicated database maintenance. In: Proceedings of the Sixth Symposium on Principles of Distributed Computing. (1987) 1–12
11. UPnP: UPnP Forum, http://www.upnp.org/. (2002)
12. W3C: Web services, http://www.w3.org/2002/ws/. (2002)
13. Cabri, G., Leonardi, L., Zambonelli, F.: XML dataspaces for the coordination of internet agents. Applied Artificial Intelligence **15(1)** (2001) 35–58
14. Moffat, D.: XML-Tuples and XML-Spaces. (1999) http://uncled.oit.unc.edu/XML/XMLSpaces.html.
15. Wycko, P.: T Spaces. IBM Systems Journal **37(3)** (1998) http://www.research.ibm.com/journal/sj/373/wycko.html.
16. Biron, P.V., Malhotra, A.: XML Schema Part 2: Datatypes, http://www.w3.org/TR/xmlschema-2/. (2001)
17. Clark, D.D., Tennenhouse, D.L.: Architectural considerations for a new generation of protocols. In: Proceedings of the ACM symposium on Communications architectures & protocols, ACM, ACM Press (1990) 200–208

18. Goos, G.: Vorlesungen ueber Informatik Band 1: Grundlagen und funktionales Programmieren. 2 edn. Springer (1997)
19. Hors, A.L., Hegaret, P.L., Wood, L., Nicol, G., Robie, J., Champion, M., Byrne, S.: Document Object Model (DOM) Level 3 Core Specifcation. W3C, http://www.w3.org/TR/2002/WD-DOM-Level-3-Core-20021022/. (2002)
20. Maltz, D.A.: On-Demand Routing Multi-hop Wireless Mobile Ad Hoc Networks. PhD thesis, Carnegie Mellon University (2001)
21. Tolksdorf, R., Glaubitz, D.: Coordinating Web-based systems with documents in XMLSpaces. Lecture Notes in Computer Science **2172** (2001) 356–?? http://link.springer-ny.com/link/service/series/0558/bibs/2172/21720356.htm; http://link.springer-ny.com/link/service/series/0558/papers/2172/21720356.pdf.
22. Carriero, N., Gelernter, D.: Linda in context. Communications of the ACM **32(4)** (1989) 444–458

A Novel Mechanism for Routing in Highly Mobile Ad Hoc Sensor Networks

Jane Tateson and Ian W. Marshall

BTexact, Adastral Park, Martlesham Heath, Ipswich IP5 3RE, UK
{jane.tateson,ian.w.marshall}@bt.com

Abstract. This paper describes a novel routing mechanism for a network of highly mobile sensor nodes that routes data over dynamically changing topologies, using only information from nearest neighbours. The preferred forwarding directions of mobile sensor nodes are modelled as vectors, and a scalar trigger is used to determine data forwarding. Simulations have demonstrated that this technique operates successfully in sparse networks, where node movements are unpredictable, and data generation by nodes is non-uniform. The application scenario is a self-configuring network of mobile nodes, floating in the sea, that is tracking the movements of a shoal of fish. The requirements of the technique in terms of memory are minimal, with very few parameters and very little code being needed, as is appropriate for the low-powered microprocessors envisaged.

1 Introduction

Scientists often want to monitor processes in the environment, where these processes could be changes in glacier size, seismic activity, or the dynamics of endangered species. Water companies have a particular interest in monitoring the movements of pollutants in rivers and reservoirs. Dredging companies are interested in the effects their activities are having on the coast-line. All these groups have typically had to rely on data gathered from a handful of strategically placed, highly expensive packages, each containing many sensors. Because the environments of interest are often hostile or difficult to access, the sensor packages have had to be housed in heavy-duty casings, to prevent damage or loss. In the case of oceanography, even collecting these sensor packages for data retrieval, by hiring a ship, costs thousands of pounds. Beyond the disadvantage of high cost, the data is only retrievable at long intervals. Even in networked sensor systems[1], localized measurements only represent a tiny fraction of the total area where environmental change is taking place.

What scientists want is near real-time access to data which has been sampled at many points in the environment of interest. In order to achieve this, a much more flexible infrastructure is needed, one that consists of a large number of nodes, each containing one or more sensors, where sensor nodes are able to self-configure dynamically, manage their own resources in the context of the monitoring experiment, and collect and forward data efficiently. This would typically involve multi-hop wireless communication amongst low-powered sensor nodes, in order to forward measurement data to a number of network sinks. The sinks are likely to be higher-

H. Karl, A. Willig, A. Wolisz (Eds.): EWSN 2004, LNCS 2920, pp. 204–217, 2004.

powered devices, with much larger memory than the sensor nodes, and able to interface, perhaps via satellite, with fixed network devices on land. The important thing is that the sensor nodes are cheap enough that they can be deployed in large numbers, and that they operate collectively as an efficient data-gathering network, in spite of node failures and varying topology.

There is relevant work in the literature that refers to dense networks of sensor nodes, and how to mange the wake-up of a subset of nodes from a sleeping state, when an event of interest occurs within the monitoring area of the sensor network. Here the sensor nodes are static, and their major problem is how to conserve power between the occurrence of interesting events, and then how to use the node battery resources fairly. For example, work by Cerpa et al.[2] refers to habitat monitoring as a driver for wireless communications technology, and focuses on power-saving by nodes outside regions where interesting changes could be observed, switching themselves off, and being triggered to switch back on only when interesting activity is detected in their vicinity. Work by Xu et al.[3] again focuses on using powered-down modes for devices to conserve power, based on whether data traffic is predicted or not, and on the number of equivalent nodes nearby that could be used for alternate routing paths. The assumption here is that the underlying routing will be based on conventional *ad hoc* routing protocols such as AODV[4]. Sensor networks, however, typically would require a lighter weight approach to routing, where decisions are based on succinct information from immediate neighbours only.

Work by Heinzelman et al.[5] has as its focus the use of clustering techniques to reduce bandwidth usage by, for example, data aggregation of similar data, and using predictable transmission times, co-ordinated by the cluster heads. This approach saves significant energy, compared with an always-on approach, but the routing side is simplistic and not fully developed. They assume that devices could all broadcast to the base station if they chose to, which would not be realistic for sensor network applications, as a general rule. Other ways to reduce energy usage include the work by Singh et al.[6], who have made a detailed study of power-conservation in ad hoc networks at the MAC and network layers. They include schemes for devices to power-down in between expected transmissions, and they take into account device load as an important factor in power consumption. Their main concern is to prevent network partitioning when gaps appear in the network as a result of devices running out of battery power.

A lot of work has been done at the University of California and the Intel Berkeley Research Lab, to develop operating systems and networks for small *ad hoc* sensor devices, known as the *Smartdust*[7] project, for which *TinyOS*[8] has been developed. Their nodes self-configure into a hierarchical structure to find shortest paths to the sinks, but this process is not power-aware.

Whilst the routing schemes referred to above are appropriate for networks of static nodes, they do not handle node mobility explicitly, and would certainly have great difficulty with highly mobile nodes. *Ad hoc* routing protocols, on the other hand, are designed to cope with node mobility. Many *ad hoc* routing protocols have been devised. Some of the most widely known are DSDV[9], TORA[10], DSR[11] and AODV[4]. DSDV[9] maintains a routing table listing the next hop for each reachable destination. Routes are tagged with sequence numbers, with the most recently determined route, with the highest sequence number, being the most favoured. There are periodic updates of routes and sequence numbers. TORA[10] discovers routes on demand and gives multiple routes to a destination. Route query and update packets are sent for each destination. Although routes are established fairly quickly, there are often routing loops, leading to dropped packets. DSR[11] uses source routing, rather than hop-by-hop routing, so each packet has a complete route, listed in its header. The protocol uses route discovery and route maintenance, with nodes maintaining caches of source routes that have been learned or overheard. AODV[4] combines route discovery and route maintenance with hop-by-hop routing. Route request packets create reverse routes for themselves back to their source nodes. Hello messages are periodically transmitted by nodes so that neighbours are aware of the state of local links. A comparison of the performance of these protocols[12] has shown widely differing results in the size of routing overhead. The total overhead is worst for TORA, and becomes unacceptable for a network size of 30 source nodes. However, the main problems with using these ad hoc network protocols for a network of mobile sensor devices is that 1) the size of processor and memory required is too large, and 2) the protocols are not energy usage aware.

Sensor networks are envisaged as consisting of very small, very cheap microprocessors, e.g.16 bit, with 32 kbytes of RAM. They will also have a finite battery supply, which will be difficult, and probably not desirable to replace. It is therefore very important that any communication protocol is energy-efficiency aware, and also pared to a minimum in communication overhead and memory usage.

2 Routing Mechanism

The application scenario is a sparsely-populated network of highly mobile, wireless sensor nodes that take measurements from the environment and send data back to one of the network sink devices. Although there has been work published on communication between sensor nodes[6-9], what is unique to this work is its ability to route data well even when nodes are moving rapidly. The method is independent of network scale.

The technique being presented here assumes that nodes know their relative positions. There are many ways that this could be achieved. Relative position can be determined if nodes have a directional antenna, a ranging mechanism, and a digital compass for a reference direction. Nodes can determine even their absolute positions themselves [13,14], using some fixed nodes or GPS-enabled nodes as reference points. And a knowledge of position is generally essential for environmental measurements to be meaningful.

Each mobile data-gathering sensor node needs a forwarding direction to send its data back to a network sink. Because the nodes are moving rapidly, even their nearest neighbours may change between data transmissions. Routing decisions must be made 'on-the-fly', using very recently gathered information. In this work, an analogy is made between the forwarding direction of a node and the co-ordinates of a *polar* bond, associated with that node, that is able to rotate to find its optimal orientation. A *polar* bond has one end that is negatively charged, and one end that is positively charged. Polar bonds will tend to align themselves with their neighbours' polarities. If a chain of real polar molecules formed, you would find alternating positive and negative polarities lined up all along the chain, just as you would find that a set of magnets, placed together in a line, would seek to have opposite poles touching : + - + - + - + - .

When a node is ready to transmit data, it can determine its forwarding direction by calculating the 'optimal' orientation of its associated polar bond. This is achieved by combining the alignment influences of the forwarding directions of neighbouring nodes.

Ideally, a node wants to do more than identify a transmission hop in the right direction, it wants to forward according to the best chance of its data getting all the way back to a sink. A complete route back to a sink is given by a chain of nodes, where each node is within forwarding range of the previous node in the chain. Such a structure may only form quite briefly, but, when it forms, it will result in stronger interactions between the 'polar bonds' along the chain. Calculating the strength of an interaction is a natural extension of the calculation of forwarding direction, and gives us a scalar trigger to determine when data forwarding should take place.

Each polar bond is modelled as two atoms, one atom is positively charged, the other atom is negatively charged. (The charges are equal, but opposite in sign). The network sinks are modelled as unit positive point charges. This results in nodes in the vicinity of a sink pointing their forwarding directions towards this sink. These forwarding directions (polar bond orientations), in turn, have knock-on effects on other nearby nodes that are too far away from the sink to forward to the sink directly. And these nearby nodes influence nodes yet further away from the sink, to aim their forwarding directions so as to make a path to the sink, via the intermediate nodes, and so on, throughout the network. The result is that, irrespective of network topology, all nodes (unless cut-off entirely from regions of the network with paths to a sink) will have a forwarding direction that is likely to result in the multi-hop transmission of data to a sink. Forwarding directions are updated dynamically, so that as soon as a link is re-established, transmission of data can re-start. If the network is very sparsely populated, most data transmissions may only occur when a node comes within direct range of a sink. In densely populated networks, much longer paths, in terms of number of hops, will be common. The important thing is that this method is flexible enough to cope with a wide range of circumstances, in terms of network topology and node speed, without such variations requiring special treatment.

The electrostatic analogy, where *electrostatic* simply refers to the interactions of charges, is a framework to enable the sum of alignment influences of neighbouring nodes to be determined quantitatively. We need to know how far away a node is, and what is its current preferred forwarding direction, but we also need to know how to combine the effects of several nodes, some of which will have conflicting (opposite) influences. Fortunately, there are well-established ways to combine such effects together, one example of which is found in the study of electrostatics. The routing mechanism makes use of these well-known relationships, so that the influences of neighbouring nodes are taken account of correctly. An illustration of how such forwarding directions / polar bond orientations are used to route data to a network sink or base station, is given in Figure 1.

This approach to routing could equally be used to route packets in multi-hop cellular networks to a nearby base station. No sensor-network-specific assumptions are made, e.g. that all nodes generate data. Extension to general *ad hoc* networks, where any node can send to any other, would involve the nodes maintaining a number of forwarding directions, for different target nodes. In order to overcome scaling difficulties, in this case, the nodes could be organised into hierarchical clusters, with nodes needing to have forwarding directions to cluster-heads and to nodes within their own clusters.

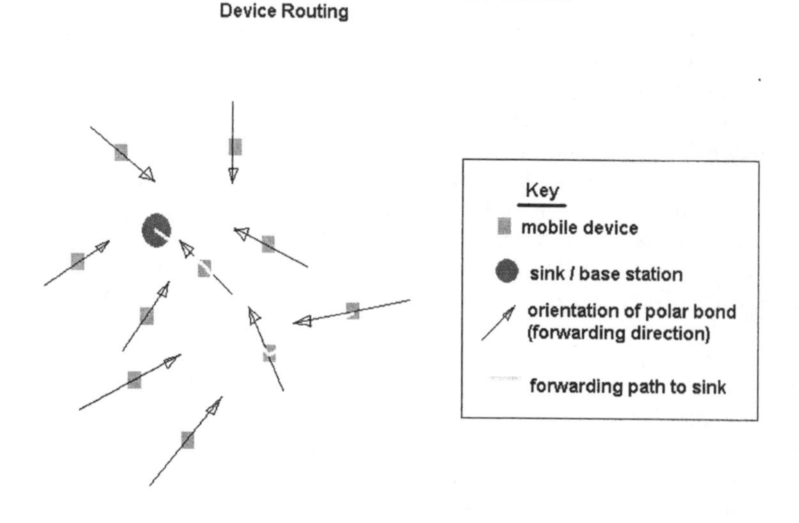

Fig. 1. Illustration of the use of polar bond orientations to give preferred forwarding directions for mobile sensor nodes, forwarding data to a network sink

Note that the type of routing reported here is, for the sake of clarity, in all cases, point-to-point. This simply means that nodes have unique identifiers, quote their identifiers when communicating with their neighbours, and ignore broadcasts that are not meant for themselves, by reading the transmitted data header. For greater assurance of packet delivery, the redundancy of multi-path routing may be desirable, but this issue is orthogonal to the main focus of this work. Also, it may be of interest to compare the hop-by-hop routing presented here to more general Distance Vector methods, for completeness. However, here - despite multiple network sinks - only a single, node-independent forwarding-direction is maintained as routing state.

Each node maintains a tuple representing its location, forwarding direction, quantity of data in its buffer, and a model parameter called the 'induced charge'. When node A has data to send, it broadcasts to see what other nodes are in the vicinity, and each neighbour replies with its tuple. Node A uses the advertised tuples to adjust its own forwarding direction, and determine its forwarding decision. This is most clearly understood as simply a novel form of route table construction that requires less state to be maintained (no addresses), and converges to "good enough" end to end routing solutions rapidly enough to allow very rapid movement to be tracked because no handshake messages are required.

Note that there are 4 different bond types used. The choice of bond is determined by the quantity of data in a device's buffer. Each polar bond has 3 properties: 1) physical reach of influence (bond length), 2) ability to influence other devices (permanent charge) and 3) ability to be influenced by other devices (*polarizability*).

We want the strongest interactions to occur between a device with a full buffer and a device with an empty buffer, for data transfer. We model empty buffers as short bonds with a strong polarising (influencing) effect, and full buffers as long bonds that are easily polarized (influenced). This model requires the devices to have permanent memory of (4 x 3) =12 values plus the forwarding threshold parameter, giving 13 values in total, which can be stored as 13 bytes.

Given the tuples from neighbouring nodes, Node A, having data to send, needs to determine its own 'optimal' forwarding direction, i.e. lowest energy polar bond orientation. Rather than finding the absolute minimum electrostatic energy orientation for the polar bond modelled on node A, this polar bond (forwarding direction) is rotated in 45 degree steps, with node A calculating the electrostatic energy 8 times (8 x 45 = 360 degrees), and choosing the orientation which has the lowest electrostatic energy. (Because the 'bonds' are symmetric about the device - equal but opposite charge on the 'atoms' - only 4 calculations are needed, with the other 4 results being simply opposite in sign.) Calculating this electrostatic energy has the form

$$U_A = \sum_{i}^{i \in A} \sum_{j}^{j \notin A} \frac{q_i q_j}{r_{ij}}, \tag{1}$$

where q_i is the permanent charge on atom i, which is on device A, and r_{ij} is the distance between atoms i and j, where j is an atom on a nearby device. Having found the lowest electrostatic energy orientation of the polar bond on device A, this orientation is adopted as the forwarding direction for node A.

As well as finding the forwarding direction, we also want a way of expressing how strong the interaction (alignment influence) is with neighbouring nodes, for a strong interaction indicates that forwarding is likely to be useful: locally favourable, in terms of node proximity and buffer capacity, and also a good chance of a long-range forwarding chain to the sink. By calculating the electric field at the negative end of the polar bond of node A, we can derive the *induced charge* on the polar bond. (Under the influence of neighbouring charges, the charge distribution of a molecule changes.) If this *induced charge* exceeds a pre-set threshold, then node A decides to forward data to the nearest device to the negative end of its polar bond, provided that this device is within transmission range. (Note that the terms referred to here: *electric field* and *charge* only have meaning in the context of the routing model, and do not refer to the battery levels or other properties of the devices themselves.)

So, the second stage of the calculation is to calculate the electric field at the negatively charged atom X, on node A, which is given by

$$E_X = \sum_{j}^{j \notin A} \frac{q_X (q_j + \delta_j)}{r_{Xj}^2}, \tag{2}$$

where δ_j is the induced charge on atom j, on a neighbouring node, and q_x is the charge on the negative atom X. From the electric field, it is simple to calculate the induced dipole moment μ_X, where this is given by $\mu_X = \alpha_X E_X$, where α_X is the atomic *polarizability* of atom X. The induced charge on the negatively charged atom of the polar bond on node *A*, δ_X, is taken as equal to μ_{C-X}/d_{C-X}, where d_{C-X} is the length of the polar bond. Note that μ_X and the bond dipole moment μ_{C-X} are treated as equivalent.

So, when the induced charge, δ_X, exceeds a fixed threshold, data is forwarded from device *A* to the nearest device to the negatively charged end of its polar bond, as long as this device is within broadcast range of *A*. If all the data has been forwarded from device *A*, the bond type on device *A* is re-set to the one appropriate for an empty buffer (short, strongly polarising), and any forwarding means that the induced charge is re-set to zero.

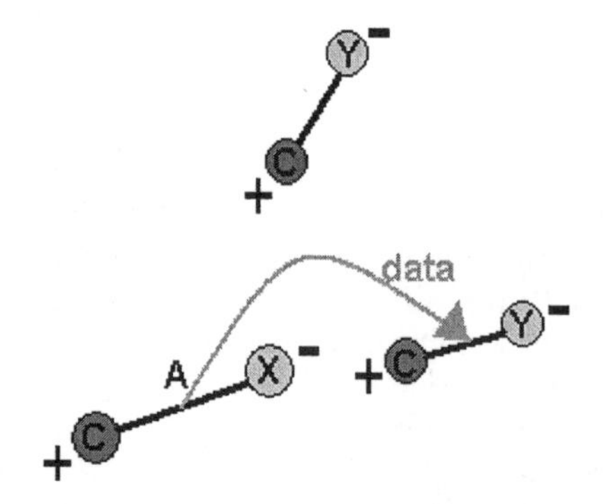

Fig. 2. X is the negatively charged end of the polar bond associated with node A, and X has been polarized beyond a fixed threshold by the 'atoms' (poles) associated with nearby nodes. Data is transferred from node A to the node which is nearest to X

Another feature of this approach is the way that nodes respond to changing battery levels. A node's broadcast range is constrained to be no greater than *range*, according to the ratio of (battery power left)/(time left), as follows:

$$\text{range} = \text{range}_{\min} + \frac{B_0 - B}{2(t_{\max} - t)}(\text{range}_{\max} - \text{range}_{\min}), \tag{3}$$

$$\text{range} \leq \text{range}_{\max}$$

where B is the current battery level and B_0 is the initial battery level, and t and t_{max} are the current time and length of the data-gathering experiment respectively. More sophisticated controls could be used, based on the quantity of data left in the device buffer, perhaps, but this simple approach has been shown to give advantage compared with a range that is independent of battery power, and is easy for the primitive devices to calculate.

Lastly, the choice of influencing neighbouring nodes is restricted to exclude the node that has just sent the forwarding device some data. This is to stop data "ping-ponging" backwards and forwards.

3 Simulation Results

The main scenario used to test and evaluate the performance of the invention is 20 mobile sensor network devices floating on the surface of the sea, and 3 network sink devices which are fixed, though this need not be the case. The simulation includes a model of water currents including a moving centre of rotation (a whirlpool or gyre) and a model of a shoal of fish with flocking behaviour. Whenever a fish comes within a certain close range of a sensor device, this generates a packet of data at the sensor device. The purpose of this scenario was to model sensor devices moving fast and unpredictably, and subject to unpredictable and unequal load. The network is sparsely populated with the *average* number of nodes within transmission range of any other node during a simulation, being less than 0.2.

There is a lot of interest in the health of fish stocks. Currently, estimates of fish numbers and locations rely on the crude approach of fishermen fishing out tagged fish from their catches and estimating where the tagged fish were found. The scenario simulated here would offer a much more comprehensive coverage of the surface area of the sea. It has been assumed that tagged fish would send out a sonar signal that would enable the floating sensor nodes to detect them.

Simulations were carried out with 20 mobile sensor devices and 3 fixed sinks, with a shoal of 30 fish. The transmission cost model used is very simple, being proportional to r^2, where r is the inter-node distance. In this work, the receive cost has been neglected; the inter-node distances are assumed to be large, on the scale of 100s of metres. The results presented are each averages of 20 simulation runs. There was a wide range of possible data to collect, between simulations, depending on the movements and interactions of sensor devices with fish. The average number of potential data packets to be collected was 4300, with a standard deviation of 3400. The mobile sensor devices were given equal, finite battery resources, some of which were entirely drained by the data gathering experiment. Results quote the percentage of potential packets of data that are recovered at the sinks by the end of the data-gathering experiment. The number of potential packets is the number of times mobile devices or sinks encountered a fish within a specified range, which would have generated a data packet. This total includes interactions between fish and devices which have exhausted their batteries and are unable to record or forward this data.

No attention has been given to the nodes conserving power by being 'asleep' during periods of inactivity, though this would be an important component of a live system.

The communication costs for exchanging the small amounts of state information needed to choose a neighbour for forwarding, have not been taken into account (except for the experiments whose results are given in Figure 8). But the requirements for the exchange of very short control messages between sensor nodes would be the same for the two methods being compared here. In summary, the technique presented here only covers a subset of the issues involved in designing an operational mobile sensor network, focusing on efficient route-finding alone; but the approach would not be in conflict with other aspects of efficient mobile sensor network design.

First, is presented in Figure 3 a graph that compares use of the power-optimising mechanism given in Equation 3, with simulations that maintain a fixed maximum broadcast range. Note that in both cases, the polarization routing mechanism was used, and the transmission range for forwarding data is variable, with power use per packet sent proportional to r^2. However, the difference is that, with the power-optimising mechanism, the *maximum* broadcast range that a node can use, even for control packets, is reduced. So, devices with lower battery power remaining are less visible, and so are less likely to be sent data, and are more limited in how far they can transmit data. Use of the power-optimising mechanism results in a greater proportion of packets being collected at the sinks, as can be seen in Figure 3.

To evaluate the polarization approach, a performance comparison has been made with a forwarding criterion that is referred to as 'distance only'. This criterion involves device A forwarding data to a neighbouring device B, if device B is the nearest device within range of A, and device B is nearer to A's nearest sink than A. This is simpler than using the concept of polarization energy, but would require a similar level of notification broadcasts of device positions. The following graphs refer to 4 sets of 20 simulations, where each set has a different forwarding cost constant $batcost$, which is used to *mutliply* the inter-device distance, so can be thought of as decreasing device density, or increasing the cost of transmissions by $batcost^2$; no additional battery resource was given. Note that the power-optimising mechanism was used for both 'polarization' and 'distance only'.

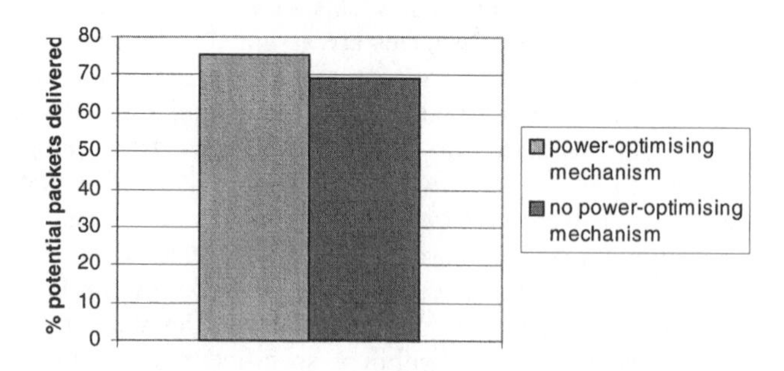

Fig. 3. % potential packets delivered on average for 20 simulations using the polarization approach for routing, with and without the 'power-optimising mechanism', which reduces node broadcast range, as relative node battery levels fall

In Figure 4, it can be seen that a significantly higher proportion of packets is successfully collected at the sinks using polarization as the criterion for forwarding data, than distance to sink alone. Similarly, in Figure 5, we see that the average cost of delivering a packet using 'distance only' is as much as 50% greater than when using the polarization forwarding criterion. This is because using the concept of an electrostatic field and polarization triggers make it more likely that data will be forwarded along routes that lead to a sink.

In Figure 6 we see how many fewer hops are needed to deliver packets when the concept of polarization is used. The use of fewer hops is an advantage as it means less processing power being used by devices (this has not been modelled in this work.) These results suggest that if *receive* energy costs were taken into account, as well as transmission costs, that the difference in performance between "polarization" and "distance only" methods would be even greater.

In Figure 7 we see a comparison of the proportion of data forwarding transmissions that occurs when there is a complete routing path to the sink. That is, when a set of devices is arranged in such a way that successive in-range transmissions could result in packets going directly to a sink. This graph shows how much more frequently this occurs when the notion of an electrostatic field is used. This shows that the success of the polarization approach is achieved by conveying long-range structural information using only local transmission interactions, so that transient data forwarding structures can be exploited.

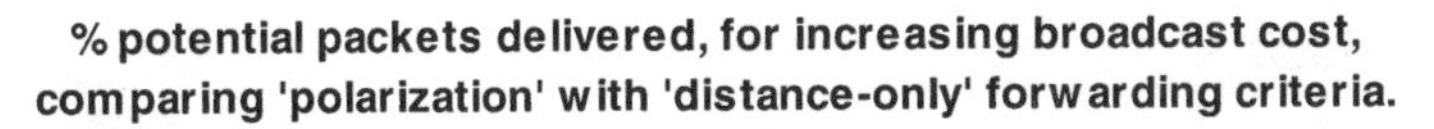

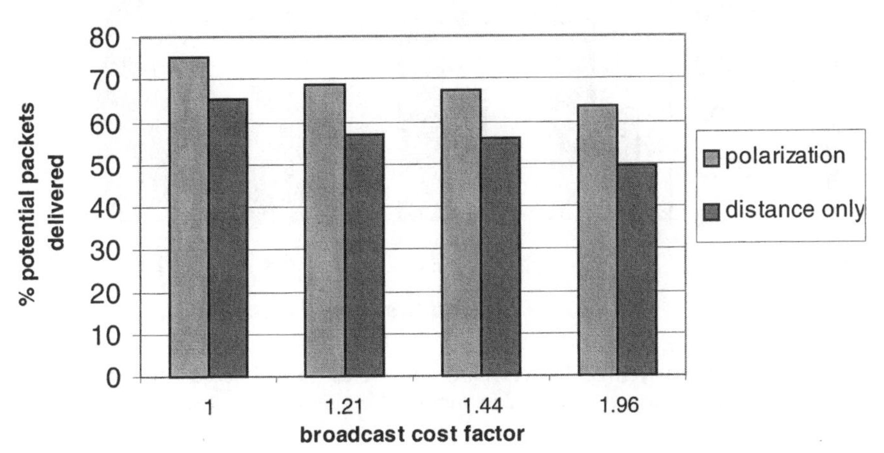

Fig. 4. A comparison of % potential packets delivered to sinks in the mobile ad hoc sensor network simulation, for two sets of forwarding rules: 'polarization' and 'distance only'. Each result quoted is the average of 20 simulations; results using 4 different broadcast cost constants are shown

Further work has shown that setting minimum *effective* interaction distances for the electrostatic calculations, improved results for the polarization method further, raising the average proportion of potential packets delivered (where the broadcast cost factor is 1.0) from 75.4%, to 77.8%. This has the effect of moderating the influence of extremely close neighbouring nodes, which would otherwise unduly dominate the alignment effects. Using this constraint also has the effect of reducing transmissions over very short distances, which would have favourable receive energy cost implications.

In addition, whereas all results quoted in figures 3-7 were generated with the sensor nodes having effectively infinite buffers, by imposing finite buffers of size 400 packets, the average potential packet delivery rate improved further from 77.8% to 82.3%. This latter improvement is the result of better load-balancing, as can be shown by the fact that the average node deaths (drained node batteries) for simulations fell from 1.25 (out of 20) to 0.5. However, this latter improvement is not specific to the polarization approach, and would be seen for all comparable routing methods.

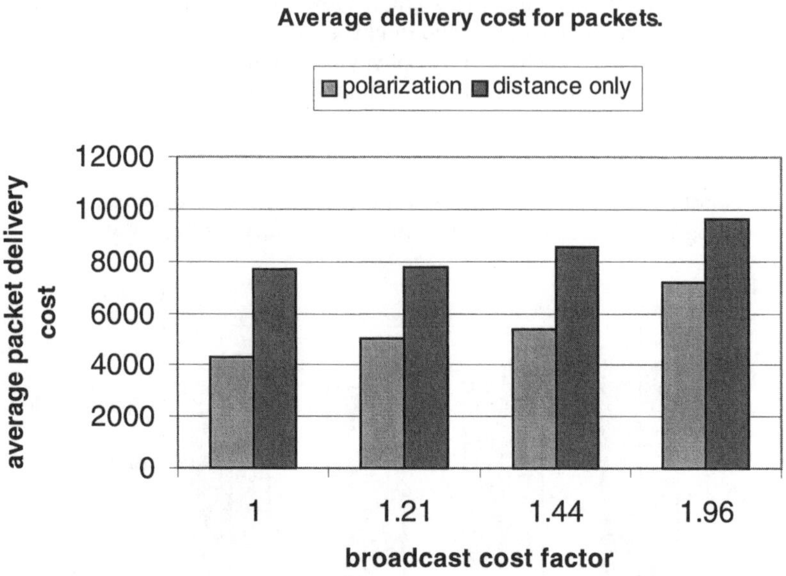

Fig. 5. The relative average delivery costs for packets in mobile ad hoc sensor network simulations, comparing the 'polarization' forwarding criterion with the 'distance only' rule

A final set of simulations was carried out to plot the performance of 'polarization' routing against average node speed. There were 48 mobile sensor nodes, moving randomly, and 2 stationary sinks. Again the transmission cost was as r^2, and receive energy cost was neglected, but control packet cost was included, being 20% of sending a data packet. The average node movement was one quarter of the maximum possible movement. Movement occurred for every node 'decision-making cycle'. 3 decision-making cycles were needed from when a node checked if it had data to forward, to the point where it had selected a receive node and was now broadcasting its data. This was to model information from neighbours becoming out of date within the timescale of routing decisions. A comparison was made with a hierarchical routing protocol, similar to the one used for *Smartdust*[7], in which the sinks are at level 0, their nearest neighbours are at level 1, and so on throughout the sensor network, so that each node's level reflects the number of shortest-path hops to its nearest sink. This protocol was implemented in the simulation with appropriate safeguards to prevent erroneous level setting. The results are given in Figure 8. There is an idiosyncrasy to these simulations: the frequency of sensor measurement increases for nodes moving over two regions of the network environment, which means that more data load is put on the network for higher average node speeds. This results in the sensor network performance deteriorating faster with node speed than it would otherwise. However, it can be seen that the 'polarization' approach outperforms the 'hierarchical' approach consistently, especially at higher average node speeds. This is partly because the 'polarization' approach means that the nodes maintain a memory of a reasonable forwarding direction, whereas the 'hierarchical' levels-based approach has nodes relying on routing levels that are changing faster than the level information can be updated.

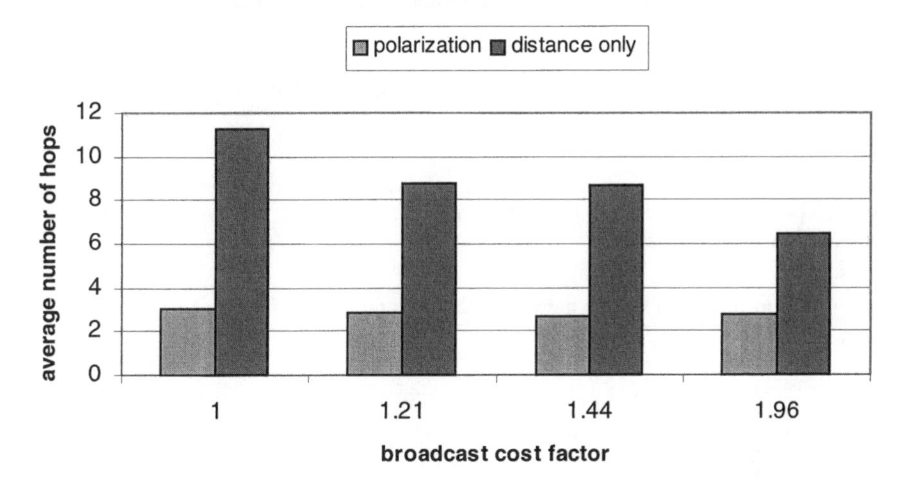

Fig. 6. The average number of hops used for packets being delivered to sinks in mobile ad hoc sensor network simulations for the two forwarding criteria: 'polarization' and 'distance only'

Also, the 'polarization' approach drops fewer packets during transmissions, as the forwarding trigger is influenced by receiving node proximity. Rather than just following a simple rule such as 'forward to nearest node at lower level', the polarization approach has a measure of how near is 'near enough'.
However, the polarization approach also has an implicit load-balancing component in that receive node buffer capacity is taken into account. (A node with a buffer that is nearly full is less attractive as a recipient of data than a node with an empty buffer.)

Forwarding is also affected by longer-range network structure. This combination of factors enables nodes using polarization routing to make more intelligent forwarding decisions, when node speed is significant.

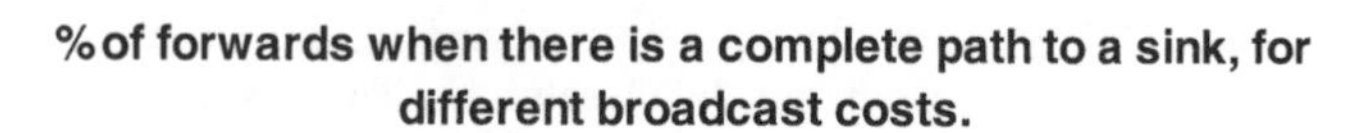

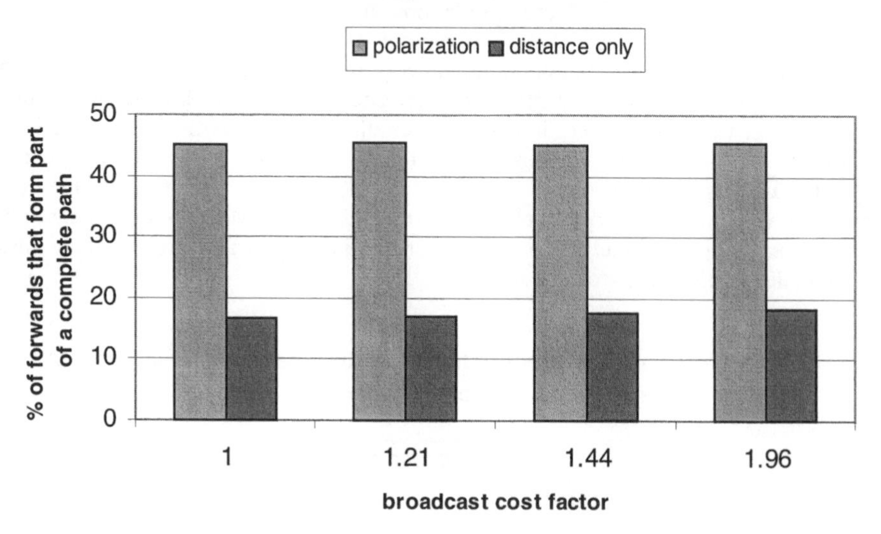

Fig. 7. A comparison of the average % of data forwarding transmissions that occur when there is a viable route all the way to a network sink, comparing the two forwarding criteria: 'polarization' and 'distance only'

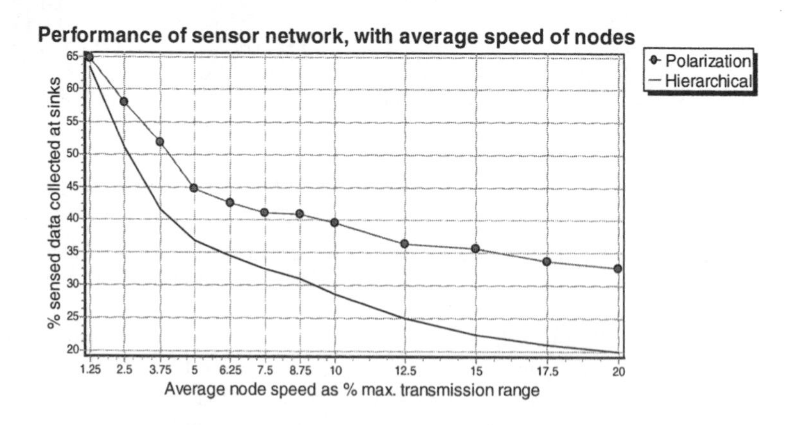

Fig. 8. Performance of sensor network, with average node speed given as percentage of maximum transmission range, per node 'decision-making cycle', for 'polarization' routing and 'hierarchical' routing

4 Conclusion

In conclusion, a novel protocol inspired by modelling the forwarding directions of mobile devices as polarizable bonds, has been used to construct dynamically-updated forwarding paths throughout a mobile device network. It has been shown that this model routes data so that less transmission energy is used for forwarding, more data is recovered, and fewer hops are needed than with a simpler approach that always chooses to forward in the direction of the nearest network sink. With increasing node speed, it has been shown that our proposed approach is more successful at routing than routing that relies on a hierarchical-levels approach. The success of the method rests on its ability to convey longer-range network information than would otherwise be available to the mobile devices, through the interactions of their preferred forwarding directions.

References

1. http://www.mysound.uconn.edu/index.html
2. Cerpa, A., Elson, J., Estrin D., Girod, L., Hamilton, M., Zhao J.: Habitat Monitoring: Application Driver for Wireless Communications Technology. ACM SIGCOMM Workshop on Data Communications in Latin America and the Caribbean (2001)
3. Xu, Y., Heidemann, J., Estrin, D.: Adaptive energy-conserving routing for multihop ad hoc networks. Tech. Rep. 527, USC/Information Sciences Institute (2000)
4. Perkins, C.: Ad Hoc On Demand Distance Vector (AODV) Routing, Internet-Draft, draft-ietf-manet-aodv-04.txt (1999)
5. Heinzelman, W.B., Chandrakasan, A.P., Balakrishnan, H.: Energy-Efficient Routing Protocols for Wireless Microsensor Networks. Proceedings of the 33rd International Conference on System Sciences (2000)
6. Singh, S., Woo, M., Raghavendra, C.: Power-Aware Routing in Mobile Ad Hoc Networks. Proceedings of the Fourth Annual ACM/IEEE International Conference on Mobile Computing and Networking (1998)
7. http://basics.eecs.berkeley.edu/sensorwebs
8. Culler, D.E., Hill, J., Buonadonna, P., Szewczyk, R., Woo, A.: A Network-Centric Approach to Embedded Software for Tiny Devices. DARPA Workshop on Embedded Software
9. Perkins, C., Bhagwat, P.: Highly Dynamic Destination-Sequenced Distance-Vector Routing (DSDV) for mobile computers. Proceedings of the SIGCOMM '94 Conference on Communications Architectures, Protocols and Applications (1994) 234–244
10. Park, V.D., Corson, M.S.: A Highly Adaptive Distributed routing Algorithm for Mobile Wireless Network. Proceedings of INFOCOM '97 (1997) 1405–1413
11. Johnson, D.B..: Routing in Ad Hoc Networks of Mobile Hosts. Proceedings of the IEEE Workshop on Mobile Computing Systems and Applications (1994) 158–163
12. Broch, J., Maltz, D.A., Johnson, D.B., Hu, Y-C.: A Performance Comparison of Multi-Hop Wireless Ad Hoc Network Routing Protocols. Proceedings of the Fourth Annual ACM/IEEE International Conference on Mobile computing and Networking, Mobicom '98 (1998) Dallas, Texas,USA
13. Robinson, D.P., Marshall, I.W.: An Iterative Approach to Location of Simple Devices in an Ad-hoc Network, Proceedings of the London Communications Symposium, LCS2002 (2002) 65–9
14. Robinson, D.P., Marshall, I.W.: Location of Simple Nodes in an Ad-hoc Network, Proceedings of Location-Based Services. LBS2002 (2002) 19/1–19/4

Building Blocks of Energy and Cost Efficient Wireless Sensor Networks

Amer Filipovic and Amitava Datta

School of Computer Science & Software Engineering,
The University of Western Australia
35 Stirling Highway, Crawley WA 6009, Australia.
{amer,datta}@csse.uwa.edu.au
http://www.csse.uwa.edu.au/

Abstract. Wireless sensor networks are limited by their energy source. The energy optimisation schemes introduced to date adapt routing techniques to achieve lower energy consumption. Little consideration is given to the underlying routing protocol itself and its energy consumption properties. This paper presents a series of simulations of wireless routing protocols and their energy consumption-related properties. We use simulations in ad-hoc networking environment to extract generic properties of wireless networks and their behaviour in different circumstances. We find that the energy consumed by the routing overhead is not indicative of the overall energy efficiency performance. We further show that two protocols can have the same energy efficiency, yet result in a considerably different network lifetimes. Topology information is shown to be very important and the protocols that operate in promiscuous mode are better equipped to handle heavy traffic loads.

1 Introduction

A wireless sensor network is a collection of inexpensive wireless devices that join together to form a network without a particular pattern. The sensor nodes organise themselves into multi-hop routed networks that operate much like ad-hoc networks; nodes are connected arbitrarily without an administrative body. The wireless property of these network systems allows them to be deployed in remote regions. They can be used for geographical surveying, monitoring weather patterns or the spread of a hazardous gas leak. The technology also has a high presence in military applications where the sensor nodes can be mounted on mobile units. Hence the sensor networks are often required to adapt to changing network topology.

Due to limited energy supply of these sensors, much of the research currently revolves around energy consumption analysis. Several studies have shown that common routing protocols can be optimised for energy efficiency. The majority of the optimisation attempts to date are based on modifying one of the common routing protocols, disregarding the properties of the protocol itself. In this paper, we test four common ad-hoc routing protocols (AODV, DSR, DSDV and

H. Karl, A. Willig, A. Wolisz (Eds.): EWSN 2004, LNCS 2920, pp. 218–233, 2004.

TORA), and use their inherent unique properties to formulate conclusions about network behaviour common to the entire family of wireless networks, specifically wireless sensor networks. We show that much of the research to date has ignored the very basic building blocks of wireless sensor networks.

Ad-hoc and wireless sensor networks are divided by a thin set of properties. Commonly, they both need to be able to reconfigure due to movements and/or failures of some nodes, to efficiently utilise the limited energy supply, and to route messages to nodes several hops away. However, the subtle differences include:

- Sensor nodes can be densely deployed with many more nodes within a small area than in an ad-hoc network [1];
- Sensor nodes are limited in computational capacity and memory [1];
- Sensor nodes may not have a global identification such as an address due to the high population [1]; and
- Sensor nodes mainly use broadcasting to communicate [1].

With the advances of technology, these differences are becoming less apparent. Sensor nodes computational power and memory are continually improving. Ad-hoc networks often broadcast public messages, therefore exhibiting similar behaviour to sensor networks. The data collected by sensor nodes is often useless unless it is known where the sensor is — which is another form of a global ID using its location. Finally the density of the sensor nodes can now be offset by reducing the power of the transceiver, effectively reducing the neighbourhood count. Hence, it seems that although the protocols for ad-hoc networks cannot be directly utilised in sensor networks, we can certainly use them to study the behaviour of wireless networks in general. Ye et al [2] outline four major sources of energy waste as collision, overhearing, control packet overhead and idle listening. Throughout our experiments we will address some of these properties in terms of energy consumption.

The novel approach of this paper is the introduction of the *topology knowledge* and *discarded control packets* as important metrics for energy consumption. Through experimental results, we show that the overall energy consumption can be only optimised by using up-to-date topology information. The routing protocols that use large amounts of energy for maintaining the routes within a network tend to have a *lower* overall energy consumption due to fewer retransmissions and optimal message paths. We further demonstrate that topology information can provide stability in networks with limited energy reserves. Our findings explain some of the results of attempts at energy optimisation using topology control [3,4].

To our knowledge this is the first publication of energy and cost efficiency analysis in terms of topology knowledge and discarded packets. The remainder of this paper is structured as follows. Section 2 outlines some of the previous work on energy analysis and how it differs from our fundamental approach. We also describe the four common routing protocols. The methodology is described in Section 3 followed by an extensive analysis of our results. Finally we conclude the paper with Section 6.

2 Related Work

2.1 Energy Optimisation

Several researchers have tackled the issue of energy consumption from different angles. Stojmenovic et al [5] proposed that the most optimal route is one that has equally spaced hops, due to the attenuation of signal in wireless medium. They provided energy consumption models for power and cost aware routing schemes. These findings coincide with recent attempts at topology control [4,3]. The aim of these schemes was to reduce the degree of individual nodes while maintaining the connectivity of the network. These techniques quickly propagated to wireless sensor networks with the availability of the variable transmission voltage Berkeley motes. However, it has also been shown that reducing the node degree does not imply better energy consumption [6].

Singh and Raghavendra [7] have managed to achieve a 40-70% reduction in energy consumption by switching off some of the inactive nodes in the network. The scheme (termed PAMAS) is performed at the MAC sublayer, and can be applied to most routing protocols. PAMAS, however, targets a different property of the network, namely the number of idle nodes. The same idea was implemented for a sensor network MAC sublayer called S-MAC by Ye et al [2]. Here, a combination of scheduled and contention based channel access is utilised to achieve higher energy efficiency.

Broch et al [8] performed some of the first protocol analyses at the packet level. Their detailed experiments generated results for reliability and the routing overhead messages and size. Broch commented on the efficiency of the protocol by referring to the reliability. Feeney [9, 10] disputed this analogy arguing that the number of routing overhead messages or their size alone do not govern the energy that the protocol will consume. To the best of our knowledge, Feeney limited her analysis to the energy consumed for routing purposes, not the overall energy consumption. Hence further research into packet-level energy consumption analysis is possible.

Jones et al [11] surveyed some of the ad-hoc routing protocols and argued that any optimisations at the lowest packet-level have already been made. In this paper we show that there is a significant deviation in performance of different protocols and that the most energy efficient routing protocol, while it may already exist, has not yet been defined. Hence there is still much to be gained by analysing packet-level operations.

2.2 Ad hoc Routing Protocols

Since the introduction of wireless networks in the 1970s, several different ad-hoc routing protocols have been developed [12,13,5,7,14,15]. This section describes the four routing protocols most commonly used in ad-hoc networks. It is important to understand how they differ in order to appreciate the results of the energy consumption analysis. While we refer to the routing protocols by name, recall that we are not aiming to measure how a particular protocol performs under different conditions, but rather how its inherent routing properties affect the network behaviour and energy consumption. Hence, the results presented in Sections 4 and 5 are applicable to all wireless networks.

Destination-Sequenced Distance Vector Routing (DSDV). DSDV is a hop-by-hop distance vector routing protocol [8]. Each node in the network maintains a routing table of paths (next hops) to every other destination node. Consistency throughout the nodes' routing tables is maintained by periodic table broadcasts. Initially, a full table broadcast is made followed by periodic incremental updates.

When a new route needs to be discovered, a *route discovery* broadcast is made with the address of the destination, best known hop-count to destination, the sequence number of the information received regarding the destination, and a new unique sequence number for the broadcast [16]. The receiving node will add its hop to the hop-count and then rebroadcast. Either the destination or a node with more recent information about the destination can reply to the *route discovery* packet.

Temporally-Ordered Routing Algorithm (TORA). TORA is a highly adaptive, distributed routing protocol based on a *link reversal* algorithm [17]. It is designed to operate in a highly dynamic environment as it provides several routes to a given destination. Optimality is only a secondary priority for TORA. All changes in the topology are contained to small local areas near the occurrence of the change. In addition, TORA only triggers updates when absolutely necessary, that is, when no route to a given destination is known.

A simple analogy for TORA is a downstream water model. Suppose that each link is a pipe capable of carrying water (messages). Water can flow along any one of the downstream pipes. If one pipe is jammed (link is broken), then the *link reversal* takes place as the water overflows into the remaining routes. If a node detects that a link has failed, it conducts erasure by broadcasting a *clear* packet throughout the network. Since the height of a node is only valid for a period of time, TORA assumes that all nodes have synchronised clocks, for example via a Global Positioning System.

Dynamic Source Routing (DSR). DSR is an on-demand routing protocol. It is topology oriented and derives its knowledge from the headers of the messages that circulate throughout the network, by operating in the promiscuous mode. This is stored in a *route cache* that is continually updated by the traffic [16].

When a source node S has a packet to send to a destination node D, it first checks the *route cache* for a valid route. If it doesn't exist, a *route request* packet is generated. Every node that receives the packet checks if it has a valid route to the destination D. If not, the node adds itself to the list of nodes that the packet has visited and rebroadcasts. When the destination node D, or an intermediate node with a valid route to D is reached, a *route reply* packet is generated and sent back to source. This packet contains the sequence of hops that the message should take to get from S to D.

Ad-hoc On-demand Distance Vector Routing (AODV). The AODV routing protocol is a variant of DSDV in the sense that it maintains a routing table to different destinations [16]. As with DSR, a *route reply* is sent back to the source S once the request reaches either the destination or an intermediate node with a valid route to the destination. However, rather than sending the entire hop sequence to the destination, the source is only advised of the next hop that the packet should take. Thus, each node's knowledge is limited to the topology of its 'one hop' neighbours. If a node wants to reach another node that is multiple hops away, it only knows in which direction to forward the message.

In order to maintain routes, AODV requires that each node transmits a control *hello* message at a pre-defined rate. Failure to receive three consecutive control messages from a neighbour indicates that the link is severed.

3 Methodology

The simulations in this paper were performed on the Network Simulator (*ns*) developed at the University of California, Berkeley [18]. *ns* is a discrete event simulator distributed with implementations of commonly used routing protocols.

Our experiments comprised 50 mobile agents moving in a flat grid area of 1500m $\times$ 300m. The experimental setup was adopted from the early work of Broch et al [8], so that our preliminary results can be verified. The simulations lasted for 900 seconds, and generated trace files up to 1GB in size. The rectangular area forces the source-destination pairs to use several intermediate nodes in their communication. This allowed to us to get a detailed view of network behaviour under different conditions.

The simulated environment was modified using the following categories:

Mobility. Each node was given a random velocity in the interval [0m/s, 20m/s]. Different mobility rates were simulated by introducing a timeout between movements. A timeout of zero seconds indicates that the node is constantly moving with its assigned velocity, whereas a timeout of 900 seconds indicates that the node is pausing 900 seconds between its movements — effectively creating a static topology.

Traffic Load. Different traffic loads were simulated by varying the number of nodes actively transmitting packets in the network. The following scenarios were used:

- 10 sending nodes — equivalent to a light traffic load, well under the network capacity;
- 20 sending nodes — equivalent to a medium traffic load, roughly the same as the network capacity; and
- 30 sending nodes — equivalent to a heavy traffic load, well above the network capacity.

As the number of sending nodes was varied, there was no need to change the packet sizes or rates of transmission. We used a constant bit rate (CBR) source with a period of 0.25 seconds. This generated 4 messages of 512 bytes every second at the sender. The constant packet generation ensures that the variable backoff mechanisms, such as the one found in TCP [19], do not mask the true behaviour of the protocol by adapting to congestion.

Energy Reserve. These experiments were aimed to reveal how the protocols react to insufficient energy reserves. The first set of each simulation was performed with infinite energies. We then varied the battery power by reducing it by 20% with every simulation. Hence we performed test runs with 100%, 80%, 60% and 40% of energy reserve. The main reason behind using percentages is that the energy consumption in the first instance (100%) revealed that the routing protocols had a significant variation in the consumed energies. In order to make this set of experiments invariant to this anomaly, and comparable to one another, we resorted to using percentages.

The energy consumption model used in this paper was developed by Feeney [9]. It is based on a linear cost function

$$C_{send/recv} = m_{send/recv} \times size + b_{send/recv} \quad (1)$$

where C denotes the total cost of sending/receiving the message, m denotes the incremental cost of sending/receiving an additional byte of traffic, and b denotes the bandwidth access cost for a send/receive event. The values used in this simulation are shown in the table below [10]:

m_{send}	1.89 μWs/byte
m_{recv}	0.42 μWs/byte
b_{send}	246 μWs/byte
b_{recv}	56.1 μWs/byte

One of the main drawbacks of *ns* is that its logging capability is quite limited. This was noted by Feeney in some earlier studies [9, 10]. Furthermore the only way of accessing the results of the simulation is through the generated trace files in post-processing. This makes extracting of additional information from the results difficult. In order to log all the information we required, we decided to implement an additional energy consumption agent for *ns* (ECANS [20]). This allowed us to monitor all the variables of interest in real-time. The main advantage of this approach is that nodes can be made to shutdown, or perform some additional task, when their energy levels are almost depleted.

4 Energy Efficiency Analysis

This section deals with the results of energy efficiency analysis — the total energy consumed in the simulations. We begin by analysing the reliability of the routing protocols to verify the results by Broch et al, and then look at the routing specific overhead and energy. The final section of these results deals with total energy consumed and its correlation to previous categories.

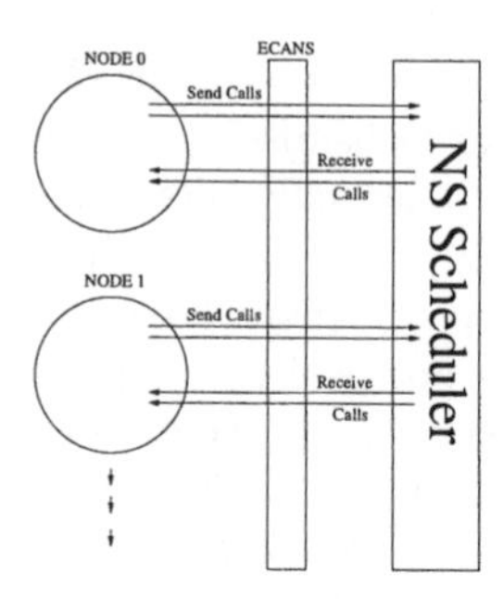

Fig. 1. Our solution to perform logging functionality. ECANS listens for any events between the scheduler and the nodes and decrements the energy as required. The advantage of performing this in real-time is that the nodes can be shutdown when their energy level reaches zero.

4.1 Reliability

The performance or reliability of the routing protocol refers to its ability to successfully deliver the data packets from a source to a destination. Comparison of the four protocols in Figure 2 shows that the on-demand AODV and DSR protocols are efficient at all degrees of mobility, with delivery rates above 95%. Since both AODV and DSR are on-demand routing protocols, the routes are discovered just prior to their use, resulting in better throughput. We only show the relevant 10 and 20 source graphs, as they are indicative of the overall performance.

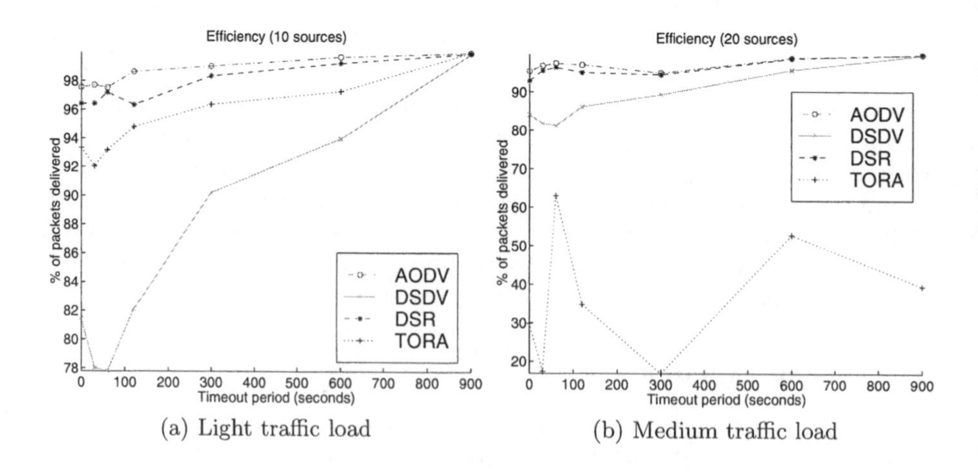

(a) Light traffic load

(b) Medium traffic load

Fig. 2. The percentage of data packets reaching their destination. As the timeout increases, the mobility decreases.

At higher level of mobility, DSDV performs poorly with its delivery rate dropping to about 78% even in light traffic load. When the nodes are consistently moving, DSDV's routing information becomes stale and hence the protocol cannot cope with the number of invalid routes. TORA produced a reasonable result of above 90% delivery rate in light traffic load, but its performance degraded to an unacceptable level (20–60%) even with optimal traffic load. TORA suffers from a degenerative condition of flooding the network with numerous control packets leaving little bandwidth for the actual data. This is why TORA performed worst of all in comparison, even lower than the periodic DSDV. We also believe that this volatile behaviour is the cause of inconsistent results at medium traffic loads shown in Figure 2(b). The periodic nature of DSDV prevented it to congest the network with route discoveries, hence resulting in a higher throughput of data packets.

4.2 Routing Overhead Analysis

Traditionally, the energy efficiency of a routing protocol is evaluated by the number of messages it sends as a part its routing strategy, or the size of the routing overhead. Figures 3 and 4, respectively, show the number of control packets generated and the size of the overhead for each routing protocol.

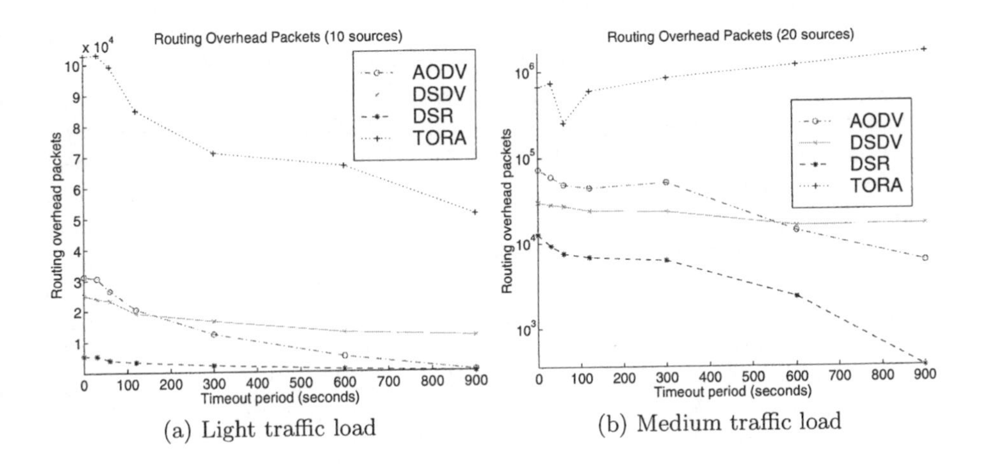

(a) Light traffic load

(b) Medium traffic load

Fig. 3. The number of control messages generated by the routing protocol. As the timeout increases, the mobility decreases.

TORA, AODV and DSR are all on-demand routing protocols, so the number of generated control (routing overhead) packets increases with mobility. This is demonstrated in Figure 3 where, as the timeout decreases, the number of packets tends to increase. In contrast, DSDV generally appears to have the same number of control packets due to its periodic broadcasts — approximately 20 and 30

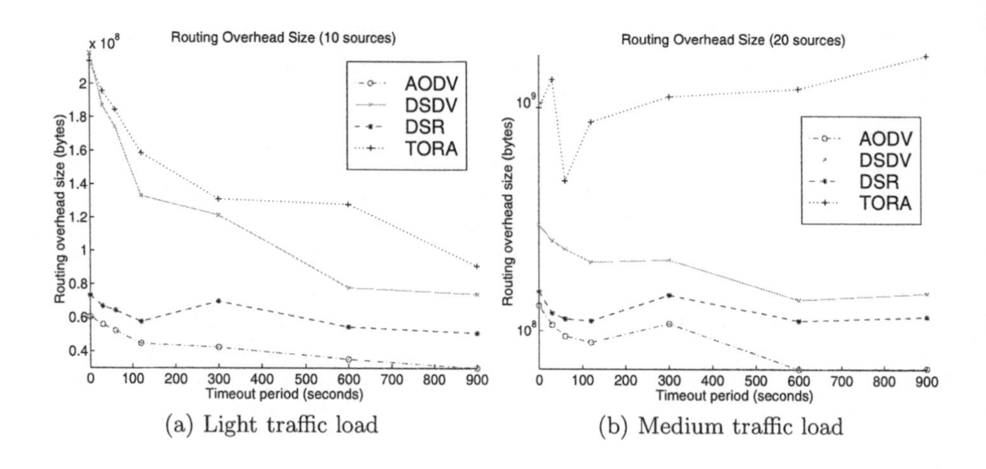

(a) Light traffic load

(b) Medium traffic load

Fig. 4. The size of the routing overhead. This includes the individual control messages and data message headers. As the timeout increases, the mobility decreases.

thousand at light and medium traffic loads respectively. TORA underperformed considerably by generating approximately 50,000 control packets in light traffic load, compared to only 35,000 data packets that it needed to deliver. This coincides with one *hello* message per second per node equalling 45,000 messages.

AODV and DSR evidently have the lowest number of control packets, both converging to almost zero in static topologies. However, at highest mobility under light traffic load, Figure 3(a), AODV generated 31,306 control packets whereas DSR only generated 5,515. The ability of DSR to learn routes from the headers of messages obliterates the need for the individual route discovery and route maintenance packets, at the expense of larger headers. Hence we must consider the *size* of the overhead as well as the number of individual routing messages.

The graphs of the overhead size in Figure 4 show AODV and DSR in reversed places. DSR's larger packet headers add up to a considerable amount causing its overhead to exceed that of AODV in all scenarios. The header size of DSR packets reached up to 200 bytes or 39% of the 512 byte data packet, whereas for AODV they rarely exceeded 10%. Even at lowest mobility, DSR generated about 60MB of overhead compared to 30MB by AODV.

Interestingly, the difference in the number of control packets and their size does not appear to be directly influencing the reliability of the protocol. Feeney recently proposed that the overhead size or the number of messages has no bearing on the energy consumed for routing purposes [9].

4.3 Energy Consumption Analysis

Two metrics are considered in energy consumption analysis: the energy consumed by the routing overhead and the overall consumed energy. These are shown in Figure 5.

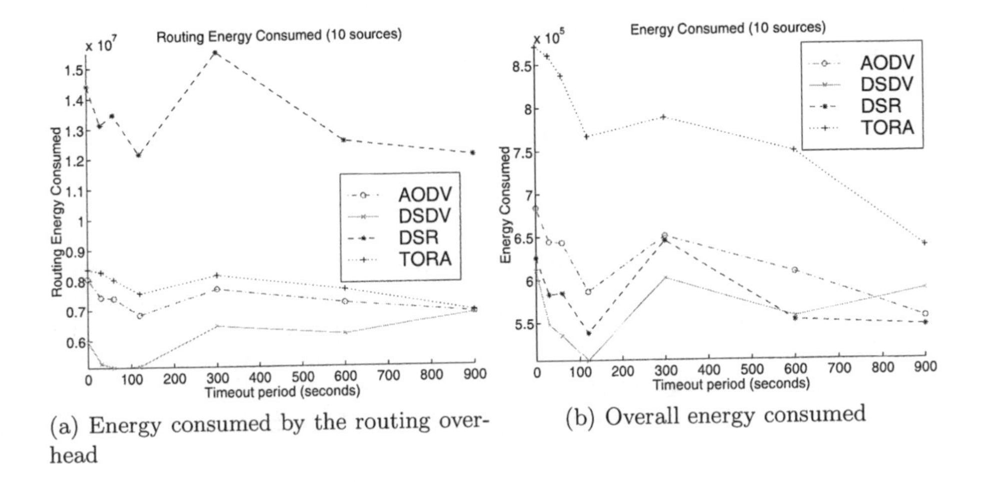

(a) Energy consumed by the routing overhead

(b) Overall energy consumed

Fig. 5. Energy consumed by the routing protocol compared to the overall energy consumed.

The DSR protocol performs very poorly as far as the energy consumed by routing is concerned. It almost doubles the energy consumed by TORA. This concurs with the results published by Feeney [9,10]. We also note that the amount is highly irregular at different mobilities. This is probably due to specific test cases in which DSR does not collect enough routing information in promiscuous mode, triggering route discoveries and therefore fueling the routing-specific energy consumption.

However, when calculating the *overall* consumed energy, DSR is quite competitive, consuming only 9% more than the best performing DSDV (Figure 5(b)). This goes against the claims that the energy consumed by routing is indicative of the overall protocol performance. The routing energy is used by DSR to gather information about the network topology that enables it to select fresh optimal routes. In thorough DSR testing, Johnson [12] concluded that DSR will, on average, select a route within a factor of 1.02 of optimal.

Despite having a lower overhead energy consumption (by approximately 50%), AODV performed worse than DSR. This again indicates that the energy consumed for routing is *not* directly related to the overall energy efficiency. The higher consumption by AODV is most likely due to a slightly better reliability result and its reliance on individual control packets to learn routing information.

DSDV protocol outperformed all others in most cases. The result is a partial trade-off with reliability, as DSDV delivers less than 80% of packets even at light traffic loads. Periodic updates also contributed to the low energy consumption as they prevented DSDV to flood the network in high mobility scenarios.

Finally, TORA performed poorly in terms of overall energy consumption. Its frequent link reversal algorithm caused a considerable inflow of control packets.

These congested the network degrading the packet delivery ratio and energy consumption.

4.4 Dropped Packets Analysis

While general packet losses cannot be avoided, a choice can be made between which packets to drop in congested conditions. Intuitively, a protocol should avoid dropping its own control packets as they carry important information about the network topology. Dropping them results in maintaining stale knowledge about the network which could induce more broken routes and dropped packets. In addition, the control packets, if dropped, are more likely to need to be retransmitted. In the worst case scenario, each node will have to see that control packet again, using up valuable energy. Our hypothesis is substantiated with the results in Figure 6.

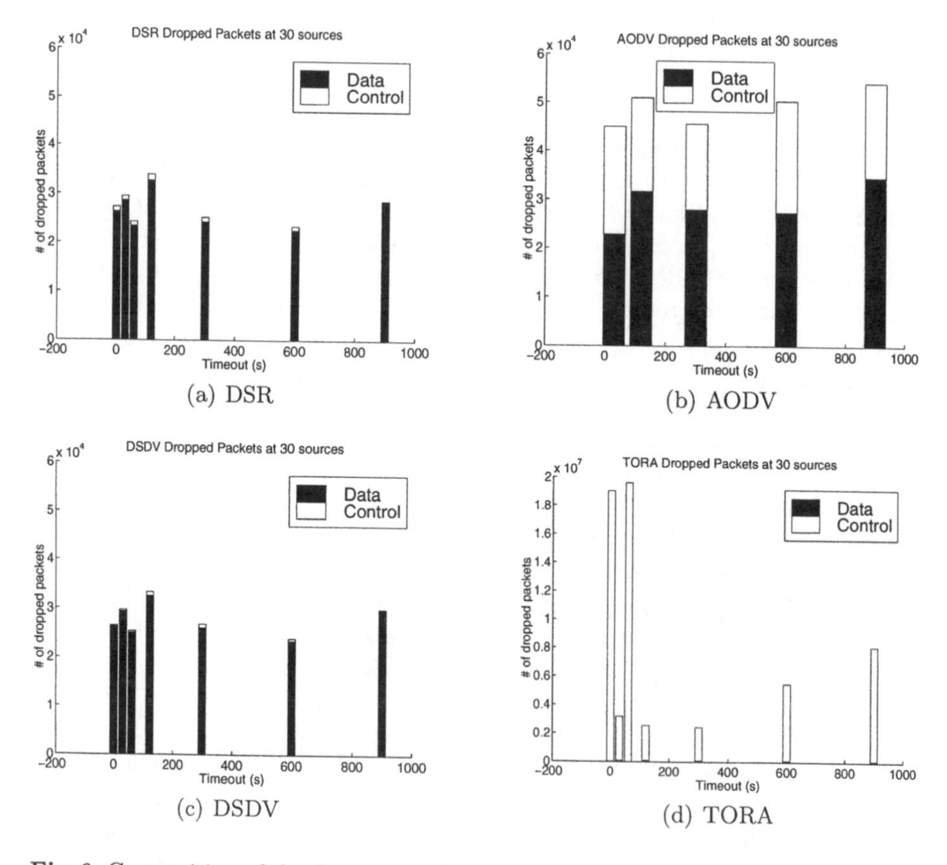

(a) DSR

(b) AODV

(c) DSDV

(d) TORA

Fig. 6. Composition of the dropped packets for each routing protocol. Increased number of dropped control packets directly influences the energy consumption. AODV did not generate results for all the simulations.

DSDV dropped the highest number of packets at 10 sources (Figure 6(c)). However, as the figure shows, almost 100% of these dropped packets are data packets. As a result, the energy efficiency of DSDV comes at a direct cost of reliability.

DSR drops very few packets. At 10 sources, it dropped well under 1,500 packets, only a few of which were control packets. For low mobility this number converges to zero; if the topology is static, DSR can manage to deliver *all* packets. Such low packet loss can be attributed to the higher amount of energy used for routing, and the consequent ability to select fresh optimal routes. The low number of dropped control packets resulted in better energy consumption.

AODV managed to drop fewer than 5,000 packets for light traffic load at all mobilities. As AODV generates many routing packets throughout its operation, most of the dropped packets were control packets as shown in Figure 6(b). This meant that the network topology information was lost. Our analysis of trace files showed that AODV ended up choosing less optimal paths than DSR — a direct result of inaccurate topology information.

We find that TORA had the highest number of dropped control packets and the worst overall performance, not just the energy consumption. At medium traffic load, TORA almost dropped 20 million packets, of which only about 34,000 (or 1.7%) were actual data packets. The problem with TORA is that it requires *hello* messages to be sent to its neighbours in addition to *route requests* and *route replies*. When a link reversal takes place, another set of *route erasures* are generated. The sheer number of control packets results in TORA literally choking itself and consuming vast amounts of energy to compensate.

5 Cost Efficiency Analysis

In this section, we analyse the performance of the two efficient routing protocols AODV and DSR in terms of the number of nodes that are completely depleted of their energy. DSDV and TORA did not generate results for all simulation. Since their performance in previous sections was lacking, we omit their cost efficiency.

The Figure 7 shows the cost efficiency of AODV against that of DSR at different mobilities, traffic loads and energy reserves. Each point on the surface corresponds to a single simulation run, resulting in a total of 168 simulations for this analysis. At light and medium traffic loads AODV and DSR are quite comparable. They both exhibit stability as no simulation ends in an unusually high number of energy depleted nodes. However, at heavy traffic loads, the surface generated by AODV clearly shows a collapse of stability. Lower energy reserves and higher mobility sees AODV deplete up to 40 nodes of their energy (80% of the total population) while DSR manages the same high reliability with only 15 energy depleted nodes.

These results are quite surprising and at first glance it is unclear as to why this would occur. Both DSR and AODV had comparable reliability and energy efficiency. AODV dropped more packets than DSR, but DSR in turn had higher routing-specific energy consumption.

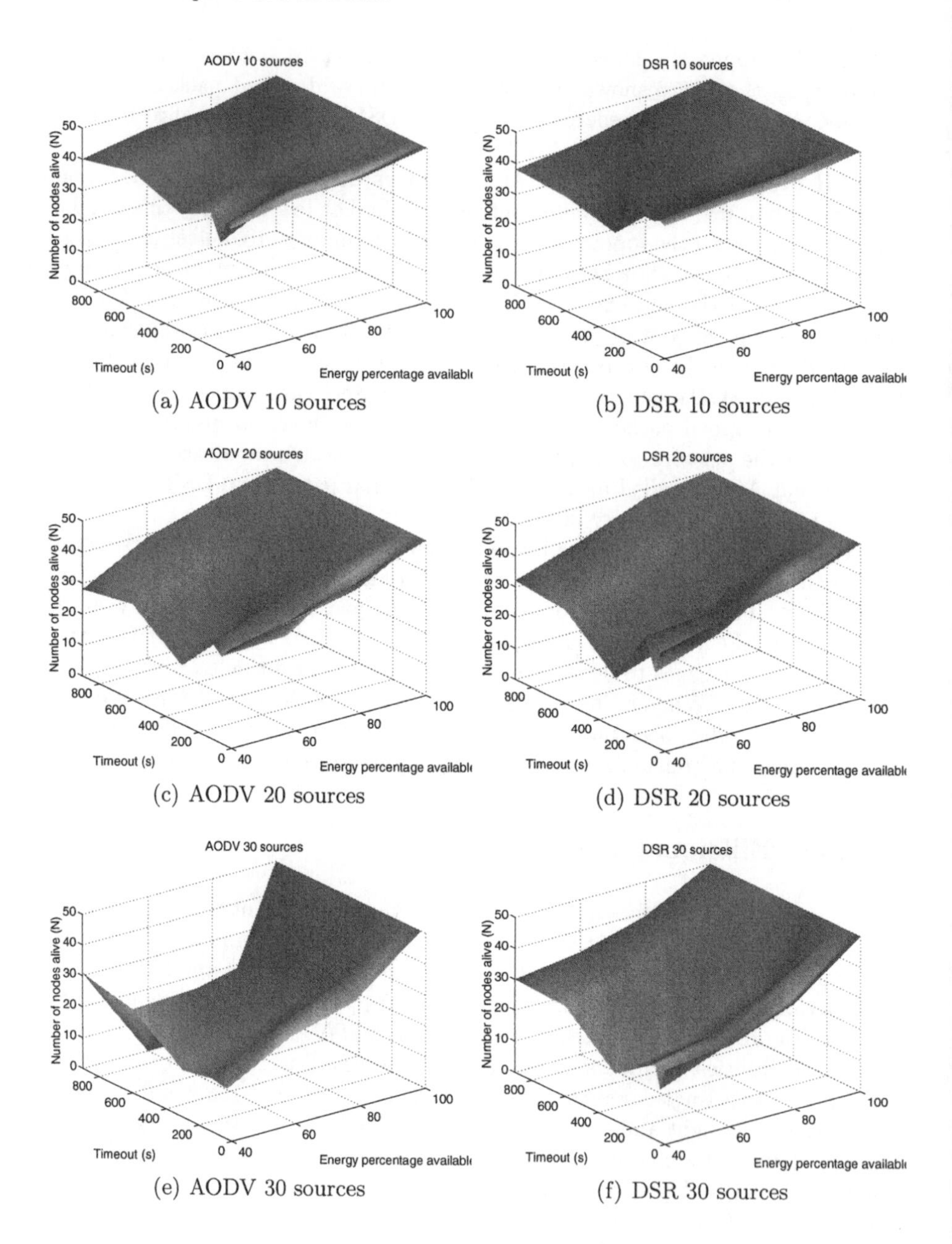

Fig. 7. AODV and DSR cost efficiency comparison.

Figure 8 shows the previously omitted graph of generated control packets. AODV generates more packets than both DSDV and DSR, with almost 140,000 control packets generated at highest mobilities. DSR has an exceptionally low reliance on the control packets, generating a mere 20,000 packets. Since there is more traffic in the network, DSR can gain even better knowledge of the network topology changes by operating in promiscuous mode and does not require individual packets. Of the 140,000 control packets, AODV drops approximately 20,000 (14%) of them. Since the information contained in these packets is vital for routing AODV attempts to rebroadcast them consuming more energy in congested areas and eventually depletes several nodes of their energy.

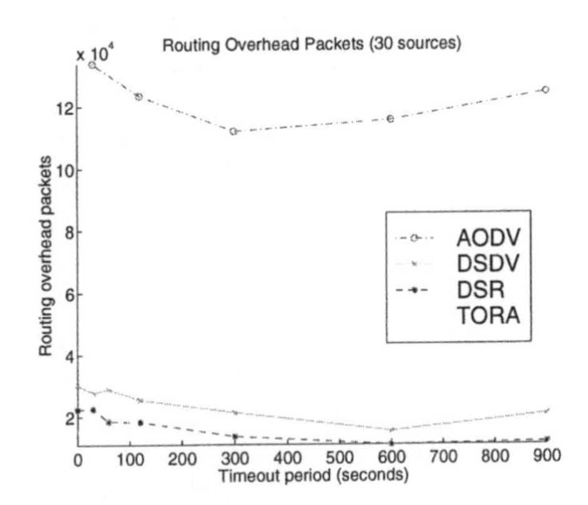

Fig. 8. The number of control packets generated by DSR, DSDV and AODV. TORA did not generate results for these simulations.

It is therefore evident that DSR's low reliance on control messages is an advantage. At high traffic loads it is able to learn a lot about the network topology, choosing optimal routes and delivering a higher percentage of packets to their destinations. Non-promiscuous routing protocols are disadvantaged at high traffic loads as their own control messages cannot get through. The source nodes resort to regenerating these control messages resulting in patches of network nodes being depleted of their energy. This results in poor cost efficiency performance.

6 Conclusions and Future Work

Performance analyses that do not take energy consumption into account are incomplete as they do not reveal potential problems of routing schemes. We have presented a detailed analysis into the energy consumption properties of four common routing protocols. This analysis revealed that current energy optimisation methods are inadequate and that some of the fundamental research into the make-up of energy efficient wireless sensor networks has been overlooked. We have demonstrated that:

- The number of control messages and their size are not an indication of the energy consumed for routing purposes;

- The energy used for routing purposes is by no means an indication of the overall energy usage, contrary to much of the current research;
- An increased number of dropped control packets tends to increase the amount of energy consumed;
- An energy efficient routing protocol will not, by default, prolong the network lifetime of the node;
- The knowledge of the network topology is one of the most important factors affecting the energy efficiency of a network and its lifetime; and
- Promiscuous mode operation, although initially costly, is highly desirable in mobile and congested environments as it provides stability and reliability.

The novel approach of this work is the consideration of the *topology knowledge* as a metric and an important factor of energy consumption. Our findings coincide with the research into various topology control schemes [3,4] that found that many of the topology control algorithms were inadequate in reducing the energy consumption. One common flaw was that by reducing the transmission radius, the knowledge of the network topology was impaired. In the end, the topology control schemes managed to optimise the metrics such as the degree and the transmission radius of each node, failed to make significant energy savings. Likewise, in our simulations, protocols that were unable to maintain a good knowledge of the network topology failed to perform well in terms of energy consumption. Limiting the nodes' energy reserves made these networks even more susceptible to the lack of topology knowledge.

Further work in this area can take many forms. We are already in the process of implementing a new topology control scheme based on these findings. Another possible direction would be to gather even more information about the network neighbourhood, such as congestion, mobility, and location, and attempt to use the acquired information to further optimise routing in wireless sensor networks.

Acknowledgements. The authors would like to thank Chris McDonald and Rachel Cardell-Oliver for their valuable discussions and feedback.

This work was supported by a Jean Rogerson Supplementary Postgraduate Award and the West Australian IVEC Doctoral Scholarship in conjunction with the Australian Postgraduate Award.

References

1. Akyildiz, I.F., Su, W., Sankarasubramaniam, Y., Cayirci, E.: Wireless sensor networks: a survey. Computer Networks **38** (2002) 393–422
2. Ye, W., Heidenmann, J., Estrin, D.: An energy-efficient mac protocol for wireless sensor networks. In: IEEE Computer and Communication Societies (INFOCOM), New York, USA (2002) 1567–1576
3. Ramanathan, R., Hain, R.: Topology control of multihop wireless networks using transmit power adjustment. In: IEEE Conference on Computer Communications (INFOCOM). Volume 2., Tel Aviv, Israel (2000) 404–413

4. Li, E.L., Halpern, J.Y., Bahl, P., Wang, Y.M., Wattenhofer, R.: Analysis of a cone-based distributed topology control algorithm for wireless multi-hop networks. ACM Symposium on Principle of Distributed Computing (PODC) (2001) 264–273
5. Stojmenovic, I., Lin, X.: Power-aware localized routing in wireless networks. In: IEEE Transactions on Parallel and Distributed Systems. Volume 12., Cacun, Mexico, USA (2001) 1122–1133
6. Filipovic, A.: Topology control problems in mobile ad hoc wireless networks. In Cardell-Oliver, R., ed.: The Twelfth University of Western Australia School of Computer Science & Software Engineering Research Conference, Yanchep, Western Australia, Uniprint (2003) 112–117
7. Singh, S., Woo, M., Raghavendra, C.S.: Power-aware routing in mobile ad hoc networks. In: ACM/IEEE International Conference on Mobile Computing and Networking (MOBICOM). Volume 4., Dallas, Texas, USA (1998) 181–190
8. Broch, J., Maltz, D.A., Johnson, D.B., Hu, Y.C., Jetcheva, J.: A performance comparison of multi-hop wireless ad hoc network routing protocols. In: ACM/IEEE International Conference on Mobile Computing and Networking (MOBICOM). Volume 4., Dallas, Texas, USA, ACM Press (1998) 85–97
9. Feeney, L.M.: An energy-consumption model for performance analysis of routing protocols for mobile ad hoc networks. ACM Journal of Mobile Networks and Applications **6** (2001) 239–249
10. Feeney, L.M., Nilsson, M.: Investigating the energy consumption of a wireless network interface in an ad hoc networking environment. In: IEEE INFOCOM. Volume 5., Anchorage, AK, USA (2001) 1548–1557
11. Jones, C.E., Sivalingam, K.M., Agrawal, P., Chen, J.C.: A survey of energy efficient network protocols for wireless networks. ACM Wireless Networks **7** (2001) 343–358
12. Johnson, D.B., Maltz, D.A.: Dynamic source routing in ad hoc wireless networks. In Imielinski, T., Korth, H., eds.: Mobile Computing. Volume 353. Kluwer Academic Publishers (1996) 153–181
13. Kravets, R., Schwan, K., Calvert, K.L.: Power-aware communication for mobile computers. In: IEEE International Workshop on Mobile Multimedia Communications (MoMuC). Volume 6., San Diego, USA (1999) 64–73
14. Perkins, C.E., Bhagwat, P.: Highly dynamic destination-sequenced distance-vector routing (DSDV) for mobile computers. In: ACM SIGCOMM Conference on Communications Architectures, Protocols and Applications. Volume 24., London, UK, ACM Press (1994) 234–244
15. Rodoplu, V., Meng, T.H.: Minimum energy mobile wireless networks. IEEE Journal on Selected Areas in Communications **17** (1998) 1333–1344
16. Royer, E.M., Toh, C.K.: A review of current routing protocols for ad hoc mobile wireless networks. IEEE Personal Communications **6** (1999) 16–55
17. Park, V.D., Corson, M.S.: A highly adaptive distributed routing algorithm for mobile wireless networks. In: IEEE Computer and Communications Societies (INFOCOM). Volume 3., Kobe, Japan (1997) 1405–1413
18. Fall, K., Varadhan, K.: ns Notes and Documentation, The VINT Project, UC Berkeley, LBL, USC/ISI, and Xerox PARC. (1997)
19. Tanenbaum, A.S.: Computer Networks. 3rd edn. Prentice Hall, Inc, Upper Saddle River, New Jersey (1996) Printed in USA.
20. Filipovic, A.: Energy efficiency of ad hoc routing protocols (2002) School of Computer Science & Software Engineering, The University of Western Australia.

Max-Min Length-Energy-Constrained Routing in Wireless Sensor Networks*

Rajgopal Kannan, Lydia Ray, Ram Kalidindi, and S.S. Iyengar

Department of Computer Science
Louisiana State University
Baton Rouge LA 70803
{rkannan,lydiaray,rkalid1,iyengar}@bit.csc.lsu.edu

Abstract. We consider the problem of inter-cluster routing between cluster heads via intermediate sensor nodes in a hierarchical sensor network. Sensor nodes having limited and unreplenishable power resources, both path length and path energy cost are important metrics affecting sensor lifetime. In this paper, we model the formation of length and energy constrained paths using a game theoretic paradigm. The Nash equilibrium of our routing game corresponds to the optimal length-energy-constrained (LEC) path. We show that this path can be computed in polynomial time in sensor networks operating under a geographic routing regime. We then define a simpler team version of this routing game and propose a distributed nearly-stateless energy efficient inter-cluster routing protocol that finds optimal routes. This protocol balances energy consumption across the network by periodically determining a new optimal path consistent with associated energy distributions. Simulation results testify to the effectiveness of the protocol in producing a longer network lifetime.

1 Introduction

A wireless sensor network is an autonomous system of numerous tiny sensor nodes equipped with integrated sensing and data processing capabilities. Sensor networks are distinguished from other wireless networks by the fundamental constraints under which they operate: a) sensor nodes are untethered and b) sensor nodes are unattended. These constraints imply that network lifetime, i.e., the time during which the network can accomplish its tasks, is finite. Therefore sensors must utilize their limited and unreplenishable energy as efficiently as possible.

Sensor network architectures can be broadly classified into two main categories: flat and hierarchical-cluster based [9]. In a hierarchical sensornet, efficient energy management is potentially easier since routing is partitioned into intra-cluster and inter-cluster, with traffic between clusters being routed through corresponding cluster heads.

* This work was supported in part by NSF grants IIS-0329738, IIS-0312632 and DARPA/AFRL grant # F30602-02-1-0198.

H. Karl, A. Willig, A. Wolisz (Eds.): EWSN 2004, LNCS 2920, pp. 234–249, 2004.

A significant amount of research has been done on hierarchical sensornets, for example, [9]. Most of these architectures are based on the assumption that cluster heads or gateway nodes can communicate directly with each other and the transmission power is adjustable at each node. In this paper, we consider a more realistic two-level architecture where cluster heads (called *leader nodes*) must use the underlying network infrastructure for communication.

The primary issue addressed in this paper is energy-constrained inter-cluster routing within the framework of the proposed architecture, i.e., leader to leader and leader to sink routing. Since sensor nodes on a route consume energy continuously by sending and receiving data packets, the longevity of a sensor node is inversely proportional to the number of routes it participates in. Network partition is therefore expedited by uneven energy distribution across sensors, resulting from improperly chosen routes. Ideally, data should be routed over a path in which participating nodes have higher energy levels relative to other non-participating nodes. Network operability will be prolonged if a critically energy deficient node can survive longer by abstaining from a route rather than taking part in a route for a small gain in overall latency. Therefore, both *path length* and *path energy cost* are critical metrics affecting sensor lifetime.

While there are several existing protocols in the literature that focus exclusively on either of these issues, there has only been some recent work that considers both aspects. For example, Shah and Rabaey [10] describe a probabilistic routing protocol where non least-energy cost paths are chosen periodically. In [8], a node attempts to balance energy across all its neighbors while finding shortest paths to the sink. However, there is no unified analytical model that explicitly considers routing under both the constraints of energy efficiency and path length. We observe that the choices of sensor nodes under these constraints are a natural fit for a game theoretic framework. In this paper, we propose a game theoretic paradigm for solving the problem of finding energy-optimal routing paths with bounded path length. We define a routing game in which sensors obtain benefits by linking to healthy nodes while paying a portion of path length Thus sensor nodes modelled as intelligent agents cooperate to find optimal routes. Our proposed model has the following benefits:

- Each player will tend to link to the healthiest possible node. Thus network partition will be delayed.
- Since each node shares the path length cost, path lengths will tend to be as small as possible. Thus delay is restricted in this model. Also smaller path lengths will prevent too many nodes from taking part in a route, thereby reducing overall energy consumption.

The Nash equilibrium of this routing game defines the optimal length-energy-constrained (LEC) path. We show that while computing this path is NP-hard in arbitrary sensor networks, it can be found in polynomial time (in a distributed manner) in sensor networks operating under a geographic routing regime.

We next propose a fully distributed and nearly stateless inter-cluster routing protocol which implements a simplified team version of the above routing game using a metric called *energy weakness*. The protocol determines the route with least weakness by combining limited global information on node energies obtained through *reverse directional flooding* (RDF) along with local geographic forwarding. While RDF involves some extra overhead and consumes node energy, global information obtained through this process ensures a significant tradeoff in terms of node energy balancing and network lifetime. Note that this protocol can be easily modified to implement a distributed version of the original game as well. The key features that distinguish our protocol from related routing protocols are summarized below.

- Unlike network lifetime maximization protocols such as [4] which require global information on current data/packet flow rates from each sensor to sink(s), our protocol utilizes very limited network state information and is thus easily implementable.
- Energy being a critical resource in sensor networks, depleted regions (i.e regions with low residual node energies) must be detected and bypassed by routing paths as quickly as possible. This is analogous to congestion in wired networks. We propose a new technique for indicating the onset of energy depletion in regions by using *energy depletion indicators*. This is used in conjunction with the energy weakness metric in our protocol to ensure energy-balanced routing.
- Geographic sensornet routing algorithms such as GPSR [6] and GEAR [8] employ elegant neighbor selection procedures using only local network information. These mechanisms enhance energy savings at individual sensor nodes and are easy to implement. However due to their predominantly local nature, there are situations in which these protocols will be slow to adapt to changing energy distributions in the network. For example, consider a region in a sensornet that is intersected by multiple routes and thus has higher energy depletion rates. While GEAR is likely to take a considerable amount of time to avoid this region through localized rerouting, our protocol, with its energy depletion indicator algorithm, will quickly detect such regions and establish new bypass routes.
- A potential drawback of protocols such as [8], [6], [11] which use local geographic forwarding is that significant backtracking is required when a hole is encountered. This situation is completely eliminated by our protocol since packets are forwarded according to the periodically updated routing table residing at each node.
- In protocols such as [8] [6] [3] [2], a single routing path (typically, the least energy path) is utilized continuously until a node's energy is completely exhausted. While the motivation behind this approach is to save energy consumption at individual sensor nodes, this might lead to unintended consequences such as the expedited partition of the network. Our protocol overcomes this drawback by selecting new length-energy-constrained routing paths periodically. ([10] also does this but in a probabilistic manner for non least-energy cost paths).
- Many of the previously proposed routing protocols are source-initiated i.e. the source predetermines the routes [3] [2]. The ad hoc infrastructure of

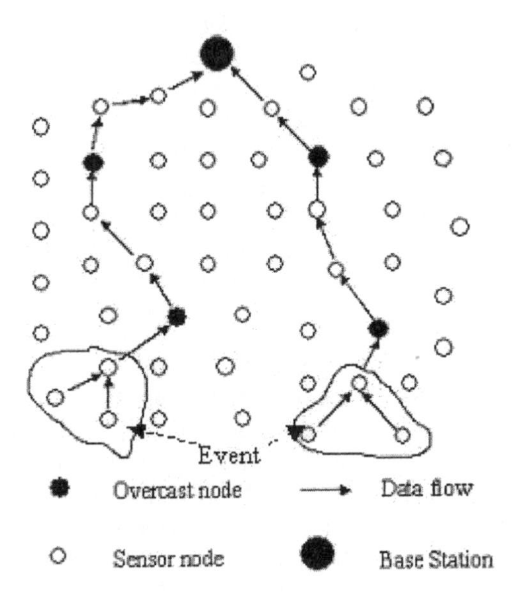

Fig. 1. Proposed Network Architecture

a sensornet makes the implementation of a centralized system extremely expensive. Our protocol reduces communication cost to a great extent by accomplishing localization of routing decisions.

We have evaluated our routing protocol using the ns-2 simulator. Simulation results indicate that compared to [6] and [8], the protocol is enormously effective in reducing energy deviation, thereby leading to equitable residual energy distribution across the sensor network. Thus the protocol should have a significant impact on sensor network survivability.

The rest of the paper is organized as follows. Section 2 describes the proposed architecture while section 3 contains the analytical model along with the proposed leader to leader routing protocol. In section 4, we describe methods for selecting routing protocol parameters. Section 5 evaluates protocol performance and conclusions are derived in section 6.

2 Network Architecture

The sensornet architecture proposed in this paper has a two-level hierarchy as shown in Figure 1. The lower level consists of a standard wireless sensor network. At the upper level, an overlay network is created among a collection of special sensor nodes (called *leader nodes*) at specific locations. Note that these leader nodes are selected from the underlying sensor network and occupy positions similar to cluster heads. In general, leader nodes are expected to be less energy-constrained than ordinary sensor nodes[1] and are responsible for aggregated leader to leader and leader to sink communication. Unlike existing clustered architectures, the proposed network is significantly different in that we do not assume direct connectivity among leader nodes. Instead, leader to leader communication utilizes underlying network infrastructure. Note that the routing responsibility of leader nodes does not preclude in any ways their functioning as sensors. Our hierarchical architecture is flexible enough for any sensor to function as a leader node. Mechanisms by which the ordinary sensors become leader nodes are not discussed in this paper. The main features of the proposed architecture are as follows:

- Sensors are assumed to be fixed (i.e. immobile).
- Each leader node is responsible for aggregating sensor data (in response to queries) from sensor nodes within a predefined neighborhood. This is similar to the clustering approach proposed in [9] but with significant differences in the way the cluster heads communicate with each other.
- Leader nodes route/communicate aggregate sensor data to other leader nodes and sinks via ordinary sensors as intermediaries. Communication of aggregate sensor data between leader nodes should result in the reduction of overall network traffic.

Since leader to leader communication has an impact on energy consumption of ordinary (intermediate) sensors too, the development of efficient protocols is of paramount importance in this architecture. In the next section, we propose an analytical model of length-constrained energy efficient leader to leader routing along with practical implementation.

3 Analytical Model of Leader to Leader Routing

A sensor network is useful as long as sensors continue to process and route data. Since sensor nodes on a route consume energy continuously by sending and receiving data packets, the longevity of a sensor node is directly proportional to the number of routing paths it participates in. Improperly chosen routing paths will lead to uneven energy consumption across sensors in the network. A highly non-uniform distribution of residual energies in sensors over the network might also expedite network partition. Therefore we must design leader to leader routing protocols that dissipate energy equitably over sensors in order to enhance network survivability.

One possible approach is to prevent low energy nodes from taking part in a route as long as they are energy-deficient relative to their neighbors. However, a route that focuses only on energy efficiency may be undesirably long since the lowest energy-cost path need not be the shortest. Conversely, if path length is measured in terms of number of hops, longer paths will result in energy depletion

[1] This is not strictly necessary for our architecture to work. However, since leader nodes have extra responsibilities, this is a reasonable expectation.

at more sensors while also increasing delay. When path length is measured in terms of Euclidian distance, longer paths lead to higher energy consumption at individual sensors since transmitting power is proportional to neighbor distance. While there are several existing protocols in the literature that focus exclusively on either of these issues, there is no unified analytical model that explicitly considers routing under both the constraints of energy efficiency and path length.

In this paper, we model sensors as intelligent agents and propose a game theoretic paradigm for solving the problem of finding energy-optimal routing paths with bounded path length. The equilibrium point(s) of the routing game define the optimal routing path. We then define a team version of this routing game and propose a distributed nearly-stateless leader to leader energy efficient routing protocol that finds the optimal route. We now formally define our analytical model and formulate the routing game.

Let $S = \{s_1, s_2, \ldots, s_n\}$ be the set of sensors in the sensor network participating in the routing game. Let L_1 and L_2 be a pair of leader nodes using sensors in S as intermediaries[2]. Data packets are to be routed from L_1 to L_2 through an optimally chosen set $S' \subset S$ of intermediate nodes by forming communication links. Note that we do not consider multicast communication between sets of leader nodes in this paper.

Strategies: Each node's strategy is a binary vector $l_i = (l_{i1}, l_{i2}, ..., l_{ii-1}, l_{ii+1}, ..., l_{in})$, where $l_{ij} = 1$ ($l_{ij} = 0$) represents sensor s_i's choice of sending/not sending a data packet to sensor s_j. Since a sensor typically relays a received data packet to only one neighbor, we assume that a node forms only one link for a given source and destination pair of leader nodes. In general, a sensor node can be modeled as having a mixed strategy [1], i.e., the l_{ij}'s are chosen from some probability distribution. However, in this paper we restrict the strategy space of sensors to only pure strategies. Furthermore, in order to eliminate some trivial equilibria, each sensor's strategy is non-empty and strategies resulting in a node linking to its ancestors (i.e. routing loops) are disallowed. Consequently, the strategy space of each sensor s_i is such that Prob. $[l_{ij} = 1] = 1$ for exactly one sensor s_j and Prob. $[l_{ij} = 1] = 0$ for all other sensors, such that no routing loops are formed.

Payoffs: Let $l = l_1 \times l_2 \times \ldots \times l_n$ be a strategy in the routing game resulting in a route $\mathcal{P}$ from source to destination leader node. Each sensor on $\mathcal{P}$ derives a payoff from participating in this route. The payoff of a sensor s_i which links to node s_j in $\mathcal{P}$ is then defined as:

$$\pi_i(l) = E_j - \xi L(\mathcal{P}) \tag{1}$$

where E_j is the residual energy level of node s_j and $L(\mathcal{P})$ the length of routing path $\mathcal{P}$. E_j represents a benefit to s_i, thus inducing it to forward data packets to higher energy neighbors. The parameter ξ represents the proportion of path length costs that are borne by sensor s_i. Choosing ξ as a positive constant or proportional to path length will inhibit the formation of longer routing paths.

[2] In general, sensors in S will be simultaneously participating in routing paths between several such pairs.

Conversely, setting ξ zero or inversely proportional to path lengths will favor the formation of paths through high-energy nodes. We choose xi as a non-zero positive constant for this routing game. Thus each sensor will forward packets to its maximal energy neighbor in such a way that the length of the path formed is bounded. This model encapsulates the process of decentralized route formation by making sensor nodes cooperate to achieve a joint goal (shorter routing paths) while optimizing their individual benefits.

A Nash equilibrium of this game corresponds to the path in which all participating sensors have chosen their best-response strategy, i.e., the one that yields the highest possible payoff given the strategies of other nodes. This equiilibrium path is the optimal Length Energy-Constrained (LEC) route in the sensor network for the given leader pair. Note that the process of determining the LEC route requires each node to determine the optimal paths formed by each of its possible successors on receiving its data. The node then selects as next neighbor that node, the optimal path through which incurs the highest payoff.

Theorem 1 Let $\hat{\mathcal{P}}$ be the optimal LEC route for a pair of source and destination leader nodes in an arbitrary sensor network. Computing $\hat{\mathcal{P}}$ is NP-Hard.

This proposition can be proved by reduction from Hamiltonian Path. For lack of space we do not include the proof in this paper.

Theorem 2 Let S be any sensor network in which sensor are restricted to following a **geographic routing** *regime. In other words, the strategy space of each sensor includes only those neighbors geographically nearer to the destination than itself. Then $\hat{\mathcal{P}}$ can be computed in polynomial time in a distributed manner.*

Again, due to space limitations, we only provide an outline of the proof. $\hat{\mathcal{P}}$ can be obtained in $O(N+E)$ steps for an arbitrary N-node geographically routed sensor network with E edges and $O(N)$ steps in a $\sqrt{N} \times \sqrt{N}$ mesh. The result hinges on the observation that in a geographically routed network, the intersection of all feasible paths from the source to a node s_i and from s_i to the destination node is empty (except for s_i). Thus one can compute the union of optimal paths from source leader to sensor and sensor to destination leader.

Next, we identify sufficient conditions under which the optimal LEC path coincides with other commonly used routing paths. For brevity, we state these results without proof.

Proposition 1: Let E^i_{max} and E^i_{min} denote the maximum and minimum neighbor node energies at sensor s_i. Then the shortest path from L_1 to L_2 will be optimal if

$$E^i_{max} - E^i_{min} < \xi(\delta S) \tag{2}$$

holds at each sensor node s_i on the shortest path, where (δS) is the difference between the shortest and second shortest paths from s_i to L_2.

Let the maximal-energy neighbor path denote the one obtained by following the maximal energy nodes from L_1 to L_2 such that a path is formed. Then we have,

Proposition 2: The maximum energy neighbor path will be optimal if

$$(\delta E) > \xi(l_h - l_s) \tag{3}$$

holds at every node s_i *on the path, where* (δE) *is the difference in energies between the maximal and second maximal neighbors of* s_i *and,* l_h *and* l_s *are the lengths of the maximal energy and shortest paths from* s_i*, respectively.*

4 Distributed Protocol Implementation

We describe a distributed implementation of the optimal LEC routing protocol in terms of a simplified 'team' version of the routing game. This team-game routing protocol can be easily modified to obtain optimal LEC paths as well and is simpler to describe. In the team LEC path heuristic, each node on a path shares the payoff of the worst-off node on it. Formally, let $\mathcal{L}$ be the set of all distinct paths from a particular source and destination leader pair. Let $E_{min}(\mathcal{P})$ be the smallest residual energy value on path $\mathcal{P}$. Then the equilibrium path of the team LEC game is defined as:

$$\hat{P} = \text{argmax}_{\mathcal{P} \in \mathcal{L}}(E_{min}(\mathcal{P}) - \xi|\mathcal{P}|) \tag{4}$$

For simplicity in the protocol description below, we set ξ to zero. However the protocol can be easily modified for non-zero ξ as well as for computing optimal LEC paths in the original LEC game. We interpret the optimal path under this condition as follows: Given any path $\mathcal{P}$, the durability of the path is inversely proportional to $E_{min}(\mathcal{P})$. A path with lower average energy but higher minimum energy should last longer than a route with the opposite attributes since the least energy node is the first to terminate and make that route obsolete. Thus the inverse of the minimum node energy on a given path reflects the energy weakness of the path. The proposed protocol will select an optimal path of bounded length with the least energy weakness. Since node energy levels are changing continuously in a sensor network due to sensing, processing and routing operations, the optimal path needs to be recomputed periodically. Thus the proposed protocol operates in two different phases: data transmission and path determination as described below:

4.1 Data Transmission Phase

During this phase, data packets are transmitted from one leader node to the other through the optimal path (with least energy weakness). Each data packet also potentially collects information about the energy consumption en route, by keeping track of residual energy levels of nodes on the path. When energy levels of a given critical number of nodes fall below a certain threshold, the data transmission phase ends and the new optimal path determination phase begins. The fundamental steps of the data transmission phase are as follows:

- Each data packet is marked by the source leader node with the geographical position of the destination node and with a threshold value th. Each data packet contains a special n-bit Energy Depletion Indicator (EDI) field, where $n << $ packet size.

- Each sensor node receiving a data packet determines whether its energy level has fallen below the threshold th. If so, and the EDI field in the data packet is not exhausted, the node sets a single bit in the EDI field. Then it forwards the packet to the best next-hop neighbor according to its routing table. We assume that before the network starts any activity, all ordinary 'non-leader' sensor nodes have the same energy level. Therefore, during the first data transmission phase, the best next-hop neighbor of a node is the one which is geographically nearest to the destination leader node. In all other phases, the routing table is updated according to the optimal LEC path calculation.
- If the receiver leader node gets a data packet with all n bits in the EDI field set to 1, it triggers a new optimal path selection procedure. Alternatively, if the destination receives n data packets with at least one bit in the EDI field set to 1 a path determination phase is triggered.

Calculation of the Threshold Value. The threshold value th plays a very important role in the data transmission phase since it is used to provide an approximate indication that the current optimal path has become obsolete. Intuitively, th must be a function of the current residual node energy levels in the network. In this paper, we use the following function:

$$th = \beta E_{min} \tag{5}$$

where $0 < \beta < 1$ and E_{min} is the minimum energy level in the current optimal path. Since E_{min} changes with time, the threshold is recalculated in each path determination phase, consistent with the current energy distribution across the network.

4.2 Path Determination Phase

This phase begins when the destination leader node receives critical EDI information and ends when the sending leader node has updated its routing table and recalculated the threshold value. The principle steps are as follows:

- The destination leader node L_2 triggers this phase by flooding the network with control packets along the geographic direction of the source leader node L_1 (Figure 2). Note that this *reverse directional flooding* occurs in the direction opposite to that of data transfer.
- Each node forwards *exactly one* control packet to all its neighbors in the geographic direction of L_1. Each control packet contains a field EM_p that indicates the maximum of the minimum energy levels of all partial paths converging at the given node, i.e., the inverse of the energy weakness of the 'strongest' partial path.
- On receiving the first control packet, each node sets a timer for a prefixed interval T. This time-period should be large enough for the node to receive future control packets from most of its neighbors (corresponding to different partial paths from the leader node terminating at this node), but not so

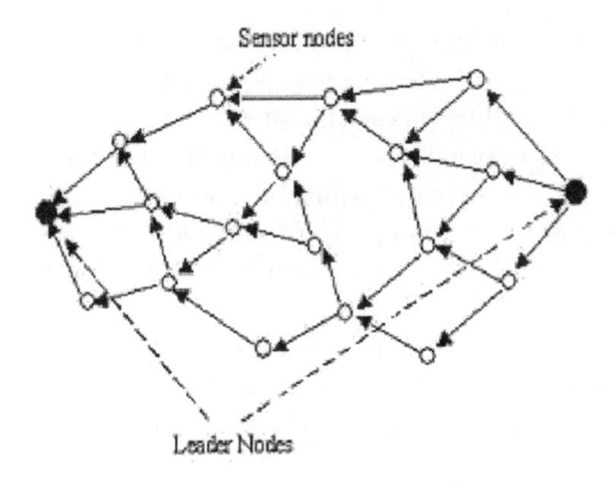

Fig. 2. Reverse Directional Flooding

large as to cause high delays. With each arriving control packet, the node updates and stores the highest EM_p value seen so far. *However, if its own energy level E_i is lower than all these values it stores E_i.* With each control packet, it also updates its routing table for destination L_2 to point to the node from which it has received the highest energy control packet. Note that this part of the protocol can be easily modified to incorporate path lengths in addition to the minimum energy computations above.

- When the timer expires, this node forwards a new control packet with EM_p field set to the stored energy value to all its neighbors in the geographic direction of L_1. Control packets arriving after the timer expires are discarded.
- Eventually, L_1 begins receiving control packets and sets its timer. Its value of T can be determined in many ways depending on the specific requirements of applications. In this paper, we calculate T to ensure that most of the paths from L_1 to L_2 are included in the optimality calculations. If (D_{max}) is the maximum transmission delay between two nodes, the value of T is determined as $(MINHOP * D_{max})$, where $MINHOP$ is an estimate of the shortest path from L_1 to L_2. This value can be estimated apriori using GPSR routing [6] before the first data transmission phase. Note that the given value of T allows control packets from paths up to twice the length of the shortest path to be forwarded to $L-1$. Also note that D_{max} is a function of the specific MAC-layer protocol being implemented in the sensor network. Finally, when the timer expires at L_1, it selects the final E_{min} value as the highest EM_p value it has received, calculates the new value of th and sets its routing table accordingly. The next data transmission phase can now begin.

5 Selection of Parameters

There are three main issues related to practical implementation efficiency of our protocol.

- To avoid unnecessary energy expenditures, control packets must be prevented from hanging around the network after the path determination phase. The parameter MINHOP serves this purpose.
- There is a tradeoff between energy consumption involved in flooding vs the gain in network lifetime due to equitable energy distribution among sensors. Therefore, frequency of invocation of the path determination phase is an important parameter. We experimentally model this using β and EDI as described later.
- Calculating the energy overhead of flooding in the path determination phase is also an important issue. We provide analytical upper bounds and discuss optimizations for further reducing this control overhead.

5.1 Selection of β

Data transmission in the proposed protocol ends when residual energy levels of at least n nodes on the current path fall below threshold th. With high β, the smaller the threshold value, and the larger the useful data transmission phase. If the traffic is fairly bursty, th should be large so that data packets in the burst can be transmitted. β is an empirical value and can be modified based on previous observations. A useful rule of thumb to set the value of β for the current period is as follows:

$$\beta_{current} = \alpha\beta_{prev1} + (1 - \alpha)\beta_{prev2} \tag{6}$$

where $0 \leq \alpha \leq 1$, β_{prev1} and β_{prev2} are the previous and previous to previous values of β correspondingly. α should be chosen according to the specific requirement i.e. whether the current value of β should be increased or decreased and to what extent.

5.2 Selection of Energy Depletion Indicator

Energy depletion indicator is an integer which indicates the maximum number of critical nodes allowed during a data transmission period. The main contribution of energy depletion indicator is to regulate the duration of data transmission phase. The higher the value of this parameter, the longer the period of data transmission. Like β, this parameter is also empirical and can be modified based on pervious observations. A rule of thumb similar to that of β can be used to modify the value of this parameter. Thus,

$$EDI_{current} = \gamma EDI_{prev1} + (1 - \gamma)EDI_{prev2} \tag{7}$$

where $0 \leq \gamma \leq 1$, EDI_{prev1} and EDI_{prev2} are the previous and previous to previous values of EDI correspondingly. γ should be chosen according to the specific requirement i.e. whether the duration of the data transmission phase should be increased or decreased and to what extent.

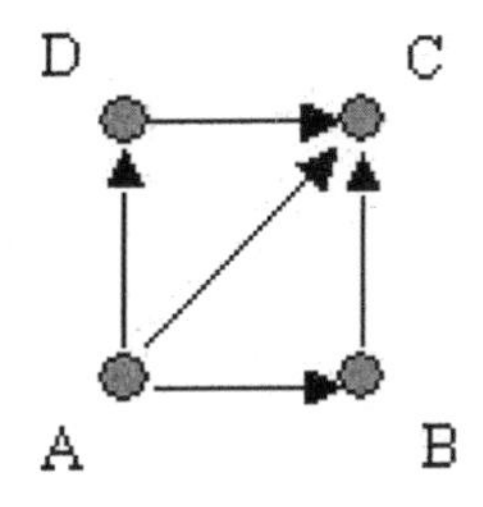

Fig. 3. Possible Control Packet Optimization

6 Overhead Due to Path Determination

The proposed protocol uses directional flooding technique to determine a new optimal path. The advantage of using directional flooding over general flooding is that the packets being forwarded only to a single direction produces less overhead. The following propositions 9stated without proof, for brevity) give an estimate of overhead due to directional flooding in terms of number of control packets.

Proposition 3: *In a general N node geographically routed wireless sensor network where a node can have at most k neighbors, the number of control packets transmitted during the path determination phase is bounded by kn.*

Proposition 4: *In a wireless sensor network as an $\sqrt{N} \times \sqrt{N}$ mesh, the number of control packets transmitted during the path determination phase is bounded by $3\sqrt{N}$.*

Note that in a mesh topology, the actual number of control packets transmitted during the path determination phase can be reduced further. Consider the following situation shown in Figure 3. In this portion of the mesh, node A sends control packets to nodes B, C and D. If B receives the packet from A before its timer expires and if the energy value on this packet is the winning value at B as well, then B does not need to send this packet to node C (which has already received a copy of the same packet from A). In this manner, the total number of control packets can be further reduced.

7 Performance Evaluation

The main objective of our protocol is to gradually balance unfair energy consumption across the network. To evaluate performance of this protocol, we use the following metrics which reflect dispersion or concentration of energy consumption across a network.

- **Variance of energy level:** The variance of the energy levels of all the nodes is the primary measure of dispersion. A high variance indicates higher energy consumption at some of the nodes compared to others.

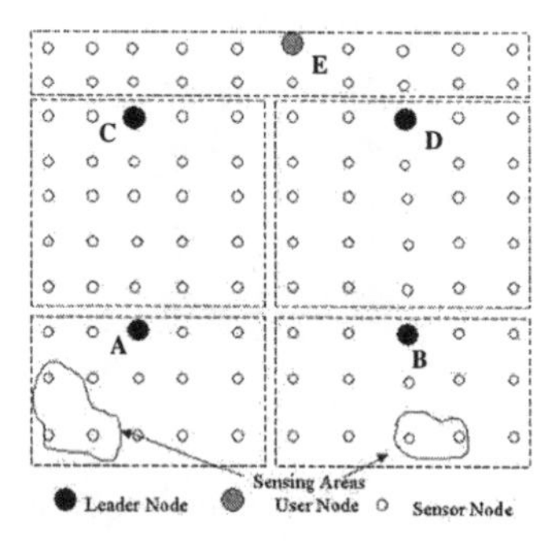

Fig. 4. Simulation Topology

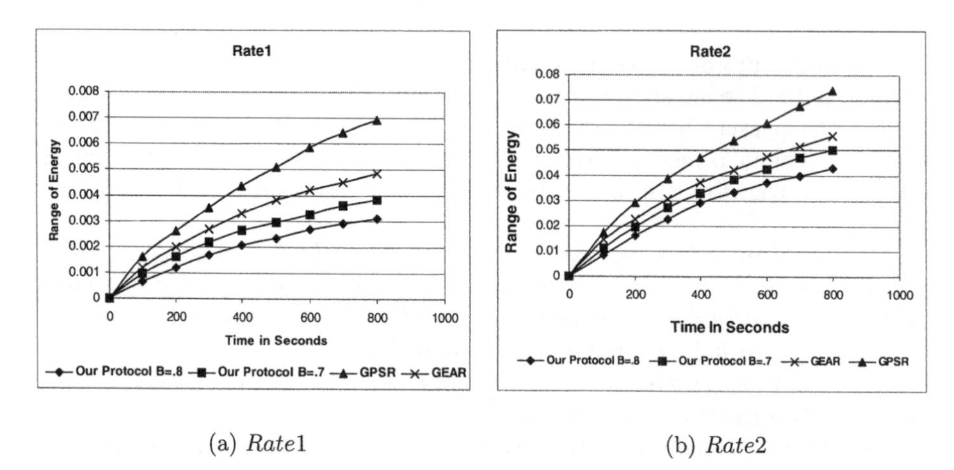

(a) *Rate*1 (b) *Rate*2

Fig. 5. Range of residual energy levels across the network with team LEC routing , GEAR and GPSR

– **Range of energy level:** This metric measures the difference between the energy levels of the maximum energy node and the minimum energy node over the whole network. A large value for this range is a result of unfair distribution of routing load among the nodes.

7.1 Experimental Setup

In our simulation we have 100 nodes in a 1000×1000 meter area, with one node at each of the positions of the 10×10 square grid. Figure 4 represents the mesh topology that we have used for evaluating our protocol. The whole network is divided into five clusters. There are two sensing areas in the regions under clusters A and B. Sensor data packets are generated from these sensing areas at a uniform rate. The leader nodes in each of the clusters A and B collect these packets and send them to the leader nodes of clusters C and D respectively

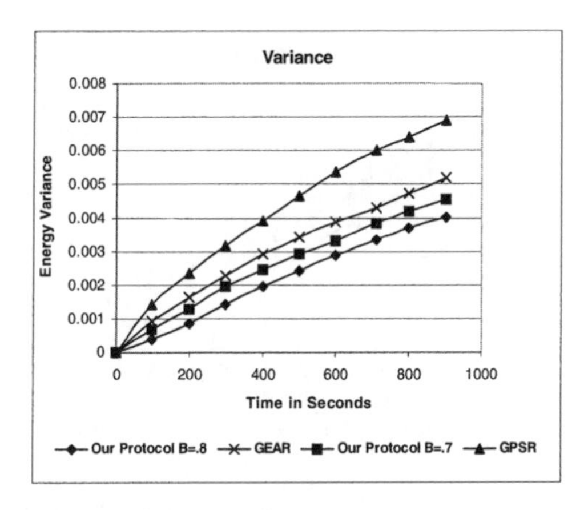

Fig. 6. Variance in residual energy levels across the network under team LEC routing with $\beta = 0.8, 0.7$, GEAR and GPSR

via intermediate sensor nodes. Leader nodes C and D forwards these packets to the sink node in cluster E. Each leader node selects the leader node which is geographically nearest to the sink for transmitting its received/sensed data. Leader to leader communication is accomplished through ordinary sensors. Reverse directional flooding is initiated when a leader node receives a sensor data packet indicating that at least *three* sensor nodes are close to the threshold th. A sender leader node sets th to the new βE_{min} obtained from the reverse flooding phase. We run the simulation for 900 seconds and compare three protocols for energy efficiency: geographic shortest path routing (GPSR [6]), GEAR [8] and the proposed team LEC protocol. In GEAR, a node N_i dynamically chooses its minimum cost neighbor for forwarding its packet, where costs are parametrically estimated as $c(N_i) = \alpha d(N_i) + (1 - \alpha)E(N_i)$. Here $d(N_i)$ and $E(N_i)$ are the neighborhood-normalized distance to destination and energy consumed at N_i respectively [8]. In our simulations, we compare the team LEC protocol for inter-leader comunication with GEAR using $\alpha = 0.5$. These experiments are carried out on our simulation test-bed which is an extension of Sensorsim [7].

7.2 Results and Analysis

We assume that before the network starts any activity, all ordinary sensor nodes have the same energy level. Therefore, in the very beginning, energy distribution is uniform across the network. When a network becomes active, the energy distribution across it gradually becomes non-uniform since nodes participating in a route inevitably consume more energy than other nodes. A protocol which uses a fixed route until one node in the route is completely drained out of its energy, ends up in producing an energy distribution with high dispersion of energy levels. On the other hand, our proposed protocol tries to adapt to the dynamically changing energy distribution and gradually uniforms the initial uneven energy

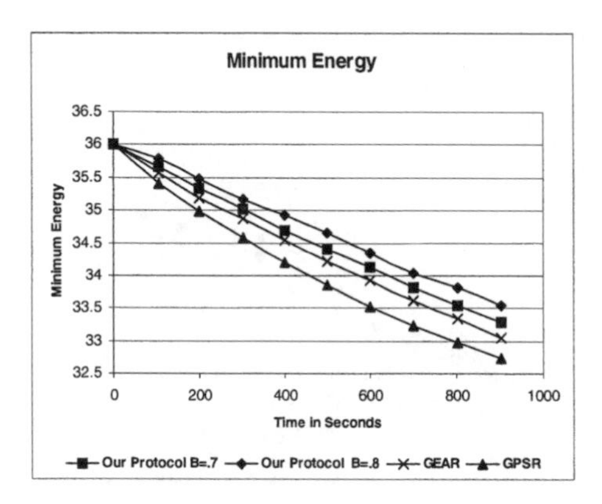

Fig. 7. Minimum residual energy levels across the network under team LEC routing with $\beta = 0.8, 0.7$, GEAR and GPSR

distribution. Therefore, it is expected that the difference between the dispersion measures produced by our protocol and those produced by any protocol with fixed routing will increase with an increasing rate with time. In this paper, we compare the performance of our protocol with that of a protocol which uses a fixed shortest path for leader to leader communication.

Results of our simulation comparing performance of our protocol with that of GEAR and GPSR reflect the outcome as expected. We use *range* (the difference between the maximum and minimum value of a distribution) and *variance* as measures of dispersion of energy distribution to evaluate our protocol. In Figure 5(a) and 5(b), we present the difference in the ranges of node energy distributions across the network over time under the three protocols with two different values of β and two different traffic rates. In both the figures, the difference of the ranges rises very sharply indicating that our protocol yields lower range of energy distribution compared to that produced by the fixed route protocol as time proceeds. Moreover, with an increased traffic rate, our protocol produces a much better result compared to that of shortest path routing. This indicates that with heavy traffic, the energy distribution across the network become more uneven in a fixed route protocol since the load is heavier on a particular route. In this case, frequent change of routing path is very much useful to bring uniformity to overall energy consumption. Note that routes being changed more frequently with a higher value of β, performance of the proposed protocol is better when $\beta = 0.8$ than that when $\beta = 0.7$.

Figure 6 represents the variance of residual energy distribution produced by our protocol with two different β values and that by GPSR and GEAR. The energy range metric does not measure the number of sensor nodes that are being treated unfairly. The high variance of the shortest path routing indicates that a significant number of sensor nodes are being treated unfairly with network traffic being concentrated at fewer nodes. This might expedite partition of the network due to energy depletion at critical nodes. With higher value of β, the proposed protocol produces lower variance due to more frequent route changes.

Figure 7 shows how the difference between the minimum energy level produced by our protocol and that by GPSR and GEAR changes with time for two different β values. In both the cases, the difference rises very sharply as time proceeds. With a higher value of β, the rise is sharper because change of route is accomplished more frequently and therefore, consumption of energy is more uniform under our protocol.

8 Conclusion

In this paper, we describe a game theoretic paradigm for inter-cluster routing in a clustered sensornet architecture in which cluster heads utilize underlying network infrastructure for communication. The max-min energy balancing inter-cluster routing protocol finds the length-energy-constrained path corresponding to the equilibrium of the routing game. This protocol balances energy consumption across the network by selecting new optimal paths periodically. The simulation results indicate effectiveness of this protocol for enhancing network survivability.

References

[1] D. Fudenberg and J. Tirole, Game Theory, MIT Press, 1991.

[2] D.B. Johnson and D.A. Maltz, "Dynamic Source Routing in Ad Hoc Wireless Sensor Networks", *Mobile Computing*, Kluwer, 1996.

[3] C. Perkins and E. Royer, "Ad Hoc On Demand Distance Vector Routing", *Proc. of 2nd IEEE Wksp Mobile Comp. Sys. and Applications*, February 1999.

[4] J. Chang and L. Tassiulas, "Energy conserving routing in wireless ad-hoc networks," *IEEE INFOCOM '2000*, March 2000.

[5] Chalermek Intanagonwiwat, Ramesh Govindan and Deborah Estrin,"Directed Diffusion: A Scalable and Robust Communication Paradigm for Sensor Networks," *In Proceedings of the Sixth Annual International Conference on Mobile Computing and Networks (MobiCOM 2000)*, August 2000, Boston, Massachusetts.

[6] Brad Karp and H. T. Kung, "GPSR: Greedy perimeter stateless routing for wireless networks," *Proc. ACM/IEEE MobiCom*, August 2000.

[7] S. Park, A. Savvides and M. B. Srivastava, "SensorSim: A Simulation Framework for Sensor Networks," *Proceedings of MSWiM 2000*, Boston, MA, August 11, 2000.

[8] Yan Yu, Ramesh Govindan and Deborah Estrin , "Geographical and Energy Aware Routing: A Recursive Data Dissemination Protocol for Wireless Sensor Networks," *UCLA Computer Science Department Technical Report UCLA/CSD-TR-01-0023*, May 2001.

[9] M.J. Handy, M. Haase and D. Timmermann, "Low Energy Adaptive Clustering Hierarchy with Deterministic Cluster-Head Selection," *4th IEEE International Conference on Mobile and Wireless Communications Networks*, Stockholm, 2002.

[10] Rahul C. Shah and Jan M. Rabaey, "Energy Aware Routing For Low Energy Ad Hoc Sensor Networks," *Proc. IEEE WCNC'02*, March 2002.

[11] R. Kannan, L. Ray, R. Kalidindi, and S.S. Iyengar, "Threshold-Energy Constrained Routing Protocol for Wireless Sensor Networks," *Sensor Letters*, Vol 1, December 2003.

Time-Synchronized Neighbor Nodes Based Redundant Robust Routing (TSN^2R^3) for Mobile Sensor Networks*

Wonjong Noh[1], Kyungsoo Lim[1], Jihoon Lee[2], and Sunshin An[1]

[1] Computer Network Lab, Dep. Of Electronics Engineering, Korea University
Sungbuk-gu, Anam-dong 5ga 1, SEOUL, KOREA, Post Code: 136-701,
Phone: +82-2-925-5377, FAX: +82-2-3290-3674
{nwj,angus,sunshin}@dsys.korea.ac.kr

[2] i-Networking Lab, Samsung Advanced Institute of Technology
San 14-1, Nongseo-Ri, Kiheung-Eup, Yongin, Kyungki-Do 449-712 Korea
vincent.lee@samsung.com

Abstract. This paper proposes a new on-demand redundant-path routing protocol considering time synchronization based path redundancy as one of route selection criteria. Path redundancy implies how many possible redundant paths may exist on a route to be built up. Our proposal aims to establish a route that contains more redundant paths toward the destination by involving intermediate nodes with relatively mode adjacent nodes in a possible route. Our approach can localize the effects of route failures, and reduce control traffic overhead and route reconfiguration time by enhancing the reachability to the destination node without source-initiated route re-discoveries at route failures. We have evaluated the performance of our routing scheme through a series of simulations using the Network Simulator 2 (ns-2).

1 Introduction

Wireless sensor networks, a collection of tiny disposable and low-power devices, are a new class of ad hoc networks that are expected to find increasing deployment in coming years[3]. Adjacent nodes communicate directly between one another over wireless channels. However, since the transmission range of nodes is limited, the nodes that are not neighboring need routing supported by intermediate nodes for communications. Due to the mobility of nodes, the topology of connections between communicating nodes may be quite dynamic. Hence, mobile wireless sensor networks require a highly adaptive routing scheme to deal with the frequent topology changes.
Traditional routing protocols [1,7,8] known from wired networks do not work efficiently because they have not been designed with the instable network topology in mobile sensor networks [5,6]. Therefore, numerous ad hoc routing protocols have been developed [2,4,9,10,11,14]. Most of existing ad hoc routing protocols are con

* This paper was supported by the Samsung Advanced Institute of Technology, under the project "Research on Core Technologies ant its Implementation for Service Mobility under Ubiquitous Computing Environment"

H. Karl, A. Willig, A. Wolisz (Eds.): EWSN 2004, LNCS 2920, pp. 250–262, 2004.

cerned about reducing control overhead and providing a stable route from source to destination and adopts on-demand routing schemes, which create routes only when desired by the source node using a route discovery process [2,4,10,14]. Dynamic source routing (DSR) [2,4] is one of typical on-demand routing protocols. DSR is based on source routing and uses the shortest path as the routing metric. Ad hoc on-demand distance vector (AODV) [10] routing protocol improves the performance characteristics of DSDV in the creation and maintenance of routing information. It minimizes the number of broadcasts by creating routes by on-demand basis and does neither maintain a complete list of routing information nor participate in periodic routing table exchanges as DSDV does. Associativity-based routing (ABR) [14] is another on-demand routing protocol that uses associativity state as a new routing metric. Associativity state implies periods of link stability and it is measured by recording the number of control beacons received by a node from its neighbors. In a dynamic ad hoc sensor network, route re-discoveries due to route failures may incur heavy control traffic through the network and cause the increase of packet transmission delay, critical QoS factor for data integration in sensor networks. Hence, it is quite required to reduce the number of route re-discoveries by maintaining multiple redundant paths, establishing alternate route promptly and localizing the effect of the failures.

This paper presents a new on-demand redundant-path sensor network routing protocol, Time-Synchronized Neighbor Nodes Based Redundant Robust Routing (TSN^2R^3), that provides dynamic and fast route reconfiguration using information about redundant paths maintained at a source and intermediate nodes on initial route. We introduce a new routing metric, 'path redundancy', which implies how many possible redundant paths on a route to be built up. TSN^2R^3 can establish a route consisting of intermediate nodes with relatively more time-synchronized neighboring nodes and hence it provides more redundant paths toward the destination. Intermediate nodes on the established route record the information of redundant links to reach downstream intermediate nodes. This protocol provides robust routing and enables efficient power usage, efficient memory and cpu time usage for its management in wireless mobile sensor networks.

The remainder of this paper is organized as follows. The following section describes detailed algorithms and procedures of TSN^2R^3. Section 3 presents the results of the analysis of TSN^2R^3. Finally, we present our conclusions in Section 4.

2 Time-Synchronized Neighbor Nodes Based Redundant Robust Routing: TSN^2R^3

2.1 Time Synchronization and Path Redundancy

TSN^2R^3 is novel in the use of 'time synchronized node redundancy' as a route selection criterion. A route's path redundancy is expressed by the sum of 'redundancy degrees' of intermediate nodes involved in the route. Each node's redundancy degree signifies the number of time synchronized redundant nodes that a node has except one incoming node and outgoing node involved in routing. The concepts of TSN^2R^3 can be expressed as follows:

$$T^e_{data,i} = Data\_Volume(bytes) \times (1+\alpha) \times Transmission\_Time\_Per\_Byte \quad (1)$$

$$T^e_{power,j} = \int_0^{\infty} P\left(T_{power,j} > t \mid T = T_c\right) dt \quad (2)$$

$$T^e_{link,i,j} = \int_0^{\infty} P\left(T_{link,i,j} > t \mid T = T_c\right) dt = \int_0^{\infty} P\left(\min\left\{ T \,\middle|\, \left| \overrightarrow{d_{ij}}(T_c) + \int_{T_c}^{T_c+T} \overrightarrow{v_{ij}}(t)\, dt \right| \geq r_i \right\} > t \right) dt \quad (3)$$

where, $\overrightarrow{d_{ij}}(T_c)$ is node(j)'s location vector with regard to node(i) at current time T_c.

$\overrightarrow{v_{ij}}(t)$ is node(j)'s relative mobility vector with regard to node(i) at time t.

r_i is transmission range of node(i)

$$T^e_{data,i} \leq \min\left\{ T^e_{link,i,j}, T^e_{power,j} \right\} \quad (4)$$

$$\sum_{i \in path(p)} \sum_{j \in nbd(i)} \beta(v_i, v_j) \cdot I_{sync,i,j} \;,\quad I_{sync,i,j} = \begin{cases} if\ j\ is\ synchronized\ to\ i, 1 \\ otherwise, 0 \end{cases} \quad (5)$$

$$\underset{path(p) \in paths(src,dst)}{\arg\max} \left\{ \sum_{i \in path(p)} \min\left\{ \sum_{j \in nbd(i)} \beta(v_i, v_j) \cdot I_{sync,i,j}, UpperLimit \right\} \right\} \quad (6)$$

Formulation(4) expresses the concept of "time synchronization". When a node(j) satisfies (4) with regard to node(i), node(j) is considered to be time synchronized to node(i). $T^e_{data,i}$ (1) is node(i)'s expected whole data transmission time, where α is a data loss coefficient. $T^e_{power,j}$ (2) is an expected power-alive time at node(j). It is determined according to node(j)'s current routing participation status. $T^e_{link,i,j}$ (3) is an expected link-alive time between neighboring two nodes, node(i) and node(j). The $T^e_{link,i,j}$ is determined by a current position and its relative mobility vector. Formulation(5) denotes the path(p)'s redundancy degree. Where, function $I_{sync,i,j}$ is a synchronization indicator function. If node(i) and node(j) is synchronized, $I_{sync,i,j}$ is one, otherwise 0. $\beta(v_i, v_j)$ is a kind of weigh function on $I_{sync,i,j}$. It is a function of each node's mobility vector, v_i and v_j. It's value ranges 0 to 1. Formulation(6) shows how to choose an optimal route from source node to destination node. To avoid the heavy storage and time computation complexity and prohibit the case of a path's redundancy degree is severely influenced by a specific node having especially large node redundancy degree, we use the upper limit on each node's redundancy degree. If several possible routes are found during initial route discovery process, a path that has the highest path redundancy is selected as an optimal route from source to destination. However, the hop distance of the route selected may be longer than in routing algorithms, such as DSR, which selects the shortest path from source to destination. To prevent a route with a relatively excessive hop distance being selected, TSN^2R^3 can choose the route that has the largest path redundancy, but is not a certain size longer than the hop distance of the shortest candidate route.

2.2 Route Establishment

Route establishment with TSN^2R^3 follows a route setup/ route reply cycle like typical on-demand ad hoc routing protocols. In this procedure, main route and redundant routes are established. TSN^2R^3's operations are based on the assumptions that wireless links between neighboring nodes are symmetric and that each node is aware of the number of nodes in its neighborhood with the help of a data link layer protocol.

2.2.1 Route Setup Process

A node initiates route establishment procedure by broadcasting a route setup message, Route Setup (RS) packet. An RS packet is flooded throughout the network as shown in Fig.1 and carries the information about redundancy degree and hop distance of nodes that it goes through. Any node that receives an RS packet does the following:
Case 1: If the node recognizes its own address as the intermediate node address, the node records the address of the neighbor node from which it received the RS packet as the upstream node. The recorded node address will be used to build a route during the route reply process. Then, it adds its own redundancy degree to that of the RS packet and broadcasts the updated packet to its neighbor nodes.

Case 2: If the node has already received the RS packet with the same identification, it records the address of the node from which it received the packet as a redundant upstream node and then drops that packet. The recorded node address will be used to build a redundant path if this node is involved in the selected route.

Case 3: If the node recognizes its own address as the destination address, it records the forwarding node address, hop count and path redundancy of the packet.

For the purpose of optimal route selection, the destination will wait for a certain number of RS packets to reach it after receiving the first RS packet. The destination node can receive several RS packets transmitted along different paths from the source node. An RS packet delivered along the shortest route will early reach the destination node and RS packets representing routes with more redundant links may come to later. The destination node adopts the RS packet that reached it later, but contained larger path redundancy per hop, and sends a Route Reply (RR) packet back to the source node via the node from which it received the RS packet.

Fig.1 illustrates how an RS packet is flooded in the entire network. An italic number under indicates each node's redundancy degree. Here we assume that $\beta(v_i, v_j)$ is 1 with regardless to v_i and v_j. The node that has the largest redundancy degree is N6. Its redundancy degree is four. Namely, N6 has four time synchronized redundant nodes. Each intermediate node does not propagate duplicate RS packets. Therefore, just two RS packets will reach the destination node, N11. One RS packet will be delivered along the path, N1-N2-N4-N8-N11, which is the shortest one in the example network. Its hop distance is four and its redundancy degree one. The other RS packet is delivered along the path, N1-N3-N6-N12-N13-N11. This path has one more hop distance and five larger route redundancy degree than the previous one. TSN^2R^3 chooses a second path as a route.

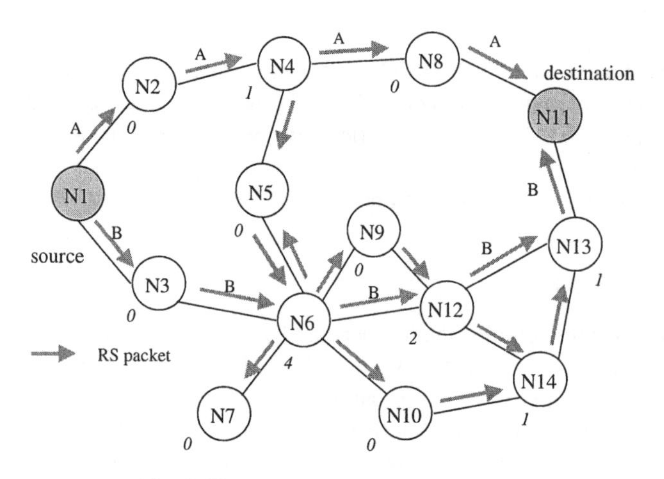

Fig. 1. The Route Setup Packet Flooding

2.2.2 Route Reply Process

A route containing redundant paths toward the destination is established during the route reply process. The destination node initiates the route reply process by sending an RR packet back to the source node via the node from which it received the corresponding RS packet. An RR packet is forwarded back along the transit nodes the RS packet was traversed. An RR packet carries the hop distance from the destination to the node that received the RR packet. The hop distance is incremented by one whenever the RR packet is forwarded at each intermediate node. Any node that receives an RR packet does the following.

Case 1: If the node recognizes itself as the target node of the received RR packet, it records the forwarding node address of the packet as the next hop for the destination. Then, the node increments the hop distance of the received packet and sends the updated packet to its upstream node, which was recorded during the route setup process.

Moreover, if the node has any redundant upstream node recorded, it generates and sends the Redundant Route Reply (RRR) packet to the redundant neighbor node(s). The hop distance of the RR packet is copied into the hop distance field of the RRR packet.

Case 2: If the node recognizes its own address as the source, it records the forwarding node address of the RR packet as the next hop for the destination in the route table.

Case 3: If the node is not targeted, the node discards the RR packet.

The RRR packet is used to setup a redundant path of a route. An RRR packet is originated from only nodes along a main route if they have redundant nodes in the upstream direction. RRR packets are forwarded at redundant nodes toward the source node. Any node that receives an RRR packet does the following:

Case 1: If the node exists along the main route, it creates the redundant route table (RRT) entry, which is a set of redundant neighbor nodes for the destination. A redundant next hop field of the RRT entry is filled with the forwarding node address of the RRR packet.

Case 2: If the node is along a redundant path and has not received the RRR packet, it records the forwarding node address of the RRR packet as a redundant next hop for the destination in the RRT entry. Then, it forwards the packet to the upstream nodes.

Case 3: If the node is along a redundant path and has already received the RRR packet with the same identification, it discards the packet. This means that a redundant path cannot have any redundant path for itself.

Case 4: If the node is not targeted, it discards the RRR packet.

Fig.2 illustrates the route reply process including redundant path setup. When the destination node N11 receives two RS packets from possible routes A and B as shown in Fig.1, it will select a more redundant, but longer route B. This route has redundant paths toward N11 at N6 and N12. N6 has two redundant links N6-N9 and N6-N10. To establish a route, N11 sends an RR packet back to the forwarding node N13 of the chosen RS packet. The RR packet is unicast to the source. Nodes receiving the RR packet increment the hop distance by one and then create or update route information for the destination. Once the RR packet reaches the source, the source begins the transmission of data packets. In this example, the data packets are transmitted along the established route N1-N3-N6-N12-N13. As for the redundant route setup, a node receiving an RR packet generates an RRR packet if it has any redundant upstream node. In Fig.2 (a), since N13 is along a main route and has a redundant upstream node N14, it generates and sends an RRR packet to N14. If N14 receives the RRR packet, it records redundant path information. The RRR packet is disseminated to other upstream nodes along a main route. In this example network, the RRR packet originating from N13 is delivered to N12 and N6. The two main nodes create the redundant route entry and maintain the redundant path information for N11. N12 has one redundant path N12-N14-N13, and N6 three redundant paths N6-N9-N12 (four hops), N6-N10-N14-N12 (five hops), N6-N10-N14-N13 (four hops). Fig.2 (b) shows the route established. In the event of a link failure between N6 and N12, N6 can forward in-transit data packets via one of three redundant paths.

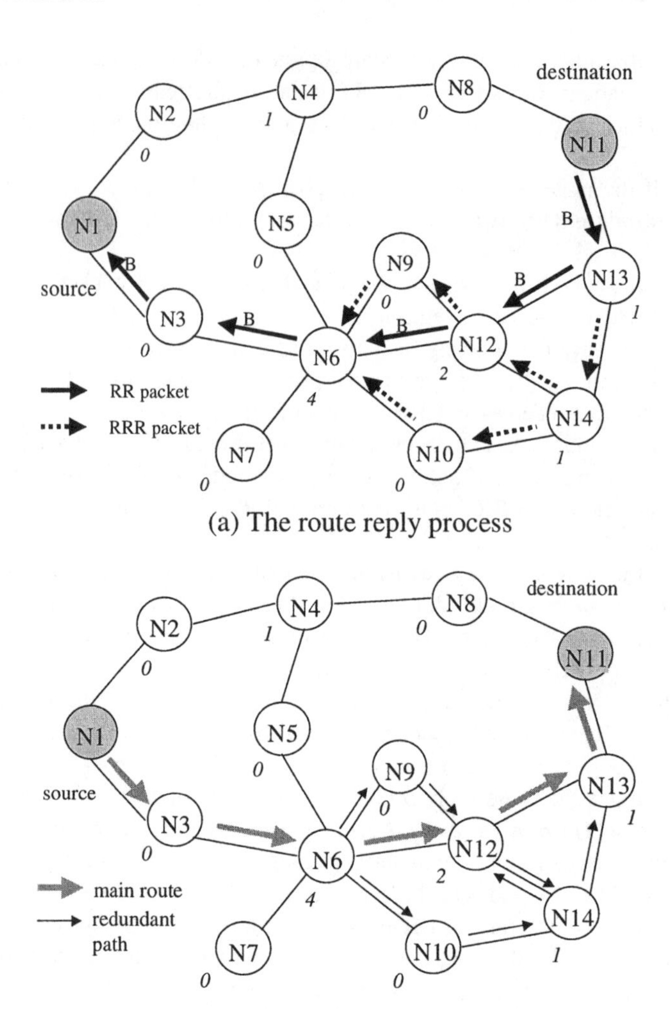

Fig. 2. Example of Route Establishment

2.3 Route Reconfiguration

2.3.1 Failure Notification

When a node detected a link failure, but did not have any redundant path for the destination. Route failure information is carried using a Failure Notification (FN) packet and stored in the failure record. Route failure information includes the information about a failure-detecting node, whether the failure-detecting node is along a main route or not, and intermediate transit nodes that an FN packet is propagated through. Any node that receives an FN packet does the following.

Case 1: If the node is along a main route (shortly, main node) and the FN packet originated from a main node, it records the failure information and finds an alternate redundant path.

Case 2: If the node is a main node and the FN packet originated from a node along a redundant path (shortly, redundant node), it removes the corresponding redundant path information from the RRT entry.

Case 3: If the node is a main node and the FN packet originated from a node along an active redundant path (shortly, active redundant node), it records the failure information and finds an alternate redundant path.

Case 4: If the node cannot find a redundant path, it adds its node address to the FN transit node list of the FN packet and propagates the updated packet. Moreover, it deletes all the information about the failed route.

Case 5: If the node is an active redundant node, it deletes the corresponding Routing Table (RT) entry and RRT entry and adds its node address to the FN transit node list of the FN packet and propagates the updated packet.

Case 6: If the node is an inactive redundant node, it deletes the corresponding RRT entry and broadcasts the packet.

2.3.2 Finding an Alternate Path

When a node is along an active main route and recognizes that it cannot forward any more data packets through its outgoing link in use, it begins the procedure to find an alternate path. In the first place, the failure-recognizing node checks if it maintains any redundant path related to the failed route. If the node doesn't have any redundant path, it tries to finds an alternative redundant path. If a node fails to find a redundant path and is the source node, it initiates a route re-discovery procedure.

3 Performance Evaluation

We have conducted a performance evaluation of TSN^{2}R^3 through simulations using CMU's wireless extensions [12] for the Network Simulator 2 (ns-2) [13]. The nodes use the IEEE 802.11 radio and MAC model provided by the CMU extensions. Radio propagation range for each node was 250 meters. Constant bit rate sources were used to generate traffic data, the size of which payload is 512 bytes. We have evaluated our routing protocol using four metrics - end-to-end delay, packet delivery ratio, control traffic overhead, and hop distance.

3.1 Comparing with Existing Single-Path Routing Mechanisms

This subsection analyzes the performance of TSN^{2}R^3, AODV and DSDV as a function of mobility speed. This simulation modeled a network of 50 mobile hosts placed randomly within a 750 meter x 750 meter area. Each simulation executed for 1000 seconds of simulation time. Fifteen sessions with randomly selected sources and destinations were simulated. Each source transmits data packets at a maximum rate of 4 packet/sec. Mobility speed varies from 3 m/s to 20 m/s across the range of simulations. Each node moves at a pause time of 10 seconds.

Fig.3 illustrates the end-to-end delay of data packets. TSN^2R^3 has shorter delay than other two protocols, and this difference becomes more obvious as the mobility speed increases. It is believed that the shorter delay is obtained from the localization of the effect of a route failure. TSN^2R^3 enables nodes constituting a route to keep redundant path information. An upstream node that detects or recognizes a link failure can promptly resume to deliver data packets via an alternate path without propagating the failure notification if it maintains any redundant path. On the other hand, AODV yields longer delay because it takes some time in reconstructing a route. When a link failure is detected, the break is propagated toward the source node. Upon receiving the failure notification, the source node restarts another route discovery process. Data packets are queued at the source node until a new route is discovered and the queuing time results in larger end-to-end delay. In DSDV, each routing table at each node lists all available destinations and the number of hops to each. That is, when a node desires to send a data packet, the route is already available. Hence there is no delay in discovering a route unlike on-demand routing schemes. This is effective at low mobility speed and in a network with light traffic as shown in Fig.3. To maintain the consistency of routing tables, each mobile node periodically transmits updates, and transmits updates immediately after a node detects a link failure. The amount of table information transmitted becomes extravagant as mobility speed increases. The end-to-end delay also increases.

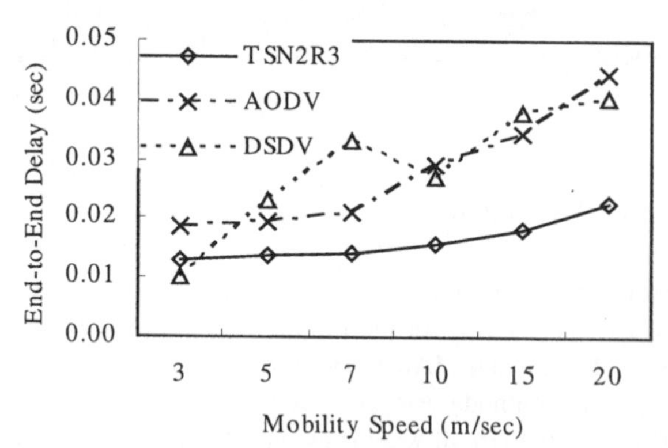

Fig. 3. End-to-End Delay

Fig.4 shows the packet delivery ratio of three routing schemes. Packets can be dropped on the way to destinations due to stale route, congestion and collisions. DSDV has very poor delivery ratio. It is believed that the main reason of DSDV's poor performance is collisions attributed to huge control traffic. Unlike DSDV, TSN^2R^3 and AODV perform well though the delivery ratio decreases slightly as the mobility speed increases. In TSN^2R^3, when a link failure happens, in-transit data packets are queued at the failure-detecting node. If the node finds a redundant route, the queued data packets are de-queued and delivered toward the destination. However, if the node does not any, the queued data packets will be dropped.

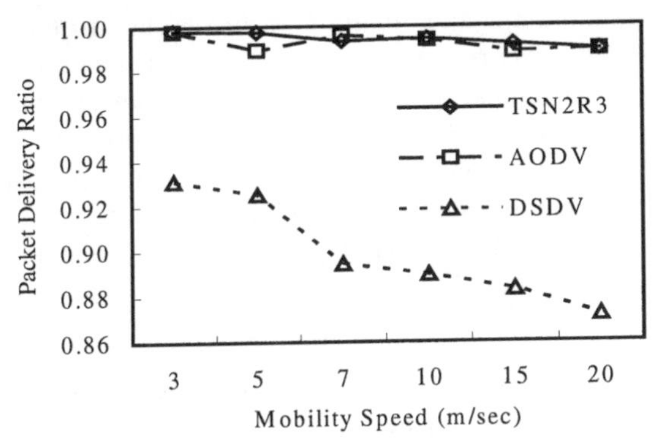

Fig. 4. Packet Delivery Ratio

Fig.5 reports the control traffic overhead of each routing scheme. As mentioned above, the result from DSDV is not out of line with our expectation. DSDV has very large control overhead even at low mobility speed 3 m/sec. TSN^{2}R^3 shows better performance than AODV.

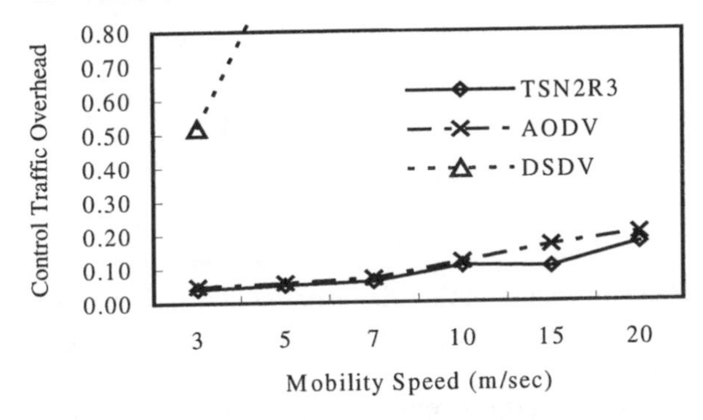

Fig. 5. Control Traffic Overhead

3.2 Comparing with Existing Multi-path Routing Mechanisms

We have compared TSN^{2}R^3 with the Split Multipath Routing (SMR) [11] and the Temporally Ordered Routing Algorithm (TORA) [11]. This simulation modeled a network of 50 mobile hosts placed randomly within a 1000 meter x 1000 meter area. Each source transmits data packets at a maximum rate of 2 packet/sec. Each node moves at a pause time of 10 seconds.

Fig.6 shows the end-to-end delay of data packets. The simulation result says that SMR has longer delay than TORA and TSN^{2}R^3. Even at low mobility speeds, the difference is conspicuously big. This is ascribed to SMR's route query message transmission policy. SMR allows duplicate transmissions of ROUTE REQUEST (RREQ) to find multiple paths maximally disjointed between source and destination. SMR produces a significant amount of RREQs flooded throughout the network and the processing of RREQs can retard the transmission of data packets. This phenomenon becomes severer as mobile nodes go beyond a certain mobility speed and the number of source-initiated route discoveries increases rapidly. The large control traffic overhead of SMR countervails the advantage of disjointed multi-path routing. On the one hand, TSN^2R^3 has shorter delay than TORA. It is thought because it takes some time for TORA to build a destination-oriented DAG in the query/reply process and route reconstruction phase.

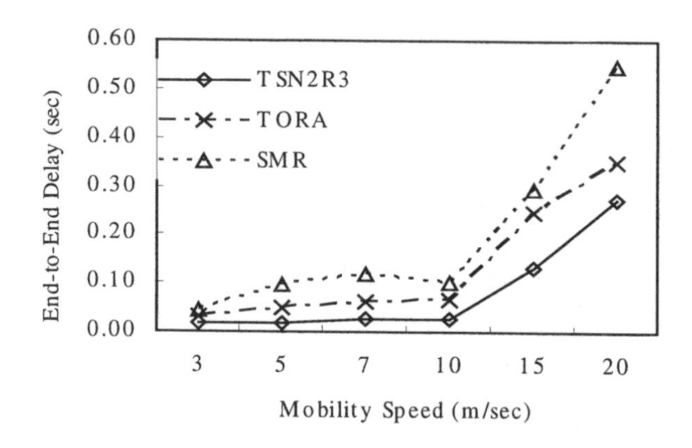

Fig. 6. End-to-End Delay

Fig.7 reveals the packet delivery ratio of each routing protocol. While TSN^2R^3 and SMR illustrate relatively robust packet delivery, TORA has shown very poor performance. It is because TORA brings about packet drops due to looping occasionally after route reconstruction.

Fig.8 presents the control traffic overhead. Control traffic overhead exhibits how many control traffic is demanded to successfully deliver data traffic from source to destination. It is said that the protocol with smaller control traffic overhead represents better protocol efficiency. According to our simulation results, SMR has larger overhead. In on-demand routing protocols using flooding method for route discovery, the proportion of route setup control traffic out of total route control traffic is huge. Hence the permission of duplicate transmissions of route query messages in SMR brings forth the result of Fig.8. TSN^2R^3 yields fewer overheads than TORA. Since TORA updates destination-oriented DAG at every link failure using UPD packets, it has more overhead than TSN^2R^3 that localizes the route reconstruction.

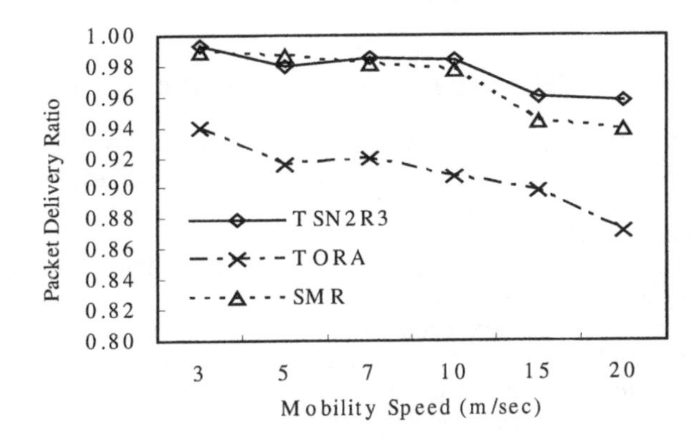

Fig. 7. Packet Delivery Ratio

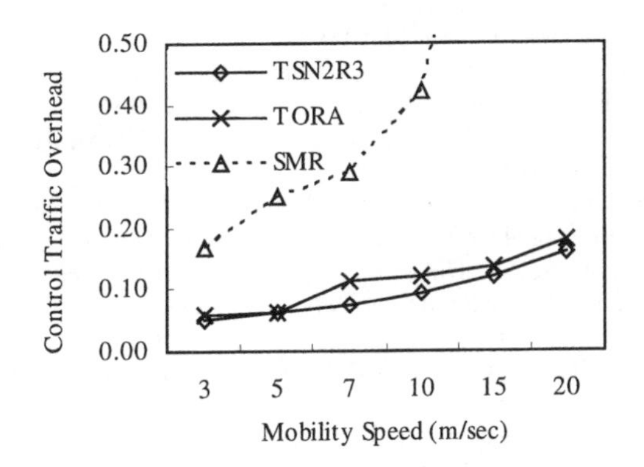

Fig. 8. Control Traffic Overhead

4 Conclusions

This paper presents a new on-demand sensor routing protocol that can accomplish dynamic and fast route reconfiguration using information about redundant paths maintained at a source node and intermediate nodes on the main route. Time-Synchronized Neighbor Node Based Redundant Robust Routing (TSN^{2}R^3) can establish a route with relatively more time synchronized redundant paths using a new routing metric-path redundancy. Our approach can localize the effects of route failures and cut down route reconfiguration time by raising the reachability to destination nodes without route re-discoveries in the event of the route failures. We conducted a performance evaluation of TSN^{2}R^3 through a simulation using Network Simulator 2. We measured TSN^{2}R^3's performance comparing with DSDV, AODV, SMR and TORA. Our results showed the effects of redundancy based routing. The simulation results say that the use of redundant paths in TSN^{2}R^3 can reduce control traffic overhead and enhance packet delivery ratio and end-to-end delay in mobile sensor networks.

References

[1] D. Bertsekas and R. Gallager, Data Networks, Prentice-Hall, New Jersey, pp. 297–333, 1987.
[2] J. Broch, D.B. Johnson, and D.A. Maltz, "The Dynamic Source Routing Protocol for Mobile Ad Hoc Networks," Internet Draft, draft-ietf-manet-dsr-00.txt, March 1998.
[3] Introduction To Wireless and Mobile Systems, D. P. Agrawal, Qing-An Zeng, Brooks/Cole, a division of Thomson Learning, 2003.
[4] D. Jhonson and D. A. Maltz, Dynamic source routing in ad hoc wireless networks, in Mobile Computing, Kluwere Academic Publishers, 1996.
[5] S. Lee and M Gerla, "A Simulation Study of Table-Driven and On-Demand Routing Protocols for Mobile Ad Hoc Networks," IEEE Network, pp. 48–54, July/Aug. 1999.
[6] B. Leiner, D. Nielson and F. Tobagi, "Issues in Packet Radio Network Design," Proceedings of the IEEE, vol. 75, no. 1, Jan. 1987.
[7] G. Malkin, "RIP Version 2 – Carrying Additional Information," Internet Draft, draft-ietf-ripv2-protocol-v2-05.txt, (work in progress), June 1998.
[8] J. Moy, "Link-State Routing," Routing in Communications Networks, edited by M.E. Steenstrup, Prentice Hall, pp. 135–157, 1995.
[9] C. Perkins and P. Bhagwat, "Highly dynamic destination-sequenced vector routing (DSDV) for mobile computers," ACM SIGCOMM, Oct. 1994.
[10] C. Perkins and E. Royer, "Ad Hoc On-Demand Distance Vector Routing," Proceeding of IEEE WMCSA, Feb. 1999.
[11] E.M. Royer and C.-K. Toh, "A Review of Current Routing Protocols for Ad-Hoc Mobile Networks," IEEE Personal Communications, vol. 6, no. 2, pp. 46–55, April 1999.
[12] The CMU Monarch Project, The CMU Monarch Project's Wireless and Mobility Extensions to ns, CS Dept., Carnegie Mellon University, Aug. 1999, http://www.monarch.cs.emu.edu/.
[13] The VINT Project, The UCB/LBNL/VINT Network Simulator-ns (version2), http://www.isi.edu/nsnam/ns.
[14] C-K Toh, "Associativity-based routing for ad hoc mobile networks," Wireless Personal Communications, vol. 4, no. 2, pp. 1–36Mar. 1997.

Design of Surveillance Sensor Grids with a Lifetime Constraint

Vivek Mhatre[1], Catherine Rosenberg[1], Daniel Kofman[2], Ravi Mazumdar[1], and Ness Shroff[1]

[1] School of Electrical and Computer Engineering, Purdue University
West Lafayette, IN-47906, USA
{mhatre,cath,mazum,shroff}@ecn.purdue.edu
[2] ENST, Paris France
{daniel.kofman}@enst.fr

Abstract. A surveillance area is to be monitored using a grid network of heterogeneous sensor nodes. There are two types of nodes; type 0 nodes which perform sensing and relaying of data within a cluster, and type 1 nodes which act as cluster heads or fusion points. A surveillance aircraft visits the area periodically, and gathers information about the activity in the area. During each data gathering cycle, the sensor nodes use multi-hopping to communicate with their respective cluster heads, while the cluster heads perform data fusion, and transmit the aggregated data directly to the aircraft. We formulate and solve a cost based optimization problem to determine the optimum number of sensor nodes (n_0), cluster head nodes (n_1) and the battery energy in each type of nodes (E_0 and E_1 respectively) to ensure at least T data gathering cycles. We observe that the number of cluster heads required, n_1, scales approximately as ${n_0}^{1-\frac{k}{4}}$ where k is the propagation loss exponent.

1 Introduction

The need for remote sensing of phenomena of interest has recently led to a surge of interest in the field of wireless sensor networks. Such networks are made up of inexpensive sensor nodes which use wireless channel to communicate with each other as well as with the base station. Sensor networks have useful applications in both military as well as civilian domains [1]. Civilian applications of sensor networks consist of smart homes, temperature control in buildings, seismic measurements, habitat monitoring etc. Military applications of sensor networks are battlefield assessment, surveillance of sensitive areas, intrusion detection etc. The application that we study in this paper is the surveillance of a sensitive area. In such applications sensor nodes are placed along the points of a grid over an area which is to be monitored. Each node has a sensing as well as communication radius of r. If the two radii are different, our analysis can be easily modified to take that into account. The nodes are organized as clusters with a single cluster head for each cluster. We call the ordinary sensor nodes to be of type 0 while the cluster head nodes to be of type 1. The number of type 0 nodes used is n_0,

H. Karl, A. Willig, A. Wolisz (Eds.): EWSN 2004, LNCS 2920, pp. 263–275, 2004.

and each has a battery energy of E_0, while the number of type 1 nodes used is E_1, and each has a battery energy of E_1. A surveillance aircraft (possibly unmanned) visits the area periodically to gather updates from the area. When the aircraft arrives, the sensor nodes send their information to their respective cluster heads using multi-hop communication. The cluster heads aggregate the received data, and send the aggregated data to the aircraft using a single hop transmission. Each such data gathering phase drains energy from the batteries of sensor nodes. Nodes are assumed to have a finite battery lifetime. For this setting, we would like to guarantee a certain minimum number of data gathering trips of the aircraft which we call the lifetime of the network. A sensor system becomes unusable when connectivity of nodes or sensing coverage of the area can no longer be ensured. We would like to minimize the cost of the entire network while ensuring node connectivity and sensing coverage of the region. We find that the required number of type 1 nodes n_1 scales approximately as ${n_0}^{1-\frac{k}{4}}$ where k is the propagation loss exponent.

This paper is organized as follows. In Section 2 we discuss some related work on similar problems. In Section 3 we formulate and solve our design problem. We present some numerical results of a case study in section 4. Finally we conclude in Section 5.

2 Related Work

In [5] the authors study a scenario in which nodes are distributed randomly over a unit area. Nodes communicate with the cluster heads using a single hop transmission. The cluster heads aggregate the collected data and and send it to the base station using a single hop transmission. Cluster heads are rotated periodically for the purpose of load balancing. For this scenario, the authors minimize the total energy spent in the network to obtain an expression for the required cluster head density. However we note that for cluster head rotation to be possible, all the nodes should be identical, and should be capable of performing data fusion as well as transmissions over long distances. This leads to additional hardware complexity in all the nodes.

In [4], the authors study a random deployment of nodes where nodes use multi-hop communication to send their data to the cluster heads. The cluster heads aggregate the received data, and use multi-hop communication to send the aggregated data to the base station. For this setting, the authors minimize the total average energy spent in the network to determine the optimum cluster head intensity. However the authors do not provide any bounds on the lifetime of the network. In [2], the authors provide upper bounds on the lifetime of a sensor network. In [3], the authors present an optimum policy to assign nodes to the cluster heads in order to improve the network lifetime.

Our work is different from all the above studies because we study a scheme which uses heterogeneous sensor nodes (two types of nodes), and take into account the costs of the two types of nodes in the overall system design problem.

3 Problem Formulation

We consider a surveillance sensor grid in which the network topology is very simple. We assume that it is possible to place the sensor nodes deterministically along grid points. While this may not always be possible, there are scenarios in military applications when we have complete control over node placement. We are interested in covering a region of unit area with sensor nodes. The nodes have a communication as well as sensing radius r, and are placed distance r apart along the grid points. This ensures coverage of the area as well as node connectivity. We assume that $r << 1$, since in most practical applications we are interested in covering a given area with a large number of nodes; each of which has a sensing radius that is much smaller than the dimensions of the area. There are two types of nodes; type 0 nodes and type 1 nodes. Type 0 nodes are the sensor nodes, while type 1 nodes act as cluster heads besides acting as sensor nodes (see Figure 1). Let n_0 denote the number of sensor nodes and n_1 denote the number of cluster heads. A surveillance aircraft visits the area periodically

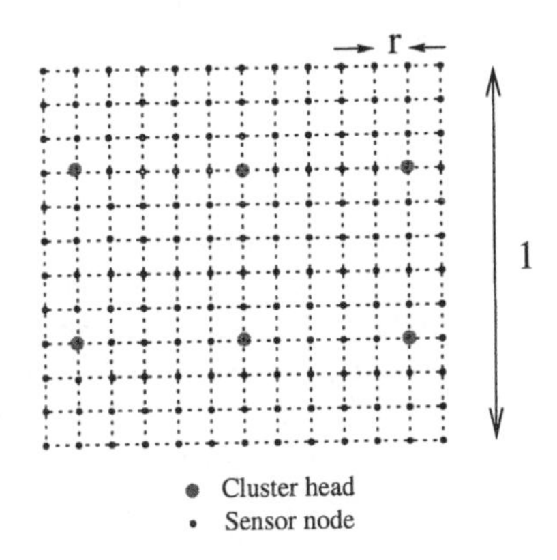

Fig. 1. A surveillance sensor grid.

and gathers information about the activity in the area. During each visit of the aircraft, each sensor node senses and sends a single data packet to its cluster head using multi-hop communication. The cluster head aggregates the received data packets into a single packet and transmits it to the aircraft. During each such data gathering phase, nodes spend a part of their battery energy for relaying of packets. Since the nodes have a finite battery energy there is an upper bound on the number of such data gathering cycles that can be sustained. Our aim is to guarantee a network lifetime of T, i.e., at least T data gathering cycles while ensuring connectivity as well as coverage of the area.

We model the cost of each node as the sum of its hardware cost and its battery cost. Hence the cost of a type i node C_i is:

$$C_i = \alpha_i + \beta E_i$$

where α_i is the hardware cost of the node, E_i is the battery energy of the node and β is the proportionality constant for the battery cost. The reason we have different cost parameters α_0, α_1 and different battery energies E_0, E_1 is that the sensor nodes have a simpler functionality as compared to the cluster head nodes. The sensor nodes perform short range transmissions and do not have to perform any data aggregation. As against this, the cluster head nodes have to perform computations for data aggregation, and also perform long range transmissions to the aircraft. The cluster heads are also responsible for managing MAC and routing in the cluster. Hence the hardware cost, as well as the battery energy of the cluster head nodes are much higher than the sensor nodes.

There are n_0 type 0 nodes each with battery energy E_0 and n_1 type 1 nodes each with battery energy E_1. If we denote $[n_0, n_1, E_0, E_1]$ by $\bar{x}$, then the total cost of the network $f(\bar{x})$ is:

$$\begin{aligned} f(\bar{x}) &= n_0(\alpha_0 + \beta E_0) + n_1(\alpha_1 + \beta E_1) \\ &= n_0\alpha_0 + n_1\alpha_1 + \beta(n_0 E_0 + n_1 E_1) \end{aligned} \tag{1}$$

Since nodes have a coverage/communication radius of r and we are interested in covering a unit area with these nodes, the condition for coverage and connectivity is:

$$n_0 + n_1 \geq \frac{1}{r^2} \tag{2}$$

We observe that since the nodes use multi-hop communication to reach the cluster heads, the type 0 nodes which are one hop away from the cluster heads have the highest burden of relaying, since every packet sent to the base station has to go through them. We call these nodes the *critical nodes*. There are 4 critical nodes in each cluster (see Figure 2).[1] We ignore the fact that the load on the critical nodes near the region boundary is likely to be less due to edge effects, and focus only on the clusters in the interior of the region.

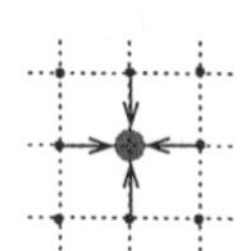

Fig. 2. Critical nodes

Since the critical nodes have the highest energy burden, these nodes are likely to drain their battery before other type 0 nodes. Besides the critical nodes, the

[1] If the sensing and communication radii are different, i.e., the communication radius (say r_c) is larger than the sensing radius (say r_s), the number of critical nodes is equal to the number of nodes that are within r_c distance of the cluster head. For simplicity, we assume $r_c = r_s$ so that there are 4 critical nodes. However the case of $r_c > r_s$ can be handled similarly. Note that $r_c < r_s$ is not feasible since no communication is possible in that case.

cluster heads also have high energy expenditure due to the long range transmissions that they have to perform to communicate with the aircraft. Thus the critical nodes and the cluster heads determine the lifetime of the network. Let P_0 and P_1 be respectively the average amounts of energy spent by a critical node and a cluster head during each cycle. Then to ensure a lifetime of T, we require:

$$\frac{E_1}{P_1} = \frac{E_0}{P_0} \geq T \tag{3}$$

where the equality of the first two terms ensures that both the critical nodes as well as the cluster heads expire at about the same time so that very little energy is wasted in the form of residual energy in the nodes when the network becomes unusable.

We assume that the nodes are asleep (with their radios turned off) in between the data gathering cycles. The nodes turn their radios on when the aircraft arrives for gathering updates. This results in a small duty cycle which in turn reduces the idle mode energy wastage. We assume that a data gathering cycle lasts for τ seconds. During these τ seconds all the nodes keep their radios in idle state. This is required because the nodes use multi-hop communication, and therefore cannot predict exactly when their neighbor nodes will send them a packet. Hence all the nodes have to keep their radios in idle state so as to be prepared to receive a packet. Since the radios are on during the entire data gathering phase, no extra energy is spent on receiving a packet. This assumption is justified by recent studies that have shown that the energy spent while receiving a packet is about the same as the energy spent in the idle mode [7]. If l be the energy spent per second in the idle mode, then the total energy spent in the radio during each cycle is simply $l\tau$. We also ignore any packet collisions, by assuming that the cluster head nodes perform scheduling to ensure perfect synchronization. Packet collisions are a potential problem for the critical nodes since they have plenty of traffic to relay, and they are close to each other. We assume that the cluster head node is in-charge of this co-ordination. We also assume a simple propagation model in which the energy spent in the RF circuitry to transmit a packet over a distance x is μx^k where k is the propagation exponent. Usually the value of k is determined through on-site measurement, and varies between 2 and 4 depending on the surrounding environment. In environments with dense vegetation, the fall off is more drastic [8]. Since the aircraft is flying at high altitude (as compared to dimensions of the antenna on the nodes), we assume that the receiver antenna is in the far-field.

Since there are n_0 sensor nodes and n_1 cluster heads, there are n_0/n_1 sensor nodes in each cluster. There are exactly 4 critical nodes in each cluster which perform the function of relaying of packets from the $n_0/n_1 - 4$ sensor nodes (see Figure 2). Thus each critical node has to relay $(n_0/n_1 - 4)/4$ packets. Besides this, the critical node also has to transmit its own packet. Let $E_0^t = \mu r^k$ be the amount of energy spent on transmission of a packet. Then the average energy spent by a critical node during each cycle is

$$\begin{aligned} P_0 &= E_0^t \left(\frac{\left(\frac{n_0}{n_1}\right) - 4}{4} \right) + E_0^t + l\tau \\ &= \frac{E_0^t}{4} \left(\frac{n_0}{n_1}\right) + l\tau \\ &= c_0 \frac{n_0}{n_1} + c_1 \end{aligned} \tag{4}$$

Each cluster head has to receive n_0/n_1 packets, aggregate them and transmit the aggregated packet to the aircraft. Let E_f be the amount of energy spent in data computations during the aggregation of each packet by the cluster head, and $E_1^t = \mu H^k$ be the amount of energy required to transmit a packet from the cluster head to the aircraft. The cluster head also has to spend $l\tau$ amount of energy in the idle mode during each cycle. Therefore the average energy spent by a cluster head during each cycle is

$$\begin{aligned} P_1 &= E_f \frac{n_0}{n_1} + E_1^t + l\tau \\ &= c_2 \frac{n_0}{n_1} + c_3 \end{aligned} \tag{5}$$

We have introduced the constants c_0, c_1, c_2 and c_3 for ease of notation.

Our objective is to minimize the cost of the network while ensuring connectivity and coverage of the area over T cycles, i.e., given r and T we would like to determine n_0, n_1, E_0 and E_1 which minimize the cost of the network. Hence we formulate the optimization problem as follows:

$$\text{minimize} \quad n_0\alpha_0 + n_1\alpha_1 + \beta(n_0E_0 + n_1E_1)$$

$$\text{subject to} \quad n_0 + n_1 \geq \frac{1}{r^2} \tag{6}$$

$$\frac{E_1}{P_1} = \frac{E_0}{P_0} \geq T \tag{7}$$

Then the first constraint can be rewritten as:

$$g_1(\bar{x}) = \frac{1}{r^2} - n_0 - n_1 \leq 0 \tag{8}$$

Substituting for P_0 and P_1 in the equality constraint of (7) from (4) and (5) we obtain

$$\begin{aligned} &E_1P_0 - E_0P_1 = 0 \\ &\Rightarrow h(\bar{x}) = E_1\left(c_0\frac{n_0}{n_1} + c_1\right) - E_0\left(c_2\frac{n_0}{n_1} + c_3\right) = 0 \end{aligned} \tag{9}$$

Similarly from the inequality constraint of (7) we obtain

$$E_1 \geq TP_1 \Rightarrow c_3 T + c_2 T \frac{n_0}{n_1} - E_1 = g_2(\bar{x}) \leq 0 \tag{10}$$

Thus from (1), (8), (9) and (10) the optimization problem can be formulated as follows

$$\begin{aligned} \text{minimize } & f(\bar{x}) \\ \text{subject to } & h(\bar{x}) = 0 \\ & g_1(\bar{x}) \leq 0 \\ & g_2(\bar{x}) \leq 0 \end{aligned}$$

This is a standard Karush-Kuhn-Tucker (KKT) optimization problem [6]. We find the solution to this problem by solving:

$$\nabla f(\bar{x}) + \lambda \nabla h(\bar{x}) + \mu_1 \nabla g_1(\bar{x}) + \mu_2 \nabla g_2(\bar{x}) = \mathbf{0} \tag{11}$$

$$\text{with } \mu_1 \geq 0, \quad \mu_2 \geq 0, \quad \mu_1 g_1(\bar{x}) + \mu_2 g_2(\bar{x}) = 0 \tag{12}$$

where λ, μ_1 and μ_2 are the constants of the KKT problem. From (1):

$$\nabla f(\bar{x}) = \begin{bmatrix} \alpha_0 + \beta E_0 \\ \alpha_1 + \beta E_1 \\ \beta n_0 \\ \beta n_1 \end{bmatrix} \tag{13}$$

From (9):

$$\nabla h(\bar{x}) = \begin{bmatrix} \frac{E_1 c_0 - E_0 c_2}{n_1} \\ -\frac{(E_1 c_0 - E_0 c_2) n_0}{{n_1}^2} \\ -c_3 - \frac{c_2 n_0}{n_1} \\ c_1 + \frac{c_0 n_0}{n_1} \end{bmatrix} \tag{14}$$

From (8):

$$\nabla g_1(\bar{x}) = \begin{bmatrix} -1 \\ -1 \\ 0 \\ 0 \end{bmatrix} \tag{15}$$

From (10):

$$\nabla g_2(\bar{x}) = \begin{bmatrix} \frac{c_2 T}{n_1} \\ -\frac{c_2 T n_0}{{n_1}^2} \\ 0 \\ -1 \end{bmatrix} \tag{16}$$

Substituting (13) to (16) in (11) and (12) we have the following four equations:

$$0 = (\alpha_0 + \beta E_0) + \lambda\left(\frac{E_1 c_0 - E_0 c_2}{n_1}\right) + \mu_1(-1) + \mu_2\left(\frac{c_2 T}{n_1}\right) \tag{17}$$

$$0 = (\alpha_1 + \beta E_1) + \lambda\left(\frac{(c_2 E_0 - c_0 E_1) n_0}{{n_1}^2}\right) + \mu_1(-1) + \mu_2\left(-\frac{c_2 T n_0}{{n_1}^2}\right) \tag{18}$$

$$0 = (\beta n_0) + \lambda\left(-c_3 - \frac{c_2 n_0}{n_1}\right) + \mu_1(0) + \mu_2(0) \tag{19}$$

$$0 = (\beta n_1) + \lambda\left(c_1 + \frac{c_0 n_0}{n_1}\right) + \mu_1(0) + \mu_2(-1) \tag{20}$$

$$0 = \mu_1\left(\frac{1}{r^2} - n_0 - n_1\right) + \mu_2\left(c_3 T + c_2 T\frac{n_0}{n_1} - E_1\right) \tag{21}$$

Assuming that a feasible solution exits, i.e., n_0, n_1, E_0, $E_1 > 0$, from (19), we get

$$\lambda = \frac{\beta n_0 n_1}{c_3 n_1 + c_2 n_0} \quad > 0 \tag{22}$$

Hence from (20), we get

$$\mu_2 = (\beta n_1) + \lambda\left(c_1 + \frac{c_0 n_0}{n_1}\right) \quad > 0 \tag{23}$$

But from the KKT theorem, this implies that the corresponding inequality (10) becomes an equality, i.e.,

$$c_2 T + c_3 T\left(\frac{n_0}{n_1}\right) - E_1 = g_2(\bar{x}) = 0 \tag{24}$$

Using the value of λ from (22) in (17),

$$\mu_1 = (\alpha_0 + \beta E_0) + \lambda\left(\frac{E_1 c_0 - E_0 c_2}{n_1}\right) + \mu_2\left(\frac{c_2 T}{n_1}\right)$$
$$= \alpha_0 + \frac{\beta(c_3 n_1 E_0 + c_0 n_1 E_1)}{c_2 n_1 + c_3 n_0} + \mu_2\left(\frac{c_2 T}{n_1}\right) \quad > 0$$

since $\mu_2 > 0$. Hence

$$\mu_1 > 0$$

Again using the KKT theorem, this implies that the inequality of (8) becomes an equality, i.e.,

$$\frac{1}{r^2} - n_0 - n_1 = g_1(\bar{x}) = 0 \tag{25}$$

So now we can re-formulate our problem as follows:

$$\begin{aligned} \text{minimize } & f(\bar{x}) \\ \text{subject to } & h(\bar{x}) = 0 \\ & g_1(\bar{x}) = 0 \\ & g_2(\bar{x}) = 0 \end{aligned}$$

Let N denote the total number of nodes of both types. Then from (8) we obtain

$$n_0 + n_1 = \frac{1}{r^2} = N \tag{26}$$

$$\Rightarrow n_0 = N - n_1 \tag{27}$$

From (10) and (9) we obtain the following expressions for E_0 and E_1

$$E_0 = T\left(c_0\frac{n_0}{n_1} + c_1\right) \tag{28}$$

$$E_1 = T\left(c_2\frac{n_0}{n_1} + c_3\right) \tag{29}$$

Eliminating n_0, E_0 and E_1 from (1) using (27), (28) and (29), and after rearranging terms we obtain

$$f(n_1) = \alpha_0 N + \beta T(c_1 + c_2 - 2Nc_0) + n_1((\alpha_1 - \alpha_0) + \beta T(c_3 + c_0 - c_1 - c_2)) + \frac{\beta T c_0 N^2}{n_1} \tag{30}$$

To minimize the cost, we differentiate (30) with respect to n_1 and obtain the following solution for n_1

$$n_1 = \sqrt{\frac{c_0 N^2}{(c_3 + c_0 - c_1 - c_2) + \left(\frac{\alpha_1 - \alpha_0}{\beta T}\right)}} \tag{31}$$

Note that the second derivative of (30) is

$$f''(n_1) = \frac{2\beta T c_0 N^2}{{n_1}^3} > 0 \quad \forall n_1 > 0$$

Hence $f(n_1)$ is convex and therefore the local minimum is also the global minimum. Re-substituting for c_0, c_1, c_2 and c_3 from (4) and (5)

$$c_0 = \frac{\mu r^k}{4}, \quad c_1 = l\tau, \quad c_2 = E_f, \quad c_3 = \mu H^k + l\tau \tag{32}$$

Since for typical practical scenarios the aircraft altitude is much larger than the communication radius of each node, i.e., $H >> r$, and usually the energy spent on communication (μH^k) is much more as compared to the energy spent on computation (E_f) for typical sensor nodes (see [5,7]), the μH^k term dominates in (31).

$$n_1 = \sqrt{\frac{\frac{\mu r^k}{4} N^2}{\mu H^k + \left(\frac{\alpha_1 - \alpha_0}{\beta T}\right)}} = \frac{N^{1-\frac{k}{4}}}{\sqrt{4H^k\left(1 + \left(\frac{\alpha_1 - \alpha_0}{\beta T \mu H^k}\right)\right)}} \tag{33}$$

We used $Nr^2 = 1$ from (26) to eliminate r. Thus we see that for a fixed H, the required number of cluster heads n_1 scales approximately as $N^{1-\frac{k}{4}}$.

In reality the sensor nodes used for surveillance have small sensing radii (10m to 100m), while the region to be monitored is usually large (a few square kilometers). Under such circumstances it is inevitable that a large number of nodes be used. To study such a scenario we consider the asymptotic behavior of n_1 as the number of nodes N becomes large (or equivalently, as r approaches zero). We note that for typical values of k (between 2 and 4), the exponent of N in (33) is always less than one. Hence n_1 scales slower than N. As a result $n_0 = N - n_1$ scales approximately as N. This is something we would expect, since the majority of the N nodes are used for sensing, while a few of them are used as cluster heads. Thus for large N, i.e., small r,

$$n_0 = N - n_1 \approx N = \frac{1}{r^2}$$

Thus we find that $n_0 \approx N$ and n_1 scales approximately as $N^{1-\frac{k}{4}}$. Hence n_1 scales approximately as ${n_0}^{1-\frac{k}{4}}$. Thus the propagation loss exponent k determines the asymptotic behavior of the required number of cluster heads.

We can also obtain expressions for the battery energies of both types of nodes using (28) and (29) and re-substituting for c_0, c_1, c_2 and c_3 from (32).

$$E_0 = T\left(\frac{\mu\sqrt{H^k\left(1 + \left(\frac{\alpha_1 - \alpha_0}{\beta T \mu}\right)\right)}}{2N^{\frac{k}{4}}} + l\tau\right) \tag{34}$$

$$E_1 = T\left(E_f N^{\frac{k}{4}}\sqrt{H^k\left(1 + \left(\frac{\alpha_1 - \alpha_0}{\beta T \mu H^k}\right)\right)} + \mu H^k + l\tau\right) \tag{35}$$

In (34) we note the dependence of E_0 on H. This implies that higher the value of H, fewer the number of cluster head nodes (see (33)), and hence larger the cluster size. As the size of the cluster is higher, the critical nodes have more relaying to do, and consequently they need more battery energy. Thus E_0 increases as H increases.

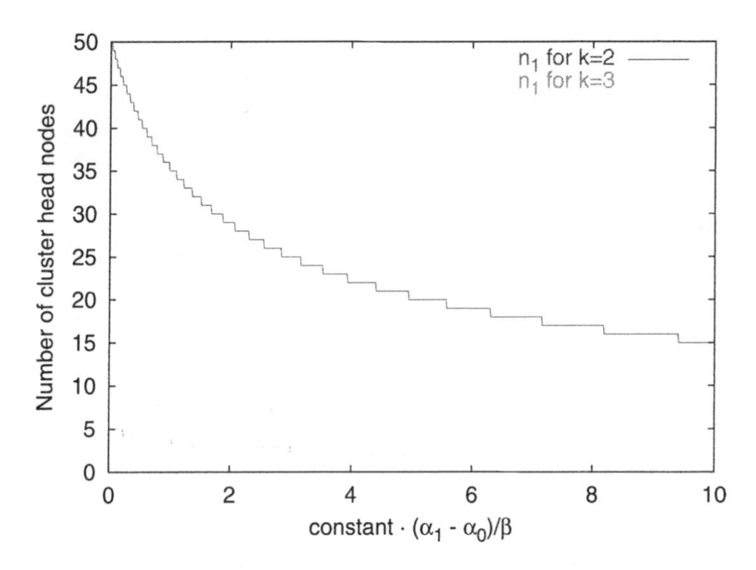

Fig. 3. Required number of cluster head as a function of hardware cost difference of two types of nodes.

4 Case Study

In our problem formulation, we have parameters α_0, α_1 and β which correspond to the hardware cost of the two types of nodes and the proportionality constant for the battery cost. In general these parameters depend on the manufacturing process of the nodes. The cluster head nodes require powerful RF amplifiers, since they have to transmit data directly to the aircraft. Typical unmanned aerial vehicles (UAV) used for surveillance fly at altitudes of a few kilometers (see [9] for more details on UAVs). On the other hand, the sensor nodes have to communicate over distances of the order of 10 to 100m (typical sensing radius). The cluster head nodes also require a more powerful CPU for data aggregation related computations and MAC and routing co-ordination. Hence we would expect $\alpha_1 >> \alpha_0$. However the exact dependence of α_0, α_1 and β on these factors are difficult to account for without knowing the manufacturing process and the components involved.

Our approach provides designers with general guidelines to choose the system parameters. For example, consider the following scenario as a case study. Assume that a region of area 10km×10km is to be covered by a sensor grid. The nodes have a sensing radius 100m. A surveillance aircraft flying at an altitude of 10km visits the region periodically. Since in our analysis we normalized the area of the region to unity, we must normalize all the distances in the case study by dividing them by 10km. Hence for this case study we have $r = 0.01$ and $H = 1$. The required number of cluster heads as a function of $(\alpha_1 - \alpha_0)/\beta T\mu$ for $k = 2$

and $k = 3$ is plotted in Figure 3. The x-axis is a measure of the hardware cost difference between the two types of nodes divided by the battery parameter.

Depending on the manufacturing process, the value of $(\alpha_1 - \alpha_0)/\beta T\mu$ can be determined. Using Figure 3 we can then determine the corresponding number of cluster head nodes required. The plots in Figure 3 have a step like shape because n_1, is rounded to the nearest integer. Note that the required number of sensor nodes is simply $1/r^2 = 10^4$. We can also obtain the required battery energy of both types of nodes using (34) and (35).

5 Conclusions and Future Work

We studied the design of surveillance sensor grids from the point of view of lifetime of the network. In our study we considered a scenario in which two types of nodes are used; type 0 nodes, i.e., the sensor nodes and type 1 nodes, i.e., the cluster heads. Nodes are placed along grid points and a surveillance aircraft visits the area periodically to collect updates about the activity in the area. Our objective was to dimension the number of cluster heads, the number of sensor nodes and their battery energies so as to provide a network lifetime of at least T cycles. During the lifetime of the network connectivity of the nodes as well as coverage of the area is ensured. We formulated and solved an optimization problem to minimize the overall network cost while attaining this objective. We found that the required number of cluster heads, n_1, scales approximately as ${n_0}^{1-\frac{k}{4}}$ where n_0 is the number of sensor nodes, and k is the propagation loss exponent.

The model of the sensor network that we consider in this work assumes a perfect grid deployment of nodes. However a more realistic scenario consists of random node deployment of (possibly) unreliable nodes. We would like to extend the results derived in this work to the random deployment scenario.

Acknowledgments. We would like to thank the annonymous reviewers for their comments and suggestions. This work was supported by a DARPA grant Contract No. MDA 972-02-1-0032.

References

1. I.F. Akyildiz, W. Su, Y. Sankarsubramaniam and E. Cayirci, "Wireless Sensor Networks: a survey," *Computer Networks*, Vol. 38, pp. 393–422, March 2002.
2. M. Bharadwaj, T. Garnett and A.P. Chandrakasan, "Upper Bounds on Lifetime of Sensor Networks," *IEEE International Conference on Communications (ICC'01)*, Helsinki Finland, June 2001.
3. C.F. Chiasserini, I. Chlamtac, P. Monti and A. Nucci, "Energy Efficient Design of Wireless Ad Hoc Networks," *Proc. European Wireless*, February 2002.
4. S. Bandyopadhyay and E. Coyle, "An Energy Efficient Hierarchical Clustering Algorithm for Wireless Sensor Networks," *proc. IEEE INFOCOM'03*, San Francisco, CA, April 2003.

5. W. Heinzelman, A. Chandrakasan and H. Balakrishnan. "An Application-Specific Protocol Architecture for Wireless Microsensor Networks," *IEEE Transactions on Wireless Communications*, Vol. 1, No. 4, October 2002.
6. E. Chong and S. Zak. *An Introduction to Optimization*, Second edition, Wiley and Sons, Inc., New York 2001.
7. G.J. Pottie and W.J. Kaiser, "Wireless Integrated Network Sensors," *Communications ACM*, 43 (5) (2000) 51–58.
8. T. S. Rappaport, Wireless Communication, Prentice-Hall, 1996.
9. NASA GSFC/Wallops Flight Facility, Unmanned Aerial Vehicles. `http://uav.wff.nasa.gov/index.html`

Design of a Secure Distributed Service Directory for Wireless Sensornetworks

Hans-Joachim Hof, Erik-Oliver Blaß, Thomas Fuhrmann, and Martina Zitterbart

Institut für Telematik
Universität Karlsruhe, Germany
{hof,blass,fuhrmann,zit}@tm.uka.de

Abstract. Sensor networks consist of a potentially huge number of very small and resource limited self-organizing devices. This paper presents the design of a general distributed service directory architecture for sensor networks which especially focuses on the security issues in sensor networks. It ensures secure construction and maintenance of the underlying storage structure, a Content Addressable Network. It also considers integrity of the distributed service directory and secures communication between service provider and inquirer using self-certifying path names. Key area of application of this architecture are gradually extendable sensor networks where sensors and actuators jointly perform various user defined tasks, e.g., in the field of an office environment.

1 Introduction

As computer miniaturisation continues and small devices become cheap, more and more of these little helpers are embedded into the user's environments to offer pervasive services. Sensor networks are special occurrences of networks build by such small devices. They are about to change the paradigm of sensors being passive devices that are connected to a more or less centralized computing engine that processes the so collected data. With the new paradigm of sensor networks, measurement and control tasks are performed inside a *sensor network* which is formed by a vast number of sensor nodes. Those nodes have low-power, are self -organising and have little computation capabilities. Such devices are called "peanut CPUs" in [9]. In the long run, sensor networks can be expected to create synergy out of the huge number and high density of devices.

In order to fully benefit from this new paradigm, sensor networks need to employ several new mechanisms that account for the special properties of their nodes: Low energy consumption, little CPU power, limited memory, regular node failures, and maintenance-free operation in heterogeneous and maybe even hostile environments.
Given these limited abilities and other characteristics, it stands to reason that traditional security methods do not work satisfyingly in the context of sensor networks. Especially asymmetric cryptography is computationally too complex for these nodes, even with elliptic curve algorithms [3]. Therefore, security architectures for sensor networks benefit of focusing on symmetric cryptography which can be used efficiently even on peanut CPUs.

H. Karl, A. Willig, A. Wolisz (Eds.): EWSN 2004, LNCS 2920, pp. 276–290, 2004.

This paper describes the construction of a secure distributed service directory which can be used as a general lookup service for sensor networks. The service directory is based on an overlay network, the Secure Content Addressable Network (S-CAN). It draws from the idea of a Content Addressable Network [7] that is extended with security features that specifically take into account the requirements of sensor networks. A typical scenario where this architecture can be advantageously deployed is an intelligent office environment with dozens or even hundreds of sensors and actuators that act jointly within a sensor network to accomplish various tasks. In order to find its respective peers for a given task, a sensor needs to discover nodes that offer the required sensor data or are able to perform specific tasks. Typically such lookups are rare as compared to the overall operation time of the network. They occur only upon addition or removal of nodes or the creation of new tasks by the user. Both can be assumed to occur only seldom.

The key idea behind the use of a CAN is to create a simple distributed system that easily scales with its growth, is considerably failsafe, and does not require any centralised component during normal operation. Hence, a user can sequentially add nodes to the network without bothering to provide enough directory capacity. Moreover, by having such a flexible lookup service, nodes can immediately benefit from other nodes already present in the network without the need for any manual configuration.

This paper is structured as follows. Chapter 2 states some goals for a distributed service directory. Chapter 3 describes Content Addressable Networks, the basis for the design of a distributed service directory. Chapter 4 presents an architecture of a sensor network platform. Chapter 5 describes the design of the proposed distributed service directory. Chapter 6 characterises a hardware prototype for further work. Chapter 7 gives a summary of this paper and an outlook on further work.

2 Goals for a Distributed Service Directory

The usual desired task of a service directory is insertion of service names and service descriptions as well as lookup of these names and descriptions. A distributed service directory spreads the load of insert and query operations on a number of nodes. A distributed solution avoids a single point of failure compared to a central implementation. Security for a distributed service directory implies:

- Secure Insertion of data
- Integrity of stored data
- Correctness of lookup answers
- Availability
- Robustness

Securing our distributed service directory starts with securing the CAN which is the basis of our directory design. See Chapter 3 for a brief description of CAN. CAN is used because it has a solid structure with non-overlapping zones each owned by exactly one node. Every zone has specific neighbours. If we could ensure that communication between neighbors in the CAN space is secure all the time, i.e. the CAN neighbors are authenticated and use encryption for communication, we can secure the structure of the CAN space. CAN uses periodic update messages to maintain neighbours states. These messages are critical for the structure of CAN and need to be secured. The best time to build trust is the joining of a device thus new devices have to be added in a secure way. In general, a secure join operation of a new device implies three steps:

1. A new device gets authenticated. Otherwise, all security is obsolete, as you can not build up trust when you do not know with whom you interact.

2. A new device joins CAN space at a random point and could not determine this point a priori. A fraudulent device *A* should not be able to absorb a specified part of the distributed hash table realized by CAN during joining by making a proper device *B* to split its zone and hand over *A* half of the zone.

3. No attacker should be able to claim to be owner of a CAN zone it does not own. If we can assure this, takeover of one or more devices has only a local effect and ensures at least functionality of those parts of the hash table not affected. It also protects joining nodes against being drawn into a fake CAN by a fraudulent node which claims to own the whole CAN space.

3 Content Addressable Network

The Secure Content Addressable Network (S-CAN) is an integral part of the architecture presented in chapter 4. S-CAN basically extends the original CAN (Content Addressable Network) [7] by security considerations. This chapter gives a short overview of CAN.

CAN utilizes a d-dimensional Cartesian coordinate space on a d-torus. We will call this virtual space CAN-space. The coordinate space is completely logical and has no relation to any physical coordinate system. CAN forms an overlay network which lies above a network layer and utilizes the communication abilities offered by it. At any time the entire CAN-space is divided into zones which are administrated by the nodes participation in the CAN. Every node "owns" a distinct zone within the overall space. The virtual coordinate space is used to store *(key,value)* pairs as follows: *key* is mapped on a point *P* in CAN-space using a hash function. The *(key,value)* pair is then stored on the node which owns the zone within which *P* lies. To retrieve an entry corresponding to a key *K*, any node can apply the same deterministic hash function on *K* to map *K* on a point *P* and then retrieve the corresponding value from the zone in which *P* lies. The request is routed through the CAN-space until it reaches the node which owns the zone including *P*. Routing is done by greedily forwarding a message to the neighbor which coordinates are closest to the destination. Each node keeps a list of neighbors that abut their own zone. This list is used for forwarding and acts as a routing table. Periodic update messages between neighbours inform about the neighbor's state and help to recognize failed devices and abandoned zones. Please note that many different paths exist between two points in the CAN-space. This means, that even if one or more of a node's neighbors crash, a node would automatically route along the next best available path. If a node loses all its neighbors, it can do an expanding ring search to get connected to the CAN again. CAN describes some repair mechanisms for restoring a consistent state. Those are of no interest for this paper and will get careful attention in further work.

As mentioned earlier CAN-space is partitioned among a number of nodes. If a new node decides to participate in CAN an existing zone is split into half and the joining device gets one of the two resulting zones. A uniform distribution of join points is eligible to maintain zones of equal size which is important because effective routing depends on a similar zone size of all zones. CAN offers some optimizations of routing and robustness. Routing can be improved using multiple CAN-spaces (called multiple realities) with different hash functions on each node. Reaching a point in CAN then translates to reaching this point in any reality. Robustness can be improved by zone overloading. This means that multiple nodes own one zone and all owners are permitted to issue answers for requests on this zone. If one owner fails, there are still other zone owners which may answer requests.

4 Architecture of Sensor Network Platform

The distributed service directory described in this paper is embedded in a general architecture for sensor networks which aims to provide a flexible, scalable and secure platform for sensor-based services. An overview of the architecture is depicted in figure 1. A detailed description of the modules of the architecture follows:

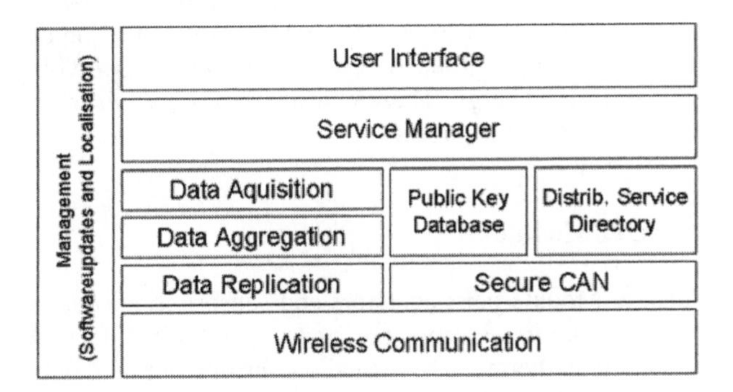

Fig. 1. Architecture of sensor network platform

4.1 Distributed Service Directory Module

The Service Directory module realises a distributed and robust service directory. There are two ways to implement a distributed service directory which are interesting for sensor networks as they take into consideration the properties of sensor networks: cluster-based or fully decentralised. We chose to implement the latter using a distributed hash-table. This table is realised by the Secure Content Addressable Network Module. This reduces the administrative overhead in contrast to a cluster-based solution. The service directory is used to store attributes about services which are available in the sensor network. Attributes include, but are not limited to, address of service provider, address of data replication point, physical location of service, validity of service, quality category, needed input to provide service, output format etc.

This paper is about the design of the service directory and S-CAN (See 4.2) on which it is based.

4.2 Secure Content Addressable Network Module

The S-CAN module realises a secure distributed hash table. A Content Addressable Network (CAN) is used as base of the secure hash table. S-CAN extends CAN by security features. CAN is the first choice, because it has a simple solid yet verifiable structure. Other overlay networks which realise a distributed hash table (like Pastry [8], Tapestry [12] or Chord [11]) miss this solid structure and are therefore not that easy to secure. Another interesting property of CAN is the low memory usage to hold neighbour information in a CAN with well distributed zones. See chapter 3 for more information on CAN. We focus on a secure construction of the CAN and on secure maintenance of its structure. S-CAN is a central service of the proposed sensor network platform and serves as the basis of the Service Directory Module. Its robustness is vital for the overall architecture.

It is not necessary that all devices of the sensor network participate in the S-CAN. Some devices could offer to issue queries to the S-CAN for devices which can not participate in the S-CAN. This is a step towards a cluster based solutions, but gives us more flexibility considering energy constraints on devices, especially in heterogeneous sensor networks. Another way to save energy is to rotate the job of S-CAN member and query server among a group of trusted devices. This involves some extra overhead but may still save energy. These considerations are out of scope of this paper and will be included in further work.

4.3 Service Manager Module

The Service Manager Module pairs actuators and sensors to execute a user task and supervises service execution. Sensors are nodes generating data. Actuators offer services other than data gathering. They usually control external devices. The pairing of actuators and sensors could be temporary or permanent, access could be simultaneous, competitive or exclusive. The Service Manager Module uses the Service Directory Module to bring actuators and sensors together. As services may require results of other services as input, the Service Manager Module may need to do cascading lookups in the service directory. Please note that the Service Manager Module has only to do lookups at the time of pairing. Later on, the paired actuators and sensors do not need any more service lookups as they know each other. The Service Manager Module also interacts with the User Interface Module and offers a simple language which can be used by the user to issue commands for actuators and sensors. User Interface Module and Service Manager Module are responsible for service provision.

4.4 Management Module

The Management Module realises some vital functions for sensor networks. This module has access to all layers as functions and informations provided by it are needed on all layers. The Management Module includes software updates and localisation information.

Software updates are vital for unattended sensor networks. They allow error correction, adaptation to unforeseen environmental conditions and recalibration. Software updates also guarantee a long lifetime of a sensor network as it is possible to redesign applications to latest needs not known when the network was deployed. Finally, software updates makes a sensor network reusable when its original aim has been achieved. It stands to reason that security is a very important matter with software updates. The goal is to ensure integrity and authenticity of updates. Another goal is to get back to a consistent state of the network as fast as possible, and, under any circumstances, must the sensors be able to recover from a failed software update. Software updates should even be possible in heterogeneous networks.

Localisation information is needed to build content aware applications. Localisation information can be acquired on different layers, starting on MAC layer with delay measurement and can go up to application layer where GPS data provides the global position on earth.

4.5 Data Replication Module

The Data Replication module is responsible for replication of sensor data. The goal is to bring frequently used data pieces as close as possibly to their inquirers to avoid large pieces of data to be multiply transported over great distances. The challenge is a timely delivery of the data and to keep several data replication points consistent. Security considerations on the integrity of replicated data are another open issue. The Data Replication Module firmly interacts with the Data Aggregation Module to reduce the overall energy usage on transmission in the whole sensor network.

4.6 Data Aggregation Module

The Data Aggregation Module is responsible for aggregation of sensor data. Transmitting messages is a very energy consuming operation in sensor networks. A considerable percentage of energy is spent on relaying messages in the sensor network thus any computation operation which reduces the data set is worth the effort as less data may be transmitted in a shorter time and therefore needs less energy. The idea is to reduce data by aggregation of data from various sensors before transmission to either get "the big picture" (which could be expressed by less data) or to eliminate redundant data. Sensor Data Aggregation is application centric as aggregation is only possible with a certain knowledge of data semantics. The Data Aggregation Module is also responsible for reducing accuracy of data. Less exactness usually means less data.

4.7 Data Acquisition Module

All data of the sensor network is acquired in the Data Acquisition Module. Data acquisition is initiated by the Service Manager Module. Data Acquisition Module also provides a part of the information needed by the Data Aggregation Module to reduce the amount of data. This information includes sensor accuracy, kind of data etc.

4.8 User Interface Module

The User Interface module enables the user to communicate with the sensor network. It gives her the possibility to issue commands to the sensor network and it provides a nice way to get results back from the sensor network. The User Interface is accountable for execution of commands and returns (aggregated) overall results of quests given to the sensor network. It represents a gateway between a the user's network (like the Internet) and an off-site sensor network.

4.9 Public Key Database Module

The Public Key Database Module realises a robust distributed public key database based on the Secure Content Addressable Network. Public keys may be necessary for the Service Manager Module to get access to confidential services etc. However, public key operations are very energy- and time-consuming so they should be kept to a minimum. The Public Key Database Module mainly uses functions of the Secure Content Addressable Network and can therefore be easily implemented.

4.10 Wireless Communication Module

The proposed architecture requires some underlying wireless communication ability. Our testbed (see chapter 6) uses Bluetooth as communication layer. However, the architecture itself is independent of the actual used communication technology and can be applied on most of the recently available sensor network platforms.

5 Design of a Distributed Service Directory

5.1 Assumptions

The design of the proposed distributed service directory is based on the following assumptions:

- Attackers are able to take over one or more nodes because nodes of our sensor network are unattended and spread in public or semi-public places. This assumption is called Big Stick principle in [9]. It implies to use tamper prove hardware. However tamper proveness and tamper resistance are way out of scope of this pa-

per. Refer to [9] for a discussion of tamper proveness and resistance. A takeover of multiple nodes is considered possible for our proposed design of a secure distributed service directory.
- Collaboration of overtaken nodes is considered possible
- Takeover of a node enables an attacker to respond on data request with fake data for the zone the overtaken node owns. Furthermore, any data which is routed through the overtaken node can be faked by the attacker.
- The service directory is used only casually and insertion of data is rare. Most communication in the sensor network is not done in CAN space but on network layer. Therefor, it is acceptable for our service directory to use overlay routing (which is not as efficient as routing on network layer) as long as it does not paralyze our sensor network for too long.
- The hardware is severely constraint, meaning that computational power, energy reserves and memory are low. This brings us in a situation where avoidance of communication, only casual use of symmetric and almost no use of asymmetric cryptography is a must.

5.2 Principles

One of the main ideas of the Secure Content Addressable Network is the use of a so called Master Device (MD). A MD is a hardware gimmick (like a ring or a key fob [15]) which is needed to bring new devices into the S-CAN. This typically involves a human interaction. It stands to reason that a part of this interaction should be authentication as this difficult action for computers is much easier for a human ("This is the device I took out of that fancy box I bought at Radio Shack"). As stated earlier it is not necessary for all devices to participate in the S-CAN. Therefor it is possible that a device joins the sensor network even if the Master Device is not available. We want the MD to be stateless so it is no Single-Point-of-Failure as the user can always keep a device constructed the same way in reserve. A stateless MD also potentiates the simultaneous use of multiple MDs. We further assume that the MD is able to perform "costly" operations like signature checking. As the MD is possessed by the user (or users) and as it is only used from time to time to bring new devices into the CAN, energy preservation is not a concern on the Master Device. E.g. it could be rechargeable. The MD is not part of our sensor network meaning it is not used as a online server of any kind. However for simplification, it has the ability to communication with the sensor network during the join of a new device. Please note that this is just for simplification and that the MD could otherwise use the communication abilities of the trusted joining device.

Another important idea is the use of zone certificates to prove the ownership of a CAN zone. Certificates are stored on every CAN node, but only the MD checks certificates when a new device is joining the CAN. This ensures, that no node of the CAN can claim possession of a CAN zone larger than the one it got assigned at the time of joining. If we combine this with the strict order for all good-natured nodes to immediately remove old certificates from their memory when they got a new one during the joining of a device, this means that a mischievous node can only claim possession of a zone of the size it had, when the attacker took over the node. Therefore it is not possible for the attacker to draw other joining nodes into a fake CAN (see protocol description in chapter 5.3 for details). This limits the impact of hostile node takeovers.

We do not want to use asymmetric cryptography, we want our MD to be stateless but we also want encrypted communication with distinguished nodes of the CAN. Our protocol is designed in a way that every communication during a join involves the Master Device. So it is not necessary for two arbitrary nodes to have a common key with each other. Our protocol presents a simple but efficient solution which only uses symmetric encryption between the MD and CAN nodes and enables the MD to calculate the needed keys on-the-fly so we do not need to store any keys on the MD (thus, it is stateless). When joining, a device gets a random join point which presents the point where the device will join into the CAN space. This point is encrypted with the private key of the MD and used as a symmetric key between the joining device and the MD. The joining device stores the key it has in common with the MD. The MD has not to keep track of all the keys it has in common with any CAN node, as it can derive the key from a join point by simply encrypting it with its private key. The join point of a device must always lie in the CAN zone the device owns. It is included in the node's zone certificate.

A Location Limited Side Channel [1] is used for the communication between the MD and the joining device. Location Limited Side Channels can be used as a secure channel as described in the paper mentioned above. Infrared or sound are examples for Location Limited Side Channels. For our prove of concept, we use simple physical contact which is another Location Limited Side Channel. This is not convenient for the user as it may take some time for the device to join the CAN and the user does definitely not want to keep physical contact between i.e. the ring on her hand and the new toy for minutes. To circumvent this, further design will exchange a short cryptographic information by physical contact or infrared and all other communication is done in the main communication channel, secured by the exchanged cryptographic information. Physical contact is preferred, because it is an intuitive way for the user to authenticate a device as stated earlier.

5.3 Protocol Description

5.3.1 Secure Join

As stated earlier, every join of a device starts with physical contact between the joining device and the Master Device. This activates any further action. The description of the proposed protocol uses the following syntax:

Channel | A -> B: Message

Channel denotes the channel a message is send in. In the present case, this can be the channel established by physical contact (*PHY*), the main communication channel like Bluetooth Scatternets in our case (*MC*) or the CAN overlay (*CAN*). *A* denotes the sender and *B* denotes the receiver of *Message*. In the case under consideration, sender and receiver can be Master Device (*MD*), joining device (*JD*), owner of the zone *JD* wants to join into (*ZO*) and neighbours of *ZO* (N_i).

Our Protocol consists of ten steps:

1.) PHY | MD -> JD: JP, E_{SK}(JP, "temp")

First, we send *JD* the join point (*JP*) at which it will join the CAN over the Location Limited Side Channel (physical contact in this case). This point is selected randomly by the Master Device. Random selection is very important as we want to achieve a equal distribution of nodes in CAN to ensure an equal zone size so that routing in the CAN is ideal. The Master Device also sends the join point encrypted with its super key (SK), which is the private key of the Master Device. The encrypted join point is used as a symmetric key by the node and will be denoted with *k* in the further description of our protocol. The string "temp" should provide an indication of the temporary character of the key. As it is important to always ensure that the join point of a device lies in its owned zone, there are cases where it is necessary for a joining device to get another join point. However it is not necessary to derive *k* from the join point. In fact, it could be any given ID as the ID is included in a certificate, issued in step 5.

2.) CAN | MD -> ZO: "who is responsible for JP", Address

Next, the Master Device sends a message into the CAN with address *JP* to find out, who owns the zone *JP* lies in. The message also includes the network layer address of the Master Device to enable the receiver to communicate with the Master Device outside the CAN. As we stated earlier, it is not necessary for the Master Device to have the ability to communication with the CAN because it could otherwise use the communication abilities of the trusted joining device. For simplification however, we assume the Master Device to be able to communication with the CAN in our further protocol description.

3.) MC | ZO -> MD: JP_{ZO}, E_k(certificate, time)

The zone owner answers the Master Device with its join point and the certificate of its zone, encrypted with the common key between the Master Device and the zone owner. The encrypted message also contains a timestamp to prevent reply attacks. The Master Device derives *k* using its private key as described earlier. It checks the zone owners certificate, compares the join point, included in the certificate, with JP_{ZO} and tests if the join point of the joining device really lies in the owned zone.

4.) PHY | MD -> JD: JP, E_{SK}(JP, "perm")

The Master Device evaluates if the joining node needs a new join point taking in consideration the join point of the zone owner and the future split line in CAN. In all cases, the joining device gets a new common symmetric key with the Master Device, which is constructed in the same way described earlier. This time, key and join point are marked as permanent. For any further interaction of the joining node with the Master Device, only the permanent key and therefore only the permanent join point is valid.

5.) PHY | MD -> JD:$key_{JD,ZO}$, E_{ZO}(*"JP,JD"*, $key_{JD,ZO}$, $certificate_{ZO}$,E_{JD}($certificate_{JD}$))

In the next step, the Master Device hands out a "ticket" to *JD* which enables it to get its new zone from *ZO*. The ticket is encrypted with *ZO*'s symmetric key and includes address (*JD*) and join point of *JD*, a symmetric key for communication between *JD* and *ZO*, a certificate for *ZO*'s zone after the split operation ($certificate_{ZO}$) and a certificate for *JD*'s new zone ($certificate_{JD}$). The certificate for *JD* is encrypted with *JD*'s symmetric key with the Master Device. If needed, step 5 is the right moment to hand *JD* a certificate for its private key and the public key of the Master Device. However, the proposed design does not (yet) make any use of asymmetric cryptography, so this step is obsolete at the moment.

6.) MC | JD -> ZO: E_{ZO}(*"JP,JD"*, $key_{JD,ZO}$, $certificate_{ZO}$,E_{JD}($certificate_{JD}$))

JD hands the ticket to *ZO*. *ZO* starts the zone split.

7.) MC | ZO -> JD: E_{key}(*data of hash table)*

ZO starts the zone split with transmission of data stored in the part of the distributed hash table, which formerly belonged to *ZO*'s zone but now belongs to *JD*. Transfer of data is encrypted with $key_{JD,ZO}$ which *JD* and *ZO* got handed out by the Master Device. This ensures, that no intermediate node is able to alter the data stored in *ZO*'s former hash table. Encryption is important to ensure integrity of the distributed hash table. All received data is acknowledged by *JD*. Those messages are not included in our protocol discussion for simplification.

8.) MC | ZO -> JD: E_{key}(E_{JD}($certificate_{JD}$))

When data transfer successfully ended, *ZO* hands out *JD*'s zone certificate. It is important that the certificate is handed over at the end of the data transfer. If the certificate would be delivered to the joining device before the successful data transmission, the joining device could simply drop any data but can prove that it is zone owner. With the certificate being handed over at the end of the data transfer, *ZO* could test some values it formerly stored. This does not prevent the joining device to drop all data anyway, but it makes cheating more difficult. From the moment of certificate handover forth, *ZO* is no longer responsible for the half of its former zone. *ZO* immediately and thoroughly destroys its old zone certificate. Extra care should be paid on this as an attacker who would be able to recover the old certificate would be capable of boarding *JD*'s zone.

9.) MC | ZO -> JD: E_{key}(*TempKeys, Neighbours)*
MC | ZO -> N_i *:* E_{Ni}(*"New Neighbour", JD, TempKey)*

As the zone split is now official in effect, *ZO* notifies all affected neighbours about the split and assigns them a temporary key for secure communication with *JD*. This messages are encrypted with the symmetric key *ZO* has in common with each of its neighbours. *ZO* sends *JD* an encrypted message containing all *JD*´s new neighbours and the temporary key which *ZO* assigned them.

10.) MC | JD -> N_i: Key construction

As the last step of our protocol, *JD* establishes a symmetric key with each of its new neighbours. This key is used for important communication between the neighbours, like periodic update messages etc.

At the end of step 10, the CAN has reached a consistent state again. *JD* joined securely the CAN and has established secrets with all of his neighbours. Every former neighbour of *ZO* has updated its routing information and *ZO* handed over its part of the distributed hash table to *JD*. This data transfer is encrypted and acknowledged. Therefore, integrity of the distributed hash table is ensured. Figure 2 summarises the protocol.

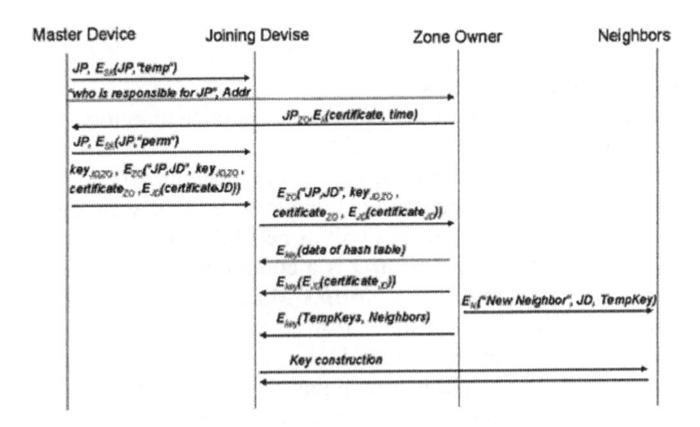

Fig. 2. Protocol overview

There are some cases where a device does not hold a valid certificate for its owned zone but is a valid owner. E.g. a node who gratefully took over an abandoned zone of a failed device has no valid certificate. Node failure can have different causes: battery outage, damage of hardware, signal jamming and environmental influences. Also, certificates of zones which are larger than a given boundary are time constraint to avoid that a hostile node takes over a large zone in the beginning of the sensor networks lifetime and is able to infiltrate a major part of the sensor network. However, the absence of a valid certificate is only grave in the case a device wants to join that zone. In this situation, the Master Device is to be considered online and can be used to issue the missing certificate. But the Master Device needs to be sure that the claiming zone owner is really the in possession of that CAN zone. To be sure, the Master Device asks the proposed zone owner to get guaranties for its ownership from its neighbours. Those guaranties include the zone limits of the neighbours, their certificates and a timestamp. The guaranties are encrypted with the neighbours' keys they have in common with the Master Device and can therefore be used to validate the zone owners claim on the zone. Protection of reply attacks is done by the timestamp. Figure 3 illustrates, how the Master Device verifies the zone claim of a device: The gray boxes illustrate zone certificates of neighbours, the dotted lines are used to boarder the zone of the claimant. Those lines are located at the edge which abuts the claimed zone. Nevertheless, the best way to avoid the loss of zone certificate is to make applications of the sensor network energy aware. An application which recognizes a power outage in the near future could hand over at least its zone certificate to a trusted device. If enough energy is available, data could be replicated, too.

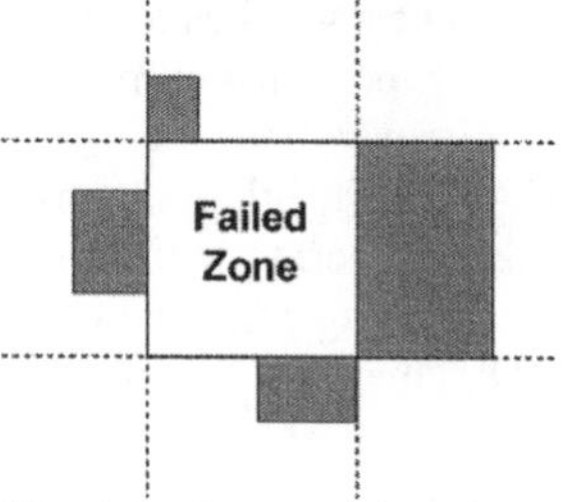

Fig. 3. Verification of zone claim after node failure

5.3.2 Self Certifying Path-Names

When registering a service in the secure distributed service directory, a node provides amongst other things its network layer address. If integrity of the proposed secure distributed service directory is ensured, how could a inquirer be sure, that she will contact the host at this address and not a malicious node? One way are Self-certifying Pathnames [6]. The idea is to store not only a network layer address in the service directory but also a *HostID* where *HostID* is a commitment to the host's public key. An entry would look like this: *Address:HostID*. Mazières [6] uses SHA-1 [4], a collision-resistant hash function, to calculate *HostID*. When the inquirer requests a service, it also asks the host to send its public key which the inquirer can verify with the corresponding *HostID*. The inquirer now holds the verified public key of the host. Inquirer and host can now establish a common symmetric key for all further communication. However, this hybrid method still involves asymmetric cryptographic operations which are very costly on peanut CPUs. As the use of public key operation is not enforced, an application can autonomously decide, if it needs that extra security for a given service. Please note that Self Certifying Path-names allow for secure communication both on network layer and in CAN.

5.3.3 Integrity

To ensure integrity of data, we use a redundancy approach like in [5] and [2]. The nodes of the CAN only store parts of the service information and artificial redundancy is added to the data, therefore not all parts of the data are needed to restore service information. Data parts are stored on different nodes. This ensures, that node failure does not affect the function of the distributed service directory, as long as there are enough nodes to restore the data. It also enables us to identify malicious or erroneous nodes, as they contribute inconsistent parts of the service information.

Another way to ensure integrity is self-certifying data: Zone owners in the CAN-space need not know the services offered by a sensor network. If a zone owner is not aware of the service names which are mapped to points in her CAN zone, we can establish data security using only hash functions. Integrity of entries is added by simply adding one new field to the entry which holds a hash of the value of the entry together with the function name. As the function name is never published into the CAN-space (hashing a service name to a point in CAN is the entry to CAN), a node not knowing available service names is not able to alter content in its hash table without the inquirer (which knows the service name) noticing. However, this approach works only if service names are long enough. Otherwise, a malicious node could use a dictionary attack to find out available service names.

6 Prototype

Further evaluation of the architecture will depend on actual prototypes of hardware like those presented by BeeCon GmbH [13]. Our prototype uses an Atmel ATMega 128 AVR Risc processor [14]. This processor comes with 128 kB of In-System reprogrammable Flash Memory, 4 kB of EEPROM and 4 kB of internal SRAM. Clock rate is 1 MHz internal and can be optionally boosted to 16 MHz using an extern clock generator. These features are responsible for severe restrictions on all operations in the sensor network. Connectivity of our testbed is realized by a robust and efficient Bluetooth stack which was developed for Ericsson Bluetooth modules in use on our current sensors and actuators. Bluetooth is used because actual hardware is available which allows us to build sensors with off-the-shelf components. There are also already some devices, which use Bluetooth (like a mobile phone). Those can be easily integrated in our sensor network. However, Bluetooth is not mandatory for the proposed architecture itself.

7 Summary and Outlook

This paper presented a work-in-progress status of an architecture for sensor networks. It showed the design of a secure distributed service directory. The design is based on an improvement of CAN, the Secure Content Addressable Network (S-CAN). Main focus of the paper is a secure join of new devices into the S-CAN. The paper showed how common secrets are constructed between S-CAN neighbours during the join of a new device without the use of public key cryptography or an infrastructure of any kind. The join process includes an intuitive and secure way to authenticate devices. The paper also presented a way to securely request services by the use of Self-Certifying Pathnames and showed how integrity protection could be done by Self-Certifying Data.

CAN in its basic implementation does not take into consideration the location of devices. This means, that eventually, communication of two physical neighbours takes plenty of hops in CAN. There are some enhancements of the basic proposal which we plan to integrate into our protocol.

With a distributed design, it is a goal to balance load to a huge number of nodes. However, hashing a service name to get a point in CAN-Space decides where a service description is stored. If there are a few services available, the load of storing service descriptions will be distributed to only a few nodes which own the zones corresponding to these points. One idea to circumvent this problem is to "spread" service names meaning to add additional information to extend the service name. Some attribute of the service description, like location, can be used. If those attributes are usually part of search queries, this procedure has an advantage as fewer entries will be returned. Another possibility is to use a hierarchy of services and to code this hierarchy in the service name string. Further research is needed in this direction.

As membership in our S-CAN is kind of energy costing, it is possible to rotate the task of S-CAN member in a closed group of trusted devices to save energy. We plan to present a robust and secure protocol for efficient rotation.

References

1. Dirk Balfanz, D. K. Smetters, Paul Stewart and H. Chi Wong: „Talking To Strangers: Authentication in Ad-Hoc Wireless Networks“, Symposium on Network and Distributed Systems Security (NDSS'02), Xerox Palo Alto Research Center, Palo Alto, USA, 2002
2. Ian Clarke, Oskar Sandberg, Brandon Wiley and Theodore W. Hong: "Freenet: A Distributed Anonymous Information Storage and Retrieval System", in Proceedings of the ICSI Workshop on Design Issues in Anonymity and Unobservability, Berkeley, USA, 2000
3. Neal Koblitz: "A course in number theory and cryptography, 2nd edition", Springer Verlag, Berlin, 1994
4. FIPS 180-1: "Secure Hash Standard", U.S. Department of Commerce/N.I.S.T., National Technical Information Service, Springfield, 1994
5. J. Kubiatowicz, et al: "OceanStore: An Architecture for Global-Scale Persistent Storage", Proceedings of ACM ASPLOS, December 2000
6. David Mazières, Michael Kaminsky, M. Frans Kaashoek and Emmett Witchel: "Separating key management from file system security", in 17th Symposium on Operating Systems Principles (SOSP'99), Kiawah Island, SC, 1999
7. Sylvia Ratnasamy, Paul Francis, Mark Handley, Richard Karp and Scott Shenker, „A Scalable Content-Addressable Network“, In Proceedings of ACM SIGCOMM 2001, August 2001
8. Antony Rowstron and Peter Druschel: "Pastry: Scalable, decentralized object location and routing for large-scale peer-to-peer systems", 18 Conference on Distributed Systems Platforms, Heidelberg, Germany, 2001
9. Frank Stajano, „Security for ubiquitous computing“, John Wiley & Sons, West Sussex, England, 2002
10. Emil Sit and Robert Morris, "Security Considerations for Peer-to-Peer Distributed Hash Tables", MIT Laboratory for Computer Science, Boston
11. Ion Stoica, Robert Morris, David Karger, Frans Kaashoek, and Hari Balakrishnan: "Chord: A Scalable Peer-to-peer Lookup Service for Internet Applications", Technical Report TR-819, Massachusetts Institute of Technology, Cambridge, USA, March 2001
12. B.Y. Zhao, K.D. Kubiatowicz and A.D. Joseph: „Tapestry: An Infrastructure for Fault-Resilient Wide-Area Location and Routing", Technical Report UCB//CSD-01-1141, Computer Science Division, U. C. Berkeley, Berkeley, USA, April 2001
13. beeCon GmbH, http://www.beecon.de/ , access on July, 12th, 2003
14. Atmel Corporation, http://www.atmel.com/dyn/resources/prod_documents/2467s.pdf, access on July, 12th, 2003
15. Dallas Semiconductor Corp, http://www.ibutton.com, access on July, 12th, 2003

Embedding Low-Cost Wireless Sensors into Universal Plug and Play Environments

Yvonne Gsottberger, Xiaolei Shi, Guido Stromberg, Thomas F. Sturm, and Werner Weber

Infineon Technologies, Corporate Research,
{Yvonne.Gsottberger,Xiaolei.Shi,Guido.Stromberg,
Thomas.Sturm,Werner.Weber}@infineon.com

Abstract. Research on Ubiquitous Computing is conducted in two opposing directions: One focuses on the seamless interaction of high-performance end devices by defining semantic mechanisms for device discovery and control. Ubiquitous computing is however also understood as the massive, unobtrusive integration of small electronic devices, in particular of small sensors and actuators, into our environment. These nodes must be cheap and energy-efficient, but are thus less flexible.
In this paper, we present an approach to harmonize these two research directions. Our objective is to integrate small sensors and actuators into one of the most established middleware platforms for distributed semantic services, namely Universal Plug and Play (UPnP). The key is to source out complex data processing from the sensor nodes to dedicated terminals which establish a proxy in the UPnP network. This allows flexible, highly-functional, UPnP compliant devices while concurrently satisfying rigid cost and power requirements.

1 Introduction

In the future, user interaction with technical appliances will become more service-oriented and less device-centric. The interface presented to the user will be harmonized and devices will learn to co-operate without user intervention. Thereby, usability is improved. The collaboration of individual appliances will open entirely new applications and will change the way technical devices are integrated into everyday life.

Besides the co-operation of existing devices, new components will be added to our environment. In particular, we envisage that numerous small sensors will be deployed to gather information on ambience such as temperature, motion, or light [1], and monitor whether e.g. doors and windows are open or closed [2]. Unobtrusive wearable sensors will monitor the health condition of ailing or elderly people and automatically place an emergency call when required. Thus, quality of life is improved. Information beacons will be installed at specific spots and provide localized, up-to-date, media-rich information using conventional smart phones or PDAs.

H. Karl, A. Willig, A. Wolisz (Eds.): EWSN 2004, LNCS 2920, pp. 291–306, 2004.

In the past, several standards have been developed to build up wireless networks of devices, such as Bluetooth [3], Wireless Local Area Networks (WLAN) IEEE 802.11x [4], Wireless Personal Area Networks (WPAN) IEEE 802.15.4 [5], or Short Range Wireless (SRW) systems [6]. These standards primarily address the physical and the lower networking layers of the wireless connection. Flexible application-layer support for autonomous device interaction is just beginning to be supported on PCs and high-tech computer appliances. One of the most established middleware platforms for this purpose is Universal Plug and Play (UPnP) [7]: UPnP defines a set of universal, open, XML-based protocols which allow the semantic description and control of various devices. Thereby it supports the flexible co-operation of these network nodes.

Unfortunately, the technical realization of the UPnP protocols is complex. Transceivers that support the entire UPnP protocol stack are thus power demanding and expensive, so that in particular many small and cheap peripheral devices will be excluded from the integration into such networks.

To this end, we have developed a distributed system architecture called Sindrion which features an effective solution for these small appliances. The basic idea is that complex operations are sourced out from the peripherals to dedicated network components, which we call terminals. Sindrion supports the wireless communication and the autonomous inter-operation of these nodes. The system is aimed at wireless sensors, information beacons as well as maintenance and control applications for white goods, for which the power consumption and a low price of the nodes are as important as the capability to embed the devices into a well-established IT infrastructure.

2 The Sindrion Concept

The basis of the Sindrion system is to set up a wireless link between peripheral devices and dedicated computing terminals. The objective of this connection is to source out complex data processing from the peripherals to the terminals. To this end, the peripheral devices contain small smart transceivers[1], the so-called Sindrion Transceivers, which are attached to embedded sensors or actuators (see Fig. 1). Typical peripheral devices are environmental sensors, small actuators like switches, or home appliances. They bear very limited or no computing power, and the embedded sensors and actuators can be controlled by simple proprietary analog or digital control lines. These are connected to the input- and output ports (I/O ports) of the Sindrion Transceivers.

The terminal is equipped with an RF transceiver which is compatible with that included in the Sindrion Transceiver. Data and protocol processing are done in the terminal, which features a virtually unlimited amount of processing power and memory compared to the Sindrion Transceiver.

---

[1] The term "smart transceiver" refers to an RF transceiver with integrated microcontroller (see Sec. 3).

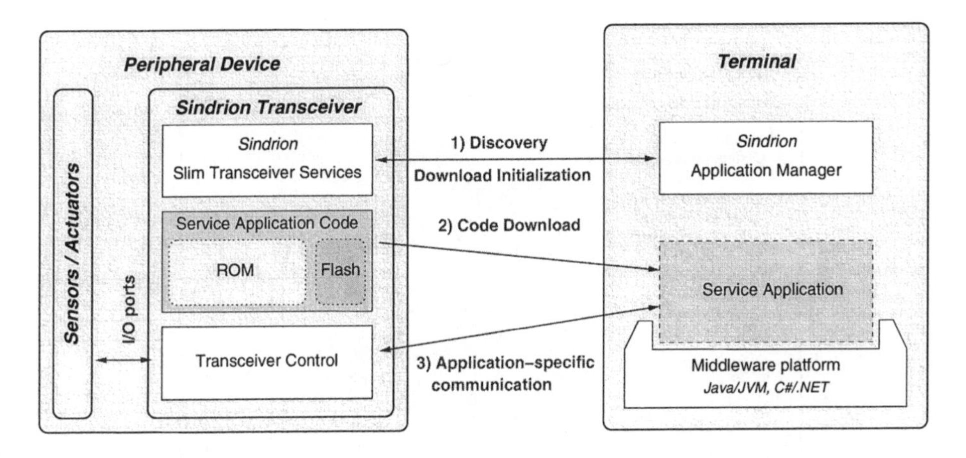

Fig. 1. Schematical Overview of the Sindrion System

Fig. 1 shows the fundamental structure establishing the communication between the terminal and a previously unknown Sindrion Transceiver. The procedure is as follows:

1. **Discovery**: The two end devices find each other in the discovery phase. To this end, the UPnP discovery protocol is used [7].
2. **Code Download**: If the terminal does not yet contain the control application for the Sindrion transceiver, the transceiver's application code is downloaded by the terminal. Preferably, this code is written for a middleware platform such as the Java Virtual Machine. This guarantees platform independence and allows for the seamless integration into various terminals. Further, high-level programming languages facilitate application development.
3. **Application-Specific Communication**: The following communication between the downloaded service application on the terminal and its counterpart - the transceiver control - on the Sindrion Transceiver may be completely application-specific and does not have to be defined by any standard.

3 Hardware Architecture of the Sensor Node

The hardware and software design of the Sindrion Transceiver[2] is the subject of this and the following section. The transceiver consists of a microcontroller which offers the necessary I/O ports, and an integrated RF transceiver (see Fig. 2). The objective of the microcontroller is to control the integrated RF transceiver, to do protocol processing to embed the transceiver into the network with the terminal, and to write and read the I/O ports according to the commands that

[2] For simplicity and where no ambiguity arises, we will usually denote the Sindrion Transceiver as *the transceiver*. Note that an RF transceiver is part of the Sindrion Transceiver as well as of the terminal.

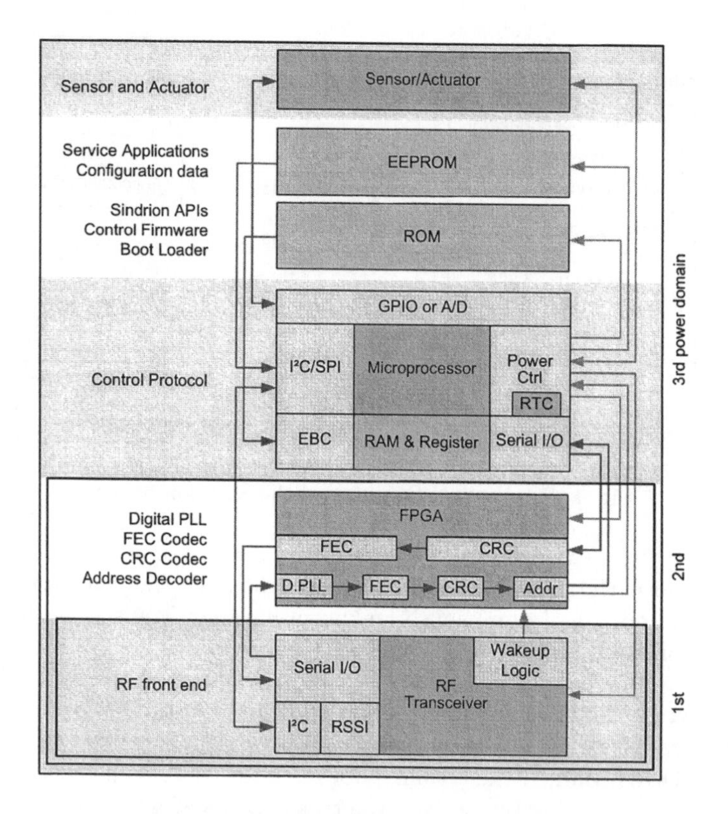

Fig. 2. Sindrion Hardware Architecture

are received via the RF link. The operation of the microcontroller is determined by the firmware of the transceiver, which is stored in non-volatile memory. The memory also contains the service application code which is executed on the terminal.

Since the Sindrion system is aimed at granting low-cost and low-power access to ambient intelligence systems, power consumption and hardware cost of the transceiver are the most rigid constraints in the Sindrion system design.

Typical application scenarios [8,9] show that there is a substantial amount of peripherals in the price range of 5$ to 20$, to which the transceiver contributes only a certain fraction. Depending on the particular application, the transceiver will be integrated with a battery or power supply, external components such as oscillator or memory, and the antenna. Additionally, integration costs must be considered. Given the price range mentioned before, we assume that the transceiver cost should not exceed 2$ - 3$. The transceiver will contain the following subcomponents:

- **RF transceiver:** To reduce the chip count, the RF transceiver should be largely self-contained. Note that wireless systems with short range are in our particular scope. In this field of application, a number of highly integrated wireless solutions are already available which require only few external components. A detailed discussion on the wireless link technology follows in Sec. 3.1.

- **Microcontroller:** The microcontroller should be small in terms of chip area and consume only little power. A simple 8-bit or 16-bit microcontroller with appropriate I/O ports suffices. As shown later, the most important characteristic is that it effectively supports different power states for sleep and idle conditions.
- **Firmware memory:** The firmware memory is a small non-volatile memory that contains the firmware (for protocol processing, control, housekeeping) for the microcontroller. The content of this memory is the same for all Sindrion transceivers. Therefore, it can be integrated as ROM.
- **Sindrion library memory:** The APIs are implemented by a fixed library, e.g. to support device control or UPnP networking. They can be used by the service application and thus depend only on the middleware platform on the terminal. Thus, the Sindrion library memory can also be stored in a ROM.
- **Service application memory:** The service application needs to be implemented by the integrator of the transceiver. Thus, the service application memory must be programmable and non-volatile. Flash or one-time-programmable (OTP) memory can be used for this purpose.

Since the individual subcomponents of the transceiver differ significantly in utilization and power consumption, the hardware design supports a number of individual power domains (see Fig. 2). In most sensor network applications, the channel occupation is very low. Thus, the entire transceiver can be operated in sleep mode for most of the time. Only when a data packet is received, the RF transceiver is powered up (see Sec. 3.2).

The bit error rate in short-range wireless networks is typically quite high. Thus, forward error correction (FEC) and error detection e.g. using CRC should be implemented in the PHY layer. Since only a fraction of the packets received over the air are destined for a specific transceiver, many packets will be discarded after address decoding. These operations of the PHY and MAC layer are regular and simple, so that they can easily be implemented in dedicated hardware. This saves a substantial amount of energy because the power-hungry microprocessor is not powered up unless an error-free packet with the correct destination address is received.

Eventually, all components except for the service application memory will be integrated in one single chip. This chip will meet the given cost requirements. Before we estimate the actual power consumption of the transceiver (see Sec. 3.3), we will detail on the two factors which have the predominant impact: The type of wireless link and the way the transceiver is networked on the MAC layer.

Table 1. Technical Parameters of some In-use Interfaces for Wireless Communication

	Data rate	Range	Memory		Power Dissipation		MIPS
			ROM	RAM	receive	idle	
SRW [10]	10-50 kbit/s	<15 m	–	–	27 mW	30 μW[1]/15 nW[2]	–
Bluetooth[11]	720 kbit/s	10 m	0.5/1 MB	32 kB	100 mW	50 mW[1]/150 μW[2]	~5
GPRS/EDGE	100-384 kbit/s	100-3000 m	>1 MB	>500 kB	>1.2 W	40 mW	300-1500
3G	>384 kbit/s	100-3000 m	>1 MB	>500 kB	>1.5 W	>50 mW	>5000
WLAN	1-54 Mbit/s	10-100 m	>500 kB	250 kB	>1 W	150 mW	>2000

[1]Sleep mode, [2]Power-down mode

3.1 Comparison of the Networking Technologies for the Sensor Network

Table 1 provides an overview of some radio link technologies that are standardized and in use today. These technologies are aimed at different target applications. Consequently, they differ in supported data rate and range.

Short range wireless (SRW) systems typically comply with certain low-level telecommunication standards, e.g. ETSI EN 300 220 [6]. The ETSI EN 300 220 specifies particular constraints on the physical data transmission such as the maximum emitted power, channel utilization (duty cycle, frequency occupation), or carrier frequencies. However, also other ETSI SRW standards exist which define higher layers for specific applications like cordless phones, audio data transmission etc.

Bluetooth also supports specific applications by defining so-called profiles [3]. A number of profiles has been specified, each of which covers a certain vertical section of the network stack from the physical layer up to the application. This guarantees compatibility for the applications which are addressed by the profiles supported by Bluetooth devices. Typical applications are wireless connections between office appliances (modems, printers, mice), or networked personal accessories (e.g. wireless headsets). Inter-operability for other applications can only be guaranteed by specifying and implementing additional profiles.

For Wireless Personal Area Networks (WPANs), the IEEE computer society has additionally founded the IEEE 802.15 working group. It is comprised of several task groups, each of which is dealing with specific target applications of WPANs. The fourth task group is working on the IEEE 802.15.4 standard for low data-rate systems which is currently in draft status [5]. It defines the PHY layer and the MAC layer of the network stack. Further, network-level security is addressed. The supported data rates are either 20 kbit/s (40 kbit/s) in the 868 MHz (915MHz) band or 250 kbit/s at 2.4 GHz.

Note that the IEEE 802.15.4 standard is also supported by the ZigBee Alliance [12]. The ZigBee Alliance is a consortium of semiconductor manufacturers, technology providers, and OEMs dedicated to providing an application and networking layer to reside on top of the 802.15.4 MAC. However, currently no commercial products are available which comply with either ZigBee or IEEE 802.15.4.

From Tab. 1, we see that the other listed transmission technologies are inappropriate to provide the basis for Sindrion Transceivers. The main reasons are power consumption and hardware complexity. The requirements regarding computing power are provided in units of million operations per second (MIPS). Compared to small networks like WPANs, this number is substantially higher for wide-area systems like GPRS or 3G due to excessive protocol processing. A significant amount of computing power is also required for data links with high bandwidth such as Wireless Local Area Networks (WLANs). Concurrently to computing power, hardware complexity increases for these systems due to requirements for larger memory.

The data rate provided by Bluetooth or the other wireless systems listed in Tab. 1 is much higher than that of SRW transceivers. However, since the control protocol is device and application specific, only very short data packets will be exchanged between the Sindrion Transceivers and the terminal to control the transceivers' operation. The only critical issue regarding the data rate is the download of the service application. Assuming an HTTP over TCP/IP download protocol according to the UPnP standard, and further assuming the MAC protocol proposed in [5], we estimate that the protocol overhead is about 9%[3].

Assuming a total data transfer rate of 50 kbit/s as shown in Table 1, this results in a net transfer rate of 45.5 kbit/s. Thus, the download of a service application of 10 kbytes would take 1.8 seconds which is a reasonably short time. However, this size has been chosen rather arbitrarily to provide an estimate on the expected latency due to the download process. The actual size of the program code firstly depends on the complexity of the application. Secondly and more significantly, it relies on the success of modularization and caching of recurring parts of the software. These parts, implementing the Sindrion API, are invariant for all transceivers and are thus only occasionally downloaded.

Note that the amount of protocol overhead is calculated for an IP packet which has the minimal recommended length that needs to be supported by hardware. A larger IP packet would concurrently increase the amount of transfer buffer. Note further that the overhead of 9% is achieved only with packets that contain the maximum amount of payload. Control messages, however, are significantly shorter in the range of approximately 10 bytes. Thus, although the data packets are very short, the relative packet overhead is intolerable. This fact emphasizes the need for header compression [14] or for a control protocol on network and transport layer which is not TCP/UDP and IP based.

[3] For large scale service applications, the overhead of the HTTP protocol is irrelevant. In total, the TCP and IP headers are typically 40 bytes long. Assuming no LLC fragmentation of IP packets, the minimal MAC payload is 576 bytes [13]. The MAC header suggested in [5] is 13 bytes, and the PHY layer requires some additional bytes for the sync field. Thus, the protocol overhead (excluding the PHY layer) will not exceed $53/(576+53) \approx 8.5\%$.

Summarizing, we conclude that the data range provided by SRW is sufficient for the Sindrion Transceiver. The low hardware complexity and power consumption suggest that this is the preferred wireless data link technology from the technical point of view.

Nevertheless, it should be taken into account that in contrast to SRW systems other systems such as Bluetooth are already widely deployed. This turns out advantageous especially regarding the availability of mobile terminals with appropriate physical communication interfaces. The propagation of other industry-wide standards such as WPANs according to IEEE 802.15.4 will lead to a diversified network infrastructure and could eventually require the support of other wireless network technologies by the Sindrion framework.

3.2 Networking Protocols for the Sensor Network

Overview on Power-Aware Technologies. Power in a sensor network can also be saved by optimizing the layers of the networking protocol stack. There are two basic approaches: One is minimizing the amount of transmitted data and transmission power, since the RF front end is one of the major power consumers in a sensor node. The other approach is switching the whole or parts of the system into sleep mode, which could save a factor of ten to a hundred in power. So far, much effort has been devoted to exploiting the power saving potential for wireless sensor networks based on these two approaches.

To reduce the payload, the highly correlated data collected by sensors can be aggregated into smaller data units [15]. To decrease the overhead, short MAC addresses can be dynamically assigned and used instead of the long global unique MAC addresses, which dramatically reduces the relative MAC overhead since the payload on a sensor network is usually very small [16]. Dynamically adjusting the transmission power for successful communication can save large amounts of power in the RF front end [17].

Besides, MAC layer duty cycle scheduling is probably the most effective way to save power, which is e.g. adopted by the IEEE802.15.4 WPAN [5] and the Bluetooth park mode [3]. The basic idea of MAC layer duty cycle scheduling is that a node sleeps for most of the time and periodically wakes up for a short time to listen to the channel for communication.

Sindrion power aware design follows different approaches in the protocol stack:

- Header compression decreases the overhead of TCP/IP.
- Dynamic short MAC addresses shorten the MAC frame header.
- FEC and proper Automatic Repeat reQuest (ARQ) protocols are used to decrease the probability of retransmission.
- Finally, MAC layer duty cycle scheduling is adopted as the most important power aware method for Sindrion.

Sindrion MAC Layer Duty Cycle Scheduling. So far, the implementation of a totally passive RF wake-up mechanism for transceivers is a big design challenge. Therefore, a wireless transceiver has to stay in receive mode to sense a possible signal on the air, which is extremely power inefficient. To this end, MAC layer duty cycle scheduling has been introduced. This reduces the time a transceiver is in receive mode. The IEEE802.15.4 standard is a typical example of a MAC layer duty cycle scheduling protocol.

In this standard, the master of a star topology periodically broadcasts a short-length beacon. All the slaves spend most of their time in sleep mode and periodically wake up to listen to the beacon. The beacon not only sends information on the beacon period, but also on the slaves the master will communicate with. If a slave has data to transmit or the beacon indicates that the master will send data to this particular slave, the slave stays awake for the rest of the cycle. Otherwise, the slave enters sleep mode until the next beacon. The duty cycle of this mechanism is the ratio between the beacon length and the cycle period: given a specific beacon length, the longer the cycle period, the smaller is the duty cycle, but the higher is the communication delay. Therefore, finding a good tradeoff between power saving and delay is a challenge. For networks with time-critical applications for which responsiveness is important, duty cycle reduction is therefore severely limited.

The Sindrion physical layer uses an SRW system which allows to detect within few bit intervals whether a certain data rate is transmitted on the channel (Data Rate Detection (DRD)) [10]. This RF transceiver also supports a so-called self-polling mode, in which the transceiver autonomously alternates between DRD mode and sleep mode. If the DRD circuitry detects that data is being transmitted, it generates an output impulse which is used to wake up the block for PHY and MAC layer processing in the Sindrion Transceiver.

We call this mechanism pseudo wake-up-by-signal, because the RF transceiver is not entirely passive but wakes up periodically. The important point is however that the time for Data Rate Detection is extremely short. Typically, it takes only the length of three bit intervals. The pseudo wake-up-by-signal mechanism is extremely power efficient since the DRD does not require support by other functional blocks like in IEEE802.15.4 in which the beacon needs to be parsed. Further, as we will show, a much smaller duty cycle can be achieved for a fixed cycle period by the pseudo wake-up-by-signal than by the IEEE802.15.4 beacon mechanism:

Assume that the data rate of the SRW transceiver is 50 kbit/s. Then, the time to check for channel occupation is 0.06 ms. Additionally, 2.6 ms are needed to switch from sleep mode to receive mode. As a result, the RF transceiver is in active receive mode for 2.66 ms each period. Therefore, to achieve a duty cycle of 1%, a cycle period of 266 ms is required. The typical length of an IEEE802.15.4 beacon is around 320-bit, which takes 6.4 ms. Including the setup time of 2.6 ms, the transceiver has to be awake for 9 ms each period, which corresponds to a duty cycle of 3.38% for the same cycle period of 266 ms.

Thus, the pseudo wake-up-by-signal is much more power efficient than the IEEE802.15.4 beacon mechanism. By comparison of duty cycles, we find that the two schemes differ by a factor of at least three in power consumption.

3.3 Power Estimation for Sensor Nodes

For a Sindrion Transceiver based on the hardware architecture shown in Fig. 2 and using an SRW system for the wireless link, the power consumption of the sensor nodes can be roughly estimated.

For a channel with light traffic, the power consumption of the sensor nodes is dominated by that of the RF transceiver[4]. If no data is received, the blocks for processing the MAC and the upper layers are in power-down mode. An off-the-shelf 16-bit single-chip microcontroller manufactured in a 3V CMOS process contributes far less than 100 μW to the overall power consumption [18]. Further assuming a duty cycle of 1%, which is reasonable to guarantee response times of less than 0.5 s (see Sec. 3.2), the power consumption of the entire transceiver is approximately 400 μW.

In active mode, the power consumed by the transceiver is also low. The required processing power for protocol processing of the lower layers is almost negligible since the data rate is low and simple access protocols will be used. The power consumption of a 16-bit microprocessor core at 10Mhz with standard peripherals is below 35 mW [18]. Thus, for a microcontroller active all the time during receive, the total system power consumption sums up to approximately 62 mW. In practice, the power consumed is significantly less since the duty cycle of the microcontroller is much lower.

We also see that even in worst-case scenarios, the power consumption of the SRW system is substantially less than that of Bluetooth. In active receive mode, the SRW system only consumes 60% of the power of a Bluetooth system. In sentry mode, the power consumption of Bluetooth exceeds that of SRW by more than two orders of magnitude.

4 Software Architecture

4.1 The UPnP Platform for Sensor Nodes

A middleware platform for the seamless interaction of end devices has to support spontaneous ad-hoc networking of components without manual installation and has to be flexible enough to integrate low-cost smart sensors and actuators.

CORBA is a universal architecture which allows a client application to call functions at a remote server. The kernel of a CORBA system is a so-called Object Request Broker (ORB) which mediates the communication between client and server. Thus, CORBA is a too heavy-weighted platform, because small sensor modules are not able to provide even a 'slim' ORB implementation.

JXTA aims for simplifying the configuration and operation of peer-to-peer networks. JXTA provides mechanisms for service discovery and an XML-based protocol framework for the network nodes. JXTA is designated for small systems

[4] The electrical characteristics for this estimation are taken from [10] as a reference implementation for SRW transceivers: The power-down power consumption is approx. 27 μW; the power consumption in active receive mode is approx. 27 mW.

although XML-based parsing requires certain computing power. In principle, JXTA offers the possibility to transfer data as well as program code between systems, but it lacks a standard for code execution.

Salutation is a protocol architecture for ad hoc networks of typical PC peripheral devices as printers or scanners. Salutation defines a set of features and data types which may be used to describe and use services. Otherwise, the possibilities for using the services are limited due to the rigid description of existing features.

The Universal Plug and Play (UPnP) Architecture [7] is one of the best established middleware platforms for Ubiquitous Computing. UPnP uses open and standardized protocols based on XML which allow to describe and control various devices. All communication is transferred over TCP/IP resp. UDP/IP and is thus hardware-independent. The most important advantages of this open concept are semantic interoperability of devices from different vendors and the simple extensibility to future devices. The price which has to be paid is the intense requirement for memory and computing power for UPnP Devices since variable XML-based protocols have to be processed. On first sight, this excludes cheap and low-power sensors and actuator nodes from participating in a UPnP environment.

On second sight, UPnP bridges resp. gateways may be used to connect a non-UPnP network to a UPnP environment. In this approach, sensor nodes have to be configured as nodes of some specific network and additionally as UPnP Devices inside the gateway. Therefore, seamless integration without user interaction and complex configuration as envisaged for ambient intelligence applications cannot be accomplished. Furthermore, if just one or few sensor nodes are added to an existing UPnP environment, an additional gateway device has to be installed as expensive overhead.

With Sindrion, we have found a third way to integrate devices into a UPnP environment which is compatible with UPnP and neither requires any gateway nor gateway configuration. For a better understanding, we take a closer look at the steps of the UPnP Device Architecture:

- **Addressing:** A UPnP Device requires a network address using a DHCP server or, if none is a available, using Auto-IP.
- **Discovery:** UPnP Devices advertise themselves by multicast messages and react according to search messages.
- **Description:** UPnP Devices and their embedded services are defined by XML description files which have to be provided for HTTP download by the device. Note that these files are static and will not change except for few lines of code.
- **Control:** UPnP Devices are controlled by the XML-based SOAP protocol [19] and therefore process XML messages and answer with dynamically created XML messages.
- **Eventing:** UPnP Devices maintain event subscriber lists, accept subscriptions and keep track of subscription durations. When an event occurs, XML-based event messages with dynamic content are sent to each subscriber.

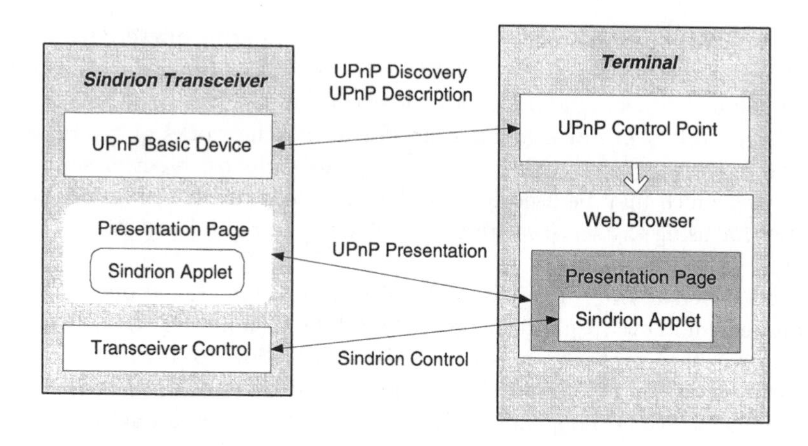

Fig. 3. Basic Device Functionality of the Sindrion Transceiver

- **Presentation:** The optional presentation page exposes an HTML-based user interface which may be used for controlling the device by embedded elements like Java applets.

The complexity of UPnP control and eventing prevents a cheap and low-power device to implement the complete UPnP Device Architecture. Nevertheless, such a device is able to do UPnP addressing, discovery, description, and presentation. Thus, it can act as a so-called UPnP Basic Device (see Sec. 4.2). Complex data processing in UPnP control and eventing can, however, also be exported to a terminal following the Sindrion approach. By this, the Sindrion Transceiver and its attached peripheral device can act as an arbitrary UPnP Device in complex Ubiquitous Computing applications (see Sec. 4.3).

4.2 The Sensor Node as UPnP Basic Devices

Due to the low-power and low-cost approach of the Sindrion sensor nodes, UPnP control and eventing capabilities will not be implemented in the Sindrion Transceivers. Sindrion Transceivers neither implement the SOAP protocol nor administrate event subscriptions.

Nevertheless, every Sindrion Transceiver will act as UPnP Basic Device, which is a special device type without embedded UPnP services. The Basic Device supports UPnP discovery, description and presentation, which need much less computational effort than UPnP control and eventing. Without prerequisites, a Sindrion Transceiver can therefore be connected to a UPnP network as displayed in Fig. 3. Following the Sindrion approach, the downloaded presentation page contains a Java applet, by which the Sindrion Transceiver and its attached peripheral device can be controlled.

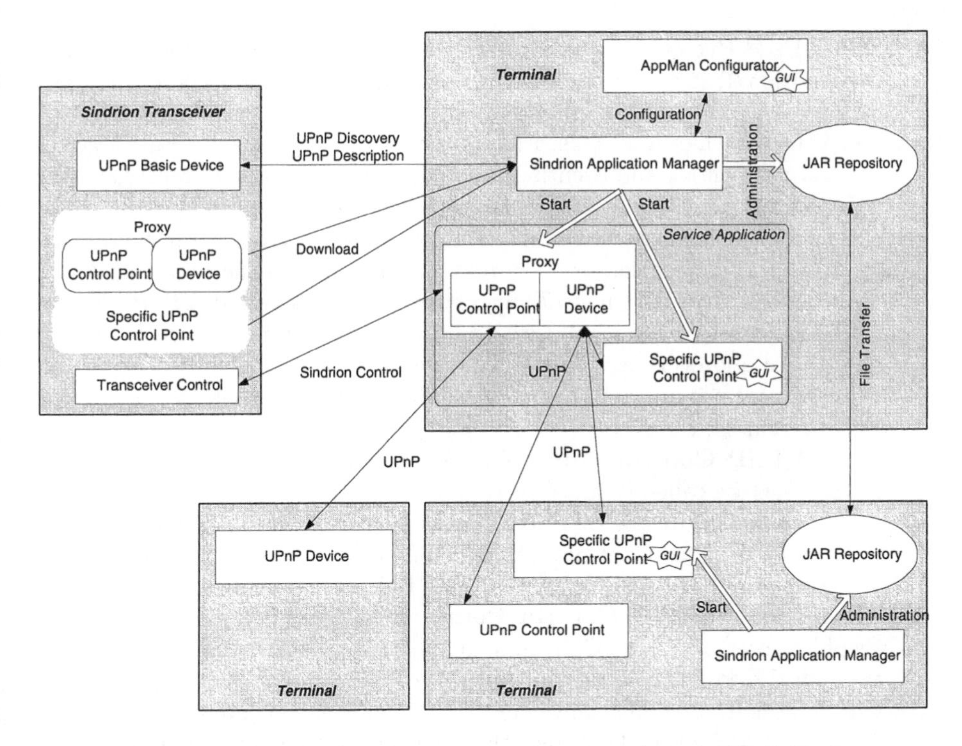

Fig. 4. The Sindrion Transceiver mapped as an Arbitrary UPnP Device

4.3 The Sensor Node in the UPnP Environment

The UPnP Basic Device of Sec. 4.2 is useful to control a Sindrion Transceiver without prerequisites on the terminal side in a point-to-point fashion. For larger control or service scenarios, an arbitrary UPnP Device with full control and eventing capabilities is needed with a longer lifetime[5] than a displayed web presentation page inside a browser.

Therefore, we envisage a Sindrion Application Manager demon running on the terminal which is used to start UPnP proxies exported from the Sindrion Transceiver. It also administrates a JAR[6] repository for the service application code. The Sindrion Application Manager discovers a Sindrion Transceiver by standard UPnP mechanisms and downloads the service application code from the transceiver. Next, the service application is started on a middleware platform like the Java Virtual Machine on the terminal and acts as a UPnP proxy for the connected Sindrion Transceiver. Thereby, it embeds the sensor node into a UPnP environment.

The detailed concept of the embedding is depicted in Fig. 4. The shown service application decomposes into two separate applications:

[5] E.g. a lighting device which is controlled by alarm equipment.

[6] Java Archive format for the Java byte code of the downloaded service application

- **The UPnP Proxy:** This application contains the exported functionality of the Sindrion Transceiver. Depending on the mirrored sensor node, the proxy may consist of two UPnP components:
  - **A UPnP Device:** This proxy provides all the needed UPnP services and full control and eventing functionality. It exclusively represents the Sindrion Transceiver and announces the represented device by UPnP discovery messages. For example, if the Sindrion Transceiver is attached to a temperature sensor, the proxy will announce a thermometer device with a service action *getTemperature*. Every interested UPnP Control Point can communicate with the announced UPnP Device by UPnP control and eventing. The proxy translates this SOAP based communication into simple application-specific messages and thereby controls the Sindrion Transceiver.
  - **A UPnP Control Point:** The members of the network which utilize services are called UPnP Control Points. In order to implement actuator functionality in the Sindrion Transceiver, the service application can contain a UPnP Control Point which maps the physical condition of the peripheral device (e.g. a switch) on UPnP control messages.
- **The specific UPnP Control Point:** The UPnP Proxy comes without graphical user interface (GUI) and communicates to the outside world only through UPnP mechanisms. The purpose of the optional *specific* UPnP Control Point is to provide a convenient GUI for direct user interaction with the Sindrion Transceiver. Typically, the GUI for the presentation page applet (see Sec. 4.2) is identical to the GUI of this application. As depicted in Fig. 4, the code for this UPnP Control Point may be exchanged from one terminal to another. Since the specific UPnP Control Point is fully UPnP compatible, it may run on any networked terminal and may be replaced by any other UPnP Control Point which is able to control the given UPnP Device.

With the given software architecture, a Sindrion Transceiver is seamlessly embedded into a UPnP environment. The underlying scheme of Fig. 4 is opaque for the other members of the UPnP network - the sensor node just appears as any fully enabled UPnP Device as shown in Fig. 5. Further, different nodes can interact by the UPnP interface to form complex services. For example, the already mentioned thermometer device (with proxy) may provide temperature data to a small LCD display (with proxy) in another room, at the same time to a PC (no proxy) to store a temperature timeline, and to an HVAC system (with proxy) for ventilation control.

5 Conclusion

In this paper, we have presented a hardware / software architecture to integrate simple, low-power and low-cost sensors and actuators into Universal Plug and Play (UPnP) environments for Ubiquitous Computing. The architecture is called Sindrion and defines interfaces for a wireless point-to-point connection between

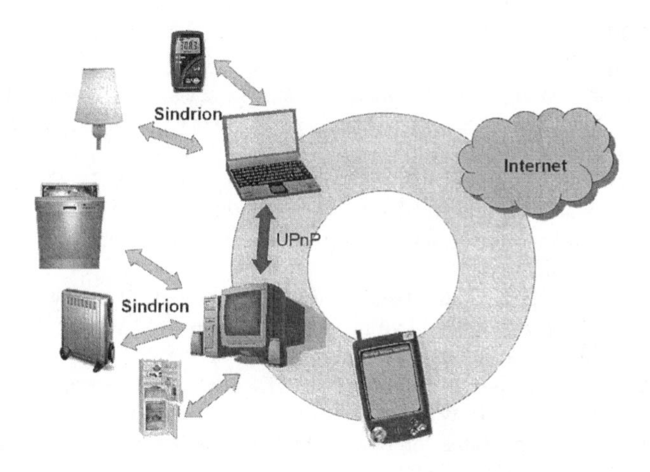

Fig. 5. Sindrion Transceivers as part of a UPnP Distributed System

so-called Sindrion Transceivers and certain computing terminals. The key principle of the Sindrion system is sourcing out complex data processing from the Sindrion Transceivers to the terminals. Thus, the transceivers are kept small, low-power and low-cost.

A Sindrion Transceiver consists of a simple microcontroller with limited computing power and memory, and of a low cost RF transceiver. Using standard input and output ports, Sindrion Transceivers are connected to arbitrary peripherals ranging from small sensors to home appliances. The comparison of several wireless networking technologies has shown that short range wireless systems are the most appropriate choice for the physical layer. Regarding networking protocols, we support short MAC addresses to reduce the amount of transmitted data. MAC Layer Duty Cycle scheduling with wake-up-by-signal is employed to manage the sleep modes of the hardware. Together with the introduction of layered power domains, we obtain a low-power and low-cost hardware system.

To connect this system to a UPnP environment, we provide the Sindrion Transceiver with a UPnP Basic Device for direct UPnP interaction. The resource-hungry UPnP control and eventing is exported to the terminal, which is a computing device with adequate computing power and memory to process XML-based protocols and to run a middleware platform like the Java Virtual Machine. Thereby, a proxy is established on the terminal which appears as UPnP Device and which uniquely represents the Sindrion Transceiver and its attached peripheral in the UPnP network.

In summary, the proposed architecture merges the two main directions of recent research on Ubiquitous Computing, namely the semantic interaction of high-performance appliances and the massive, unobtrusive deployment of low-cost and low-power devices.

References

1. Rabaey, J.M., Ammer, J., Karalar, T., Li, S., Otis, B., Sheets, M., Tuan, T.: PicoRadios for Wireless Sensor Networks: The Next Challenge in Ultra-Low Power Design. In: ISSCC. (2002)
2. http://www.emt.ei.tum.de/~heimauto/Projekte/IWOBAY/IWOBAY.htm: Project IWO-BAY: Intelligent House Instrumentation (2003)
3. Bluetooth SIG, Inc. http://www.bluetooth.org/specifications.htm: The Bluetooth V1.1 Core Specification. (2001)
4. LAN/MAN Standards Committee of the IEEE Computer Society: IEEE Standard for Part 11: Wireless LAN Medium Access Control (MAC) and Physical Layer (PHY) Specifications (1999)
5. LAN/MAN Standards Committee of the IEEE Computer Society: Draft: IEEE Standard for Part 15.4: Wireless Medium Access Control (MAC) and Physical Layer (PHY) specifications for Low Rate Wireless Personal Area Networks (LR-WPANs), P802.15.4/D18 (2003)
6. European Telecommunications Standards Institute http://www.etsi.org: ETSI EN 300 220-1,2. V1.3.1 edn. (2000)
7. Microsoft Corporation: Universal Plug and Play Device Architecture 1.0. Microsoft Corporation, Redmond, USA (2000) http://www.upnp.org.
8. Ducatel, K., Bogdanowicz, M., Scapolo, F., Leijten, J., Burgelman, J.C.: Scenarios for Ambient Intelligence in 2010. Draft final report, Information Society Technologies Advisory Group (ISTAG) (2001)
9. Neff, A.J., Chung, T.: iAppliances 2001: You've Got the Whole World in Your Hand. Bear Stearns Equity Research, Technology (2001)
10. Infineon Technologies AG: TDA5255 E1 ASK/FSK 434 MHz Wireless Transceiver. Data sheet, Infineon Technologies AG, München (2002) Preliminary Specification, Version 1.1.
11. CSR: Bluecore™2-Flash Single Chip Bluetooth™System. Data sheet, CSR, Cambridge, United Kingdom (2003) Advance Information Data Sheet for BC215159A, BC219159A.
12. ZigBee™ Alliance: ZigBee™ Alliance. http://www.zigbee.org (2003)
13. USC/Information Sciences Institute: Internet Protocol, DARPA Internet Program, Protocol Specification. The Internet Engineering Task Force, http://www.ietf.org/ (1981) http://www.ietf.org/rfc/rfc791.txt?number=791.
14. Jacobson, V.: RFC 1144: Compressing TCP/IP headers for low-speed serial links (1990) Status: PROPOSED STANDARD.
15. Heinzelman, W.R., Sinha, A., Wang, A., Chandrakasan, A.P.: Energy-Scalable Algorithms and Protocols for Wireless Microsensor Networks. In: ICASSP. (2000)
16. Schurgers, C., Kulkarni, G., Srivastava, M.B.: Distributed Assignment of Encoded MAC Addresses in Sensor Networks. In: MobiHoc. (2001)
17. Monks, J.P., Bharghavan, V., Hwu, W.E.: A Power Controlled Multiple Access Protocol for Wireless Packet Networks. In: INFOCOM. (2001)
18. Infineon Technologies AG: C161 PI Microcontrollers. Data sheet, Infineon Technologies AG, München (1999) Preliminary.
19. W3C: SOAP version 1.2 W3C working draft. http://www.w3c.org/TR/2001/WD-soap12-20010709 (2001)

Lessons from a Sensor Network Expedition

Robert Szewczyk[1], Joseph Polastre[1], Alan Mainwaring[2], and David Culler[1,2]

[1] University of California, Berkeley
Berkeley CA 94720
{szewczyk,polastre,culler}@cs.berkeley.edu
[2] Intel Research Laboratory at Berkeley
Berkeley CA 94704
{amm,dculler}@intel-research.net

Abstract. Habitat monitoring is an important driving application for wireless sensor networks (WSNs). Although researchers anticipate some challenges arising in the real-world deployments of sensor networks, a number of problems can be discovered only through experience. This paper evaluates a sensor network system described in an earlier work and presents a set of experiences from a four month long deployment on a remote island off the coast of Maine. We present an in-depth analysis of the environmental and node health data. The close integration of WSNs with their environment provides biological data at densities previous impossible; however, we show that the sensor data is also useful for predicting system operation and network failures. Based on over one million data and health readings, we analyze the node and network design and develop network reliability profiles and failure models.

1 Introduction

Application-driven research has been the foundation of excellent science contributions from the computer science community. This research philosophy is essential for the wireless sensor network (WSN) community. Integrated closely with the physical environment, WSN functionality is affected by many environmental factors not foreseen by developers nor detectable by simulators. WSNs are much more exposed to the environment than the traditional systems. This allows WSNs to densely gather environment data. Researchers can study the relationships between collected environmental data and sensor network behavior. If a particular aspect of sensor functionality is correlated to a set of environmental conditions, network designers can optimize the behavior of the network to exploit a beneficial relationship or mitigate a detrimental one.

Habitat monitoring is widely accepted as a driving application for wireless sensor network research. Many sensor network services are useful for habitat monitoring: localization [1], tracking [3,14,16], data aggregation [9,15,17], and, of course, energy efficient multihop routing [5,13,25]. Ultimately the data collected needs to be meaningful to disciplinary scientists, so sensor design [19] and in-the-field calibration systems are crucial [2,24]. Since such applications need to run unattended, diagnostic and monitoring tools are essential [26].

H. Karl, A. Willig, A. Wolisz (Eds.): EWSN 2004, LNCS 2920, pp. 307–322, 2004.

While these services are an active area of research, few studies have been done using wireless sensor networks in long-term field applications. During the summer of 2002, we deployed an outdoor habitat monitoring application that ran unattended for four months. Outdoor applications present an additional set of challenges not seen in indoor experiments. While we made many simplifying assumptions and engineered out the need for many complex services, we were able to collect a large set of environmental and node diagnostic data. Even though the collected data was not useful for making scientific conclusions, the fidelity of the sensor data yields important observations about sensor network behavior. The data analysis discussed in this paper yields many insights applicable to most wireless sensor deployments. We utilize traditional quality of service metrics such as packet loss; however the sensor data combined with network metrics provide a deeper understanding of failure modes. We anticipate that with system evolution comes higher fidelity sensor readings that will give researchers an even better understanding of sensor network behavior.

The paper is organized as follows: Section 2 provides a detailed overview of the application, Section 3 analyzes the network behaviors that can be deduced from the sensor data, Section 4 contains the analysis of the node-level data. Section 5 contains related work and Section 6 concludes.

2 Application Overview

In the summer of 2002, we deployed a 43 node sensor network for habitat monitoring on an uninhabited island 15km off the coast of Maine, USA. Biologists have seasonal field studies with an emphasis on the ecology of the Leach's Storm Petrel [12]. The ability to densely instrument this habitat with sensor networks represents a significant advancement over traditional instrumentation. Monitoring habitats at scale of the organism was previously impossible using standalone data loggers. To assist the biologists, we developed a complete sensor network system, deployed it on the island and monitored its operation for over four months. We used this case study to deepen our understanding of the research and engineering challenges facing the system designers, while providing data that has been previously unavailable to the biologists. The biological background, the key questions of interest, and core system requirements are discussed in depth in [18,20].

In order to study the Leach's Storm Petrel's nesting habits, nodes were deployed in underground nesting burrows and outside burrow entrances above ground. Nodes monitor typical weather data including humidity, pressure, temperature, and ambient light level. Nodes in burrows also monitored infrared radiation to detect the presence of a petrel. The specific components of the WSN and network architecture are described in this section.

2.1 Application Software

Our approach was to simplify the problem wherever possible, to minimize engineering and development efforts, to leverage existing sensor network platforms and components, and to use off-the-shelf products. Our attention was focused on the sensor network operation. We used the Mica mote developed by UC Berkeley [10] running the TinyOS operating system [11].

In order to analyze the long term operation of a WSN, each node executed a simple periodic application that met the biologists requirements in [20]. Every 70 seconds, each node sampled each of its sensors. Data readings were timestamped with 32-bit sequence numbers kept in flash memory. The readings were transmitted in a single 36 byte data packet using the TinyOS radio stack. We relied on the underlying carrier sense MAC layer protocol to prevent against packet collisions. After successfully transmitting the packet of data, the motes entered their lowest power state for the next 70 seconds. The motes were transmit-only devices and the expected duty cycle of the application was 1.7%. The motes were powered by two AA batteries with an estimated 2200mAh capacity.

2.2 Sensor Board Design

We designed and built a microclimate sensor board for monitoring the Leach's Storm Petrel. We decided to use surface mount components because they are smaller and operate at lower voltages than probe-based sensors. Burrow tunnels are small, about the size of a fist. A mote was required to fit into the tunnel and not obstruct the passage. Above ground, size constraints were relaxed. To fit the node into a burrow, the sensor board integrated all sensors into a single package to minimize size and complexity. To monitor the petrel's habitat, the sensor board included a photoresistive light sensor, digital temperature sensor, capacitive humidity sensor, digital barometric pressure sensor, and passive infrared detector (thermopile with thermistor). One consequence of an integrated sensor board is the amount of shared fate between sensors; a failure of one sensor likely affects all other sensors. The design did not consider fault isolation among independent sensors or controlling the effects of malfunctioning sensors on shared hardware resources.

2.3 Packaging Strategy

In-situ instrumentation experiences diverse weather conditions including dense fog with pH readings of less than 3, dew, rain, and flooding. Waterproofing the mote and its sensors is essential for prolonged operation.

Sealing electronics from the environment could be done with conformal coating, packaging, or combinations of the two. Since our sensors were surface mounted and needed to be exposed to the environment, we sealed the entire mote with parylene leaving the sensor elements exposed. We tested the sealant with a coated mote that ran submerged in a coffee cup of water for days.

Our survey of off-the-shelf enclosures found many that were slightly too small for the mote or too large for tunnels. Custom enclosures were too costly. Above ground motes were placed in ventilated acrylic enclosures. In burrows, sealed motes were deployed without enclosures.

(a) Sealed block (b) Cylinder with vents and drainage

Fig. 1. Acrylic enclosures used at different outdoor applications.

Of primary concern for the packaging was the effect it has on RF propagation. We decided to use board-mounted miniature whip antennas and build enclosures out of acrylic. There were significant questions about RF propagation from motes inside burrows, above ground on the surface, within inches of granite rocks, tree roots and low, dense vegetation. When we deployed the motes we noted the ground substrate, distance into the burrow, and geographic location of each mote to assist in the analysis of the RF propagation for each mote. Enclosures built for this deployment can be seen in Fig. 1.

2.4 Application Realization

The network architecture had a multi-level structure as shown in Fig. 2. The first level consisted of motes with sensors. In general, sensor nodes perform general-purpose computing and networking, as well as application-specific tasks. The *gateway* is responsible for transmitting packets from the *sensor patch* through a local transit network to one or more *base stations*. The base stations in the third level provide database services as well as Internet connectivity. The fourth and final level consists of remote servers supporting analysis, visualization and web content. Mobile devices may interact with any of the networks–whether it is used in the field or across the world connected to a database replica.

The sensor patch consisted of the motes with their sensor boards in a single hop network. The single hop network was chosen not only for simplicity but also to evaluate the characteristics of a single radio cell without interfering cells. The gateway was implemented with a relay mote connected to a high-gain Yagi antenna to retransmit data from the sensor patch over a 350 foot link to the base station. The relay node ran at a 100% duty cycle powered by a solar cell and rechargeable battery. The data was logged by a laptop into a Postgres database and then replicated through an Internet satellite connection.

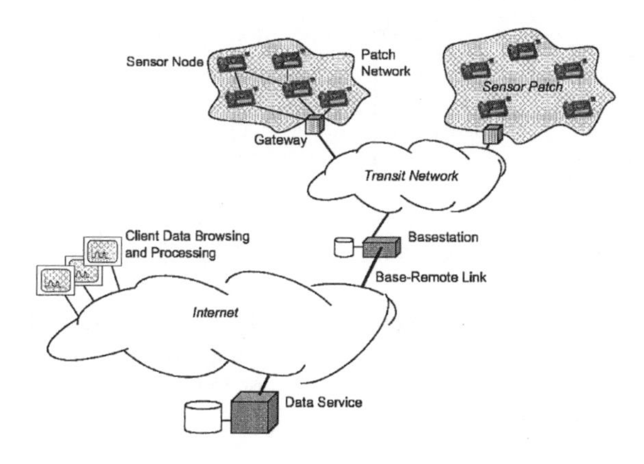

Fig. 2. System architecture for habitat monitoring

2.5 Experimental Goals

Since our deployment was the first long term use of the mote platform, we were interested in how the system would perform. Specifically, this deployment served to prove the feasibility of using a miniature low-power wireless sensor network for long term deployments. We set out to evaluate the efficacy of the sealant, the radio performance in and out of burrows, the usefulness of the data for biologists including the occupancy detector, and the system and network longevity. Since each hardware and software component was relatively simple, our goal was to draw significant conclusions about the behavior of wireless sensor networks from the resulting data.

After 123 days of the experiment, we logged over 1.1 million readings. During this period, we noticed abnormal operation among the node population. Some nodes produced sensor readings out of their operating range, others had erratic packet delivery, and some failed. We sought to understand why these events had occurred. By evaluating these abnormalities, future applications may be designed to isolate problems and provide notifications or perform self-healing. The next two sections analyze node operation and identify the causes of abnormal behavior.

3 Network Analysis

We need evaluate the behavior of the sensor network to establish convincing evidence that the system is operating correctly. Our application was implemented as a single hop network, however the behavior in a single hop is equivalent to what occurs in any WSN radio cell. We begin by examining WSN operation and its performance over time in order to evaluate network cell characteristics.

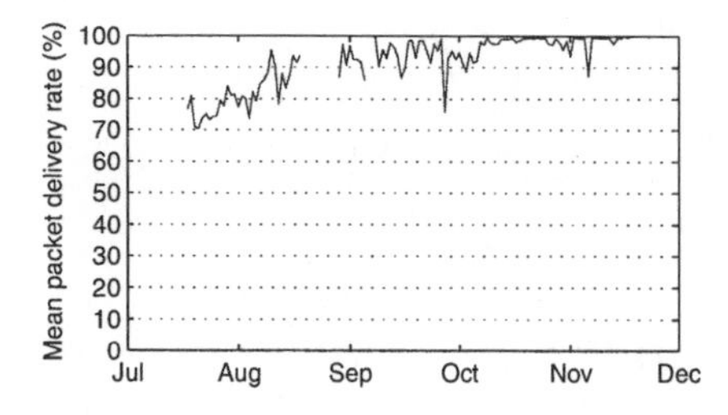

Fig. 3. Average daily losses in the network throughout the deployment. The gap in the second part of August corresponds to a database crash.

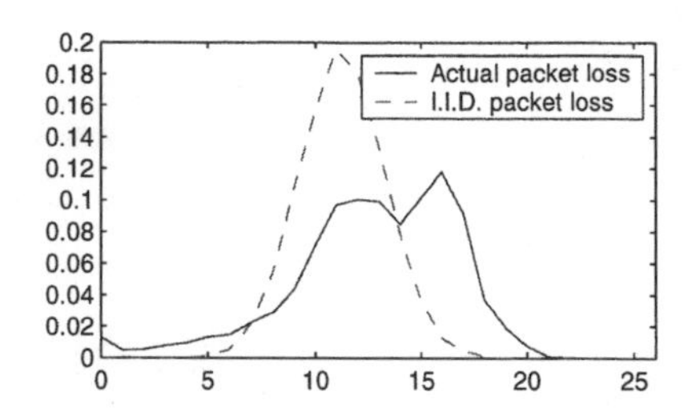

Fig. 4. Distribution of packet losses in a time slot. Statistically, the losses are not independently distributed.

3.1 Packet Loss

A primary metric of network performance is packet loss in the network over time. Packet loss is a quality-of-service metric that indicates the effective end-to-end application throughput [4]. The average daily packet loss is shown in Fig. 3. Two features of the packet loss plot demand explanation: (1) why was the initial loss rate high and (2) why does the network improve with time? Note that the size of the sensor network is declining over time due to node failures. Either motes with poor packet reception die quicker or the radio channel experiences less contention and packet collisions as the number of nodes decreases. To identify the cause, we examine whether a packet loss at a particular node is dependent on losses from other nodes.

The periodic nature of the application allows us to assign virtual time slots to each data packet corresponding with a particular sequence number from each node. After splitting the data into time slices, we can analyze patterns of loss within each time slot. Fig. 5 shows packet loss patterns within the network during the first week of August 2002. A black line in a slot indicates that a packet expected to arrive was lost, a white line means a packet was successfully received. If all packet loss was distributed independently, the graph would contain a random placement of black and white bars appearing as a gray square. We note that 23 nodes do not start to transmit until the morning of August 6; that reflects the additional mote deployment that day. Visual inspection reveals patterns of loss: several black horizontal lines emerge, spanning almost all nodes, *e.g.* midday on August 6, 7, and 8. Looking closer at the packet loss on August 7, we note it is the only time in the sample window when motes 45 and 49 transmit packets successfully; however, heavy packet loss occurs at most other nodes. Sequence numbers received from these sensors reveal they transmitted data during every sample period since they were deployed even though those packets were not received.

More systematically, Fig. 4 compares the empirical distribution of packet loss in a slot to an independent distribution. The hypothesis that the two distributions are the same is rejected by both parametric (χ^2 test yields 10^8) and non-

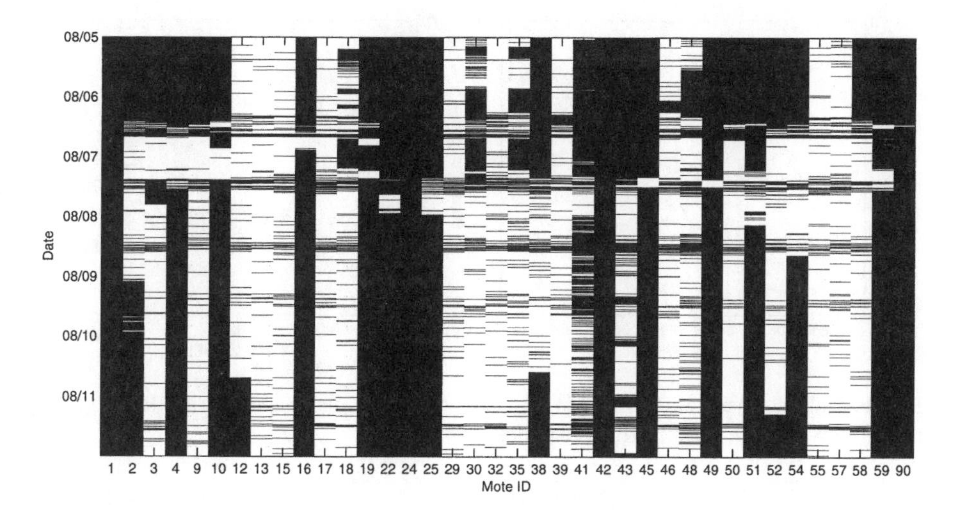

Fig. 5. Packet loss patterns within the deployed network during a week in August. Y-axis represents time divided into virtual packet slots (note: time increases downwards). A black line in the slot indicates that a packet expected to arrive in this time slot was missed, a white line means that a packet was successfully received.

parametric techniques (rank test rejects it with 99% confidence). The empirical distribution appears a superposition of two Gaussians: this is not particularly surprising, since we record packet loss at the end of the path (recall network architecture, Section 2). This loss is a combination of potential losses along two hops in the network. Additionally, packets share the channel that varies with the environmental conditions, and sensor nodes are likely to have similar battery levels. Finally, there is a possibility of packet collisions at the relay nodes.

3.2 Network Dynamics

Given that the expected network utilization is very low (less than 5%) we would not expect collisions to play a significant role. Conversely, the behavior of motes 45 and 49 implies otherwise: their packets are only received when most packets from other nodes are lost. Such behavior is possible in a periodic application: in the absence of any backoff, the nodes will collide repeatedly. In our application, the backoff was provided by the CSMA MAC layer. If the MAC worked as expected, each node would backoff until it found a clear slot; at that point, we would expect the channel to be clear. Clock skew and channel variations might force a slot reallocation, but such behavior should be infrequent.

Looking at the timestamps of the received packets, we can compute the phase of each node, relative to the 70 second sampling period. Fig. 6 plots the phase of selected nodes from Fig. 5. The slope of the phase corresponds to a drift as a percentage of the 70-second cycle. In the absence of clock drift and MAC delays, each mote would always occupy the same time slot cycle and would appear as a

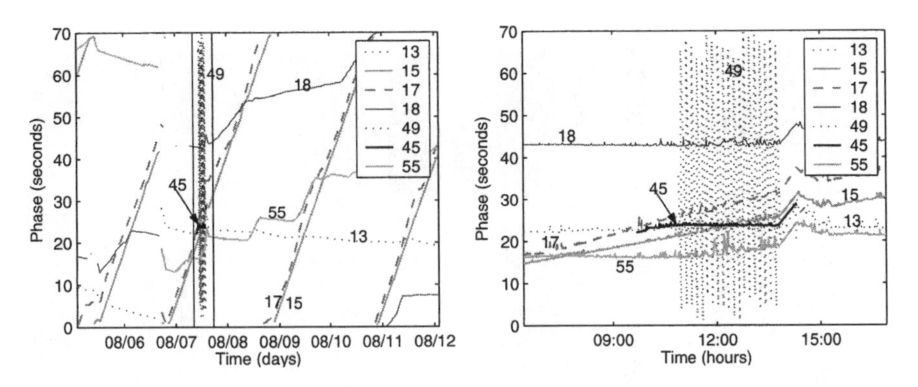

Fig. 6. Packet phase as a function of time; the right figure shows the detail of the region between the lines in the left figure.

horizontal line in the graph. A 5 ppm oscillator drift would result in gently sloped lines, advancing or retreating by 1 second every 2.3 days. In this representation, the potential for collisions exists only at the intersections of the lines.

Several nodes display the expected characteristics: motes 13, 18, and 55 hold their phase fairly constant for different periods, ranging from a few hours to a few days. Other nodes, *e.g.* 15 and 17 appear to delay the phase, losing 70 seconds every 2 days. The delay can come only from the MAC layer; on average they lose 28 msec, which corresponds to a single packet MAC backoff. We hypothesize that this is a result of the RF automatic gain control circuits: in the RF silence of the island, the node may adjust the gain such that it detects radio noise and interprets it as a packet. Correcting this problem may be done by incorporating a signal strength meter into the MAC that uses a combination of digital radio output and analog signal strength. This additional backoff seems to capture otherwise stable nodes: *e.g.* mote 55 on August 9 transmits in a fixed phase until it comes close to the phase of 15 and 17. At that point, mote 55 starts backing off before every transmission. This may be caused by implicit synchronization between nodes caused by the transit network.

We note that potential for collisions does exist: the phases of different nodes do cross on several occasions. When the phases collide, the nodes back off as expected, *e.g.* mote 55 on August 9 backs off to allow 17 to transmit. Next we turn to motes 45 and 49 from Fig. 5. Mote 45 can collide with motes 13 and 15; collisions with other nodes, on the other hand, seem impossible. In contrast, mote 49, does not display any potential for collisions; instead it shows a very rapid phase change. Such behavior can be explained either though a clock drift, or through the misinterpretation of the carrier sense (*e.g.* a mote determines it needs to wait a few seconds to acquire a channel). We associate such behavior with faulty nodes, and return to it in Section 4.

4 Node-Level Analysis

Nodes in outdoor WSNs are exposed to closely monitor and sense their environment. Their performance and reliability depend on a number of environmental factors. Fortunately, the nodes have a local knowledge of these factors, and they may exploit that knowledge to adjust their operation. Appropriate notifications from the system would allow the end user to pro-actively fix the WSN. Ideally, the network could request proactive maintenance, or self-heal. We examine the link between sensor and node performance. Although the particular analysis is specific to this deployment, we believe that other systems will be benefit from similar analyses: identifying outlier readings or loss of expected sensing patterns, across time, space or sensing modality. Additionally, since battery state is an important part of a node's self-monitoring capability [26], we also examine battery voltage readings to analyze the performance of our power management implementation.

4.1 Sensor Analysis

The suite of sensors on each node provided analog light, humidity, digital temperature, pressure, and passive infrared readings. The sensor board used a separate 12-bit ADC to maximize the resolution and minimize analog noise. We examine the readings from each sensor.

Light readings: The light sensor used for this application was a photoresistor that we had significant experience with in the past. It served as a confidence building tool and ADC test. In an outdoor setting during the day, the light value saturated at the maximum ADC value, and at night the values were zero. Knowing the saturation characteristics, not much work was invested in characterizing its response to known intensities of light. The simplicity of this sensor combined with an *a priori* knowledge of the expected response provided a valuable baseline for establishing the proper functioning of the sensor board. As expected, the sensors deployed above ground showed periodic patterns of day and night and burrows showed near to total darkness. Fig. 7 shows light and temperature readings and average light and temperature readings during the experiment.

The light sensor operated most reliably of the sensors. The only behavior identifiable as failure was disappearance of diurnal patterns replaced by high value readings. Such behavior is observed in 7 nodes out of 43, and in 6 cases it is accompanied by anomalous readings from other sensors, such as a 0^oC temperature or analog humidity values of zero.

Temperature readings: A Maxim 6633 digital temperature sensor provided the temperature measurements While the sensor's resolution is 0.0625^oC, in our deployment it only provided a 2^oC resolution: the hardware always supplied readings with the low-order bits zeroed out. The enclosure was IR transparent to assist the thermopile sensor; consequently, the IR radiation from direct sunlight would enter the enclosure and heat up the mote. As a result, temperatures measured inside the enclosures were significantly higher than the ambient

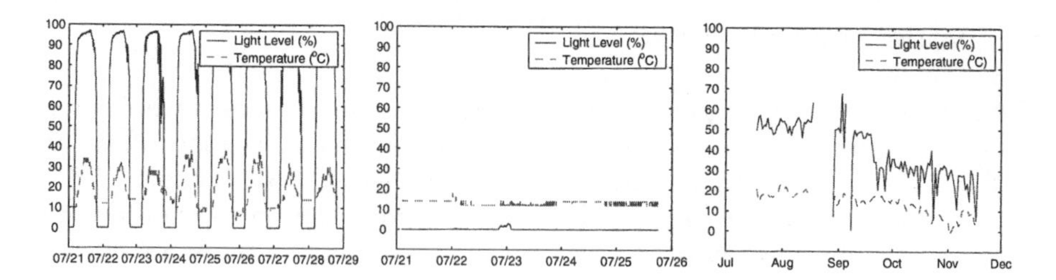

Fig. 7. Light and temperature time series from the network. From left: outside, inside, and daily average outside burrows.

temperatures measured by traditional weather stations. On cloudy days the temperature readings corresponded closely with the data from nearby weather buoys operated by NOAA.

Even though motes were coated with parylene, sensor elements were left exposed to the environment to preserve their sensing ability. In the case of the temperature sensor, a layer of parylene was permissible. Nevertheless the sensor failed when it came in direct contact with water. The failure manifested itself in a persistent reading of 0°C. Of 43 nodes, 22 recorded a faulty temperature reading and 14 of those recorded their first bad reading during storms on August 6. The failure of temperature sensor is highly correlated with the failure of the humidity sensor: of 22 failure events, in two cases the humidity sensor failed first and in two cases the temperature sensor failed first. In remaining 18 cases, the two sensors failed simultaneously. In all but two cases, the sensor did not recover.

Humidity readings: The relative humidity sensor was a capacitive sensor: its capacitance was proportional to the humidity. In the packaging process, the sensor needed to be exposed; it was masked out during the parylene sealing process, and we relied on the enclosure to provide adequate air circulation while keeping the sensor dry. Our measurements have shown up to 15% error in the interchangeability of this sensor across sensor boards. Tests in a controlled environment have shown the sensor produces readings with 5% variation due to analog noise. Prior to deployment, we did not perform individual calibration; instead we applied the reference conversion function to convert the readings into SI units.

In the field, the protection afforded by our enclosure proved to be inadequate. When wet, the sensor would create a low-resistance path between the power supply terminals. Such behavior would manifest itself in either abnormally large (more than 150%) or very small humidity readings (raw readings of 0V). Fig. 8 shows the humidity and voltage readings as well as the packet reception rates of selected nodes during both rainy and dry days in early August. Nodes 17 and 29 experienced a large drop in voltage while recording an abnormally high humidity readings on Aug. 5 and 6. We attribute the voltage drop to excessive load on the batteries caused by the wet sensor. Node 18 shows an more severe effect of rain: on Aug. 5, it crashes just as the other sensors register a rise in the humidity

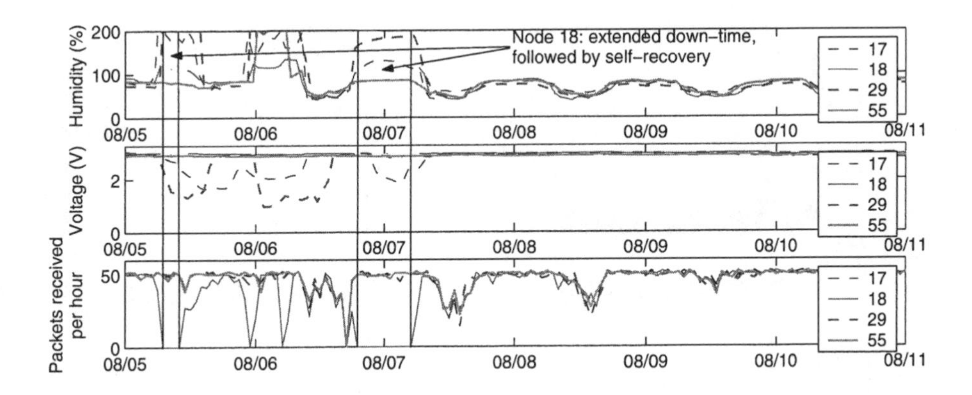

Fig. 8. Sensor behavior during the rain. Nodes 17 and 29 experience substantial drop in voltage, while node 55 crashes. When the humidity sensor recovers, the nodes recover.

readings. Node 18, on the other hand, seems to be well protected: it registers high humidity readings on Aug. 6, and its voltage and packet delivery rates are not correlated with the humidity readings. Nodes that experienced the high humidity readings typically recover when they dried up; nodes with the unusually low readings would fail quickly. While we do not have a definite explanation for such behavior, we evaluate that characteristics of the sensor board as a failure indicator below.

Thermopile readings: The data from the thermopile sensor proved difficult to analyze. The sensor measures two quantities: the ambient temperature and the infrared radiation incident on the element. The sum of thermopile and thermistor readings yields the object surface temperature, *e.g.* a bird. We would expect that the temperature readings from the thermistor and from the infrared temperature sensor would closely track each other most of the time. By analyzing spikes in the IR readings, we should be able to deduce the bird activity.

The readings from the thermistor do, in fact, track closely with the temperature readings. Fig. 9 compares the analog thermistor with the digital maxim temperature sensor. The readings are closely correlated although different on an absolute scale. A best linear fit of the temperature data to the thermistor readings on a per sensor per day basis yields a mean error of less than 0.9^oC, within the half step resolution of the digital sensor. The best fit coefficient varies substantially across the nodes.

Assigning biological significance to the infrared data is a difficult task. The absolute readings often do not fall in the expected range. The data exhibits a lack of any periodic daily patterns (assuming that burrow occupancy would exhibit them), and the sensor output appears to settle quickly in one of the two extreme readings. In the absence of any ground truth information, *e.g.* infrared camera images corresponding to the changes in the IR reading, the data is inconclusive.

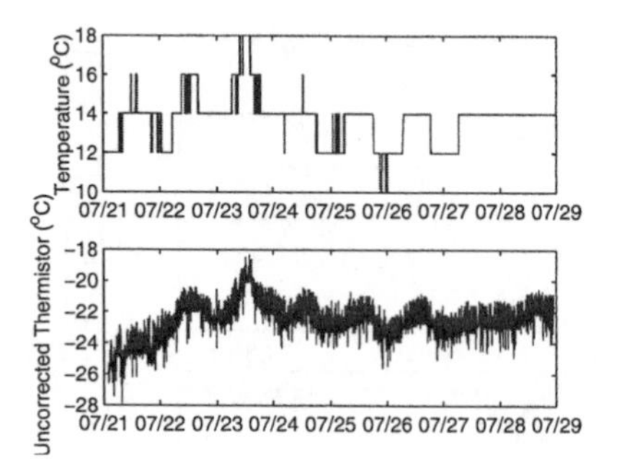

Fig. 9. The temperature sensor (top) and thermistor (bottom), though very different on the absolute scale, are closely correlated: a linear fit yields a mean error of less than 0.8°C.

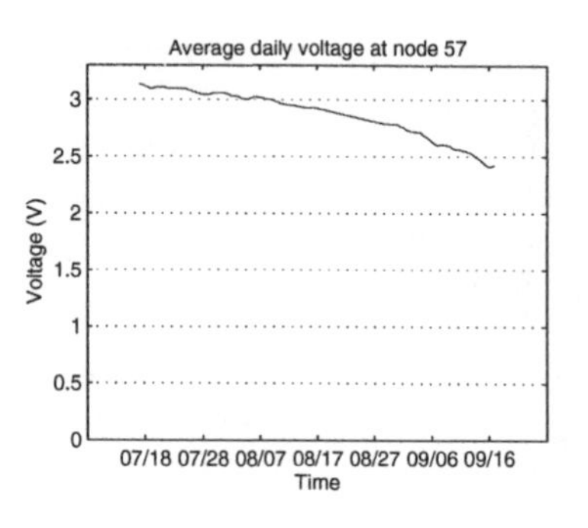

Fig. 10. Voltage readings from node 57. Node 57 operates until the voltage falls below 2.3V; at this point the alkaline cells can not supply enough current to the boost converter.

4.2 Power Management

As mentioned in Section 2, one of the main challenges was sensor node power management. We evaluate the power management in the context of the few nodes that did not exhibit other failures. Motes do not have a direct way of measuring the energy they consumed, instead we use battery voltage as an indirect measure. The analysis of the aggregate population is somewhat complicated by in-the-field battery replacements, failed voltage indicators, failed sensors and gaps in the data caused by the database crashes. Only 5 nodes out of 43 have clearly exhausted their original battery supply. This limited sample makes it difficult to perform a thorough statistical analysis. Instead we examine the battery voltage of a single node without other failures. Fig. 10 shows the battery voltage of a node as a function of time. The battery is unable to supply enough current to power the node once the voltage drops below 2.30V. The boost converter on the Mica mote is able to extract only 15% more energy from the battery once the voltage drops below 2.5V (the lowest operating voltage for the platform without the voltage regulation). This fell far short of our expectations of being able to drain the batteries down to 1.6V, which represents an extra 40% of energy stored in a cell [6]. The periodic, constant power load presented to the batteries is ill suited to extract the maximum capacity. For this class of devices, a better solution would use batteries with stable voltage, *e.g.* some of the lithium-based chemistries. We advocate future platforms eliminate the use of a boost converter.

4.3 Node Failure Indicators

In the course of data analysis we have identified a number of anomalous behaviors: erroneous sensor readings and application phase skew. The humidity sensor seemed to be a good indicator of node health. It exhibited 2 kinds of erroneous behaviors: very high and very low readings. The high humidity spikes,

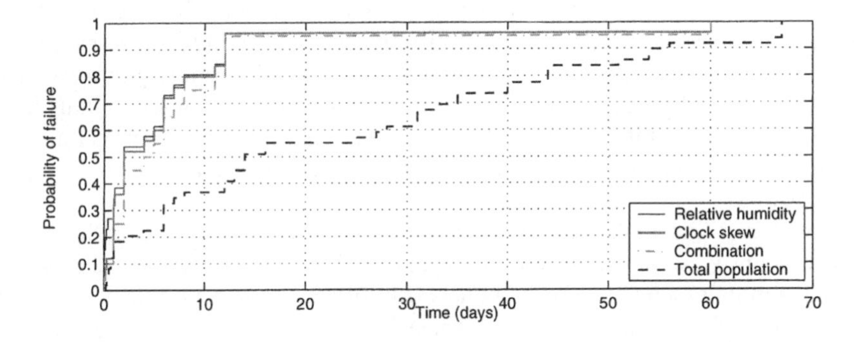

Fig. 11. Cumulative probability of node failure in the presence of clock skew and anomalous humidity readings compared with the entire population of nodes.

even though they drained the mote's batteries, correlated with recoverable mote crashes. The humidity readings corresponding to a raw voltage of 0V correlated with permanent mote outage: 55% of the nodes with excessively low humidity readings failed within two days. In the course of packet phase analysis we noted some motes with slower than usual clocks. This behavior also correlates well with the node failure: 52% of nodes with such behavior fail within two days.

These behaviors have a very low false positive detection rate: only a single node exhibiting the low humidity and two nodes exhibiting clock skew (out of 43) exhausted their battery supply instead of failing prematurely. Fig. 11 compares the longevity of motes that have exhibited either the clock skew or a faulty humidity sensor against the survival curve of mote population as a whole. We note that 50% of motes with these behaviors become inoperable within 4 days.

5 Related Work

Traditional data loggers for habitat monitoring are typically large in size and expensive. They require that intrusive probes and corresponding equipment immediately adjacent. They are typically used since they are commercially available, supported, and provide a variety of sensors. One such data logger is the Hobo Data Logger [19]. Due to size, price, and organism disturbance, using these systems for fine-grained habitat monitoring is inappropriate.

Other habitat monitoring studies install one or a few sophisticated weather stations an "insignificant distance" from the area of interest. With this method, biologists cannot gauge whether the weather station actually monitors a different microclimate due to its distance from the organism being studied. Using the readings from the weather station, biologists make generalizations through coarse measurements and sparsely deployed weather stations. Instead, we strive to provide biologists the ability to monitor the environment on the scale of the organism, not on the scale of the biologist [8,21].

Habitat monitoring for WSNs has been studied by a variety of other research groups. Cerpa et. al. [3] propose a multi-tiered architecture for habitat monitoring. The architecture focuses primarily on wildlife tracking instead of habitat monitoring. A PC104 hardware platform was used for the implementation with future work involving porting the software to motes. Experimentation using a hybrid PC104 and mote network has been done to analyze acoustic signals [23], but no long term results or reliability data has been published. Wang et. al. [22] implement a method to acoustically identify animals using a hybrid iPaq and mote network.

ZebraNet [14] is a WSN for monitoring and tracking wildlife. ZebraNet nodes are significantly larger and heavier than motes. The architecture is designed for an always mobile, multi-hop wireless network. In many respects, this design does not fit with monitoring the Leach's Storm Petrel at static positions (burrows). ZebraNet, at the time of this writing, has not yet had a full long-term deployment.

The number of deployed wireless sensor network systems is extremely low. The Center for Embedded Network Sensing (CENS) has deployed their Extensible Sensing System [7] at the James Mountain Reserve in California. Their architecture is similar to ours with a variety of sensor patches connected via a transit network that is tiered. Intel Research has recently deployed a network to monitor Redwood canopies in Northern California and a second network to monitor vineyards in Oregon. We deployed a second generation multihop habitat monitoring network on Great Duck Island, Maine in the summer of 2003. These networks are in their infancy but analysis may yield the benefits of various approaches to deploying habitat monitoring systems.

6 Conclusion

We have analyzed the environmental data from one of the first outdoor deployments of WSNs. While the deployment exhibited very high node failure rates, it yielded valuable insight into WSN operation that could not have been obtained in simulation or in an indoor deployment. We have identified sensor features that predict a 50% node failure within 4 days. We analyzed the application-level data to show complex behaviors in low levels of the system, such as MAC-layer synchronization of nodes. We have shown that great care must be taken when deploying WSNs such that the MAC implementation, network topology, synchronization, sensing modalities, sensor design, and packaging be designed with awareness of each other.

Sensor networks do not exist in isolation from their environment; they are embedded within it and greatly affected by it. This work shows that the anomalies in sensor readings can predict node failures with high confidence. Prediction enables pro-active maintenance and node self-maintenance. This insight will be very important in the development of self-organizing and self-healing WSNs.

Notes

Data from the wireless sensor network deployment on Great Duck Island can be view graphically at `http://www.greatduckisland.net`. Our website also includes the raw data for researchers in both computer science and the biological sciences to download and analyze.

This work was supported by the Intel Research Laboratory at Berkeley, DARPA grant F33615-01-C1895 (Network Embedded Systems Technology), the National Science Foundation, and the Center for Information Technology Research in the Interest of Society (CITRIS).

References

1. Bulusu, N., Bychkovskiy, V., Estrin, D., and Heidemann, J. Scalable, ad hoc deployable, RF-based localization. In *Proceedings of the Grace Hopper Conference on Celebration of Women in Computing* (Vancouver, Canada, Oct. 2002).
2. Bychkovskiy, V., Megerian, S., Estrin, D., and Potkonjak, M. Colibration: A collaborative approach to in-place sensor calibration. In *Proceedings of the 2nd International Workshop on Information Processing in Sensor Networks (IPSN'03)* (Palo Alto, CA, USA, Apr. 2003).
3. Cerpa, A., Elson, J., Estrin, D., Girod, L., Hamilton, M., and Zhao, J. Habitat monitoring: Application driver for wireless communications technology. In *2001 ACM SIGCOMM Workshop on Data Communications in Latin America and the Caribbean* (San Jose, Costa Rica, Apr. 2001).
4. Chalmers, D., and Sloman, M. A survey of Quality of Service in mobile computing environments. *IEEE Communications Surveys 2*, 2 (1992).
5. Chen, B., Jamieson, K., Balakrishnan, H., and Morris, R. Span: An energy-efficient coordination algorithm for topology maintenance in ad hoc wireless networks. In *Proceedings of the 7th Annual International Conference on Mobile Computing and Networking* (Rome, Italy, July 2001), ACM Press, pp. 85–96.
6. Eveready Battery Company. Energizer no. x91 datasheet. `http://data.energizer.com/datasheets/library/primary/alkaline/energizer_e2/x91.pdf`.
7. Hamilton, M., Allen, M., Estrin, D., Rottenberry, J., Rundel, P., Srivastava, M., and Soatto, S. Extensible sensing system: An advanced network design for microclimate sensing. `http://www.cens.ucla.edu`, June 2003.
8. Happold, D. C. The subalpine climate at smiggin holes, Kosciusko National Park, Australia, and its influence on the biology of small mammals. *Arctic & Alpine Research 30* (1998), 241–251.
9. He, T., Blum, B., Stankovic, J., and Abdelzaher, T. AIDA: Adaptive Application Independant Data Aggregation in Wireless Sensor Networks. *ACM Transactions in Embedded Computing Systems (TECS), Special issue on Dynamically Adaptable Embedded Systems* (2003).
10. Hill, J., and Culler, D. Mica: a wireless platform for deeply embedded networks. *IEEE Micro 22*, 6 (November/December 2002), 12–24.
11. Hill, J., Szewczyk, R., Woo, A., Hollar, S., Culler, D., and Pister, K. System architecture directions for networked sensors. In *Proceedings of the 9th International Conference on Architectural Support for Programming Languages and Operating Systems (ASPLOS-IX)* (Cambridge, MA, USA, Nov. 2000), ACM Press, pp. 93–104.
12. Huntington, C., Butler, R., and Mauck, R. *Leach's Storm Petrel (Oceanodroma leucorhoa)*, vol. 233 of *Birds of North America*. The Academy of Natural Sciences, Philadelphia and the American Orinthologist's Union, Washington D.C., 1996.

13. INTANAGONWIWAT, C., GOVINDAN, R., AND ESTRIN, D. Directed diffusion: a scalable and robust communication paradigm for sensor networks. In *Proceedings of the 6th Annual International Conference on Mobile Computing and Networking* (Boston, MA, USA, Aug. 2000), ACM Press, pp. 56–67.
14. JUANG, P., OKI, H., WANG, Y., MARTONOSI, M., PEH, L.-S., AND RUBENSTEIN, D. Energy-efficient computing for wildlife tracking: Design tradeoffs and early experiences with ZebraNet. In *Proceedings of the 10th International Conference on Architectural Support for Programming Languages and Operating Systems (ASPLOS-X)* (San Jose, CA, USA, Oct. 2002), ACM Press, pp. 96–107.
15. KRISHANAMACHARI, B., ESTRIN, D., AND WICKER, S. The impact of data aggregation in wireless sensor networks. In *Proceedings of International Workshop of Distributed Event Based Systems (DEBS)* (Vienna, Austria, July 2002).
16. LIU, J., CHEUNG, P., GUIBAS, L., AND ZHAO, F. A dual-space approach to tracking and sensor management in wireless sensor networks. In *Proceedings of the 1st ACM International Workshop on Wireless Sensor Networks and Applications* (Atlanta, GA, USA, Sept. 2002), ACM Press, pp. 131–139.
17. MADDEN, S., FRANKLIN, M., HELLERSTEIN, J., AND HONG, W. TAG: a Tiny AGgregation service for ad-hoc sensor networks. In *Proceedings of the 5th USENIX Symposium on Operating Systems Design and Implementation (OSDI '02)* (Boston, MA, USA, Dec. 2002).
18. MAINWARING, A., POLASTRE, J., SZEWCZYK, R., CULLER, D., AND ANDERSON, J. Wireless sensor networks for habitat monitoring. In *Proceedings of the 1st ACM International Workshop on Wireless Sensor Networks and Applications* (Atlanta, GA, USA, Sept. 2002), ACM Press, pp. 88–97.
19. ONSET COMPUTER CORPORATION. HOBO weather station. `http://www.onsetcomp.com`.
20. POLASTRE, J. Design and implementation of wireless sensor networks for habitat monitoring. Master's thesis, University of California at Berkeley, 2003.
21. TOAPANTA, M., FUNDERBURK, J., AND CHELLEMI, D. Development of Frankliniella species (Thysanoptera: Thripidae) in relation to microclimatic temperatures in vetch. *Journal of Entomological Science 36* (2001), 426–437.
22. WANG, H., ELSON, J., GIROD, L., ESTRIN, D., AND YAO, K. Target classification and localization in habitat monitoring. In *Proceedings of IEEE International Conference on Acoustics, Speech, and Signal Processing (ICASSP 2003)* (Hong Kong, China, Apr. 2003).
23. WANG, H., ESTRIN, D., AND GIROD, L. Preprocessing in a tiered sensor network for habitat monitoring. *EURASIP JASP Special Issue on Sensor Networks 2003*, 4 (Mar. 2003), 392–401.
24. WHITEHOUSE, K., AND CULLER, D. Calibration as parameter estimation in sensor networks. In *Proceedings of the 1st ACM International Workshop on Wireless Sensor Networks and Applications* (Atlanta, GA, USA, Sept. 2002), ACM Press.
25. XU, Y., HEIDEMANN, J., AND ESTRIN, D. Geography-informed energy conservation for ad hoc routing. In *Proceedings of the 7th Annual International Conference on Mobile Computing and Networking* (Rome, Italy, July 2001), ACM Press.
26. ZHAO, J., GOVINDAN, R., AND ESTRIN, D. Computing aggregates for monitoring wireless sensor networks. In *Proceedings of the 1st IEEE International Workshop on Sensor Network Protocols and Applications* (Anchorage, AK, USA, May 2003).

Prototyping Wireless Sensor Network Applications with BTnodes

Jan Beutel[1], Oliver Kasten[2], Friedemann Mattern[2], Kay Römer[2], Frank Siegemund[2], and Lothar Thiele[1]

[1] Computer Engineering and Networks Laboratory
Department of Information Technology and Electrical Engineering
ETH Zurich, Switzerland
{beutel,thiele}@tik.ee.ethz.ch
[2] Institute for Pervasive Computing
Department of Computer Science
ETH Zurich, Switzerland
{kasten,mattern,roemer,siegemun}@inf.ethz.ch

Abstract. We present a hardware and software platform for rapid prototyping of augmented sensor network systems, which may be temporarily connected to a backend infrastructure for data storage and user interaction, and which may also make use of actuators or devices with rich computing resources that perform complex signal processing tasks. The use of Bluetooth as the wireless networking technology provides us with a rich palette of Bluetooth-enabled commodity devices, which can be used as actuators, infrastructure gateways, or user interfaces. Our platform consists of a Bluetooth-based sensor node hardware (the BTnode), a portable operating system component, and a set of system services. This paper gives a detailed motivation of our platform and a description of the platform components. Though using Bluetooth in wireless sensor networks may seem counter-intuitive at first, we argue that the BTnode platform is indeed well suited for prototyping applications in this domain. As a proof of concept, we describe two prototype applications that have been realized using the BTnodes.

1 Introduction

Recent advances in wireless communication and micro system technology allow the construction of so-called "sensor nodes". Such sensor nodes combine means for sensing environmental parameters, processors, wireless communication capabilities, and an autonomous power supply in a single tiny device that can be fabricated in large numbers at low cost. Large and dense networks of these untethered devices can then be deployed unobtrusively in the physical environment in order to monitor a wide variety of real-world phenomena with unprecedented quality and scale while only marginally disturbing the observed physical processes [1,9].

In other words, Wireless Sensor Networks (WSNs) provide the technological foundation for performing many "experiments" in their natural environment instead of using an artificial laboratory setting, thus eliminating many fundamental limitations of the latter. It is anticipated that a wide range of application domains can substantially benefit

H. Karl, A. Willig, A. Wolisz (Eds.): EWSN 2004, LNCS 2920, pp. 323–338, 2004.

from such a technological foundation. Biologists, for example, want to monitor the behavior of animals in their natural habitats; environmental research needs better means for monitoring environmental pollution; agriculture can profit from better means for observing soil quality and other parameters that influence plant growth; geologists need better support for monitoring seismic activity and its influences on the structural integrity of buildings; and of course the military is interested in monitoring activities in inaccessible areas.

These exemplary applications domains demonstrate one important property of WSNs: their inherent and close integration with the real world, with data being captured and processed automatically in the physical environment, often in real time. This is in contrast to traditional computing systems, which are typically mostly decoupled from the real world. This paradigmatic change comes with a number of important implications, also with respect to the development and deployment of WSNs.

One such implication is that sensor networks are hard to simulate, since their exact behavior very much depends on many parameters of the physical environment. Already rather subtle differences of the simulation models and the real world can make the difference between a working and a broken implementation [14]. For example, many wireless radio simulations assume a spherical communication range. However, measurements in a network of about 100 sensor nodes revealed that the radio communication range is far from spherical in practice [12]. This can lead to asymmetric or even unidirectional communication links, which can easily break algorithms that assume symmetrical links.

Hence, experimentation platforms for performing real experiments are of particular importance in the sensor network domain. As part of various research projects, a number of such sensor network platforms were developed [3,16]. These platforms are typically optimized for energy efficiency, providing rather restricted processing capabilities and a low-power radio platform. While this leads to sensor networks with improved longevity, the chosen RF-communication technology also imposes a number of limitations. This is especially true in experimental settings, where the purpose is the development and evaluation of algorithms and prototypical application scenarios.

We have explored a different point of the design space and have developed a sensor node – the BTnode – which is optimized for programming comfort and interoperability, using a more powerful processor, more memory, and a Bluetooth radio. Our choice was motivated by the need for a functional, versatile fast prototyping platform that is certainly not the ultimate in low-power device technology but rather available today, to an extent energy aware and low cost. Additionally, we were inspired by the observation that sensor networks are almost always part of larger systems consisting of heterogeneous devices, whose collaboration requires an interoperable networking technology.

In the remainder of this paper, we present the BTnode platform, including the hardware, the operating system software, and the service infrastructure. We begin with motivating our design decisions and the use of Bluetooth as a networking technology for sensor networks. Sect. 3 describes the BTnode hardware and its operating system while Sect. 4 characterizes the system services we designed for building complex applications. As a proof of concept, we present two such applications that have been developed using our platform in Sect. 5. Power consumption issues are described in Sect. 6. Related work and conclusions are presented in Sect. 7 and 8, respectively.

2 Motivation

Our sensor network platform uses Bluetooth as a communication means. Many features of the Bluetooth specification and its implementations make the use of Bluetooth in sensor networks inadequate at first glance. But on the other hand, Bluetooth has become an interoperable wireless networking standard which is implemented in a number of consumer devices (e.g., PDAs, laptops, mobile phones, digital cameras) or hardware extensions (e.g., USB sticks, PCMCIA cards), and is supported by all major operating systems. This interoperability with a large variety of devices makes Bluetooth valuable in the sensor network domain, in particular for prototyping of applications now while other technologies are still under development. Below we outline some of the more interesting Bluetooth features and sketch concrete applications where these come in handy.

Bluetooth Features. When compared with "traditional" communication approaches for wireless sensor networks, Bluetooth has a rather high power consumption, suffers from somewhat long connection setup times, and has a lower degree of freedom with respect to possible network topologies [17].

On the other hand, Bluetooth is a connection-oriented, wireless communication medium that assures interoperability between different devices and enables application development through a standardized interface. It offers a significantly higher bandwidth (about 1 megabit per second) compared to many low-power radios (about 50 kilobit per second). Furthermore, the standardized Host Controller Interface (HCI) provides a high level interface that requires no knowledge of the underlying baseband and media access layers and their respective processing. This abstraction offers built-in high-level link-layer functionality such as isochronous and asynchronous communication, link multiplexing, QoS, integrated audio, forward error correction, automatic packet retransmission, user authentication using link keys, and encryption. Bluetooth also offers service discovery, serial port emulation, and IP connectivity.

Infrastructure Integration. Sensor networks are typically connected to some back-end infrastructure for tasking the network, as well as for storage and evaluation of sensing results. Also, since limited resources preclude the execution of many services (e.g., for complex data processing) in the sensor network, such services might be provided by an external infrastructure possessing sufficient resources.

For these reasons it is often necessary to connect a sensor network to an external network, such as the Internet. In the Great Duck Island habitat monitoring experiment [20], for example, a number of isolated sensor network patches are linked to a base station, which in turn is connected to the Internet via a satellite link. A database system executing on the Internet is used to store and query sensing results.

To implement such solutions, typically proprietary gateway solutions are required to bridge between the sensor network and the Internet. However, if the sensor nodes support Bluetooth, off-the-shelf devices can be used to implement such solutions. A Bluetooth-enabled laptop computer, for example, can be used to bridge between Bluetooth and WLAN. Instead of a satellite link, a mobile phone can be used as a gateway to connect a sensor network to the Internet.

User Interaction. Many Bluetooth-enabled devices can also be used for direct interaction with isolated sensor networks that are not connected to a background infrastructure. Interaction with a user might be required to specify the sensing task, or to alert the user of sensing results. A Bluetooth-enabled mobile phone, for example, can be used to alert a nearby user. By placing a phone in the vicinity of the sensor network, the sensor network can even send short text messages (SMS) to a user who is not currently present. It is advantageous to be able to use a widely spread class of devices with a well acquainted user interface for such interactions.

Actuators. In some sensor network applications, just storing measurement results for later evaluation is not sufficient but an immediate action is needed upon detecting certain phenomena – either by means of automated actuation or by notifying a user. Consider for example an animal habitat monitoring application, where it might be useful to notify a biologist or to take a picture in case a certain animal is present.

If the sensor network uses Bluetooth, we can choose among a wide variety of commercial off-the-shelf Bluetooth-enabled consumer devices to implement such actuation. For example, the sensor network can trigger a Bluetooth-enabled camera to take a picture when it suspects something interesting is happening.

Debugging. The development of algorithms and applications for sensor networks is non-trivial for various reasons. Firstly, sensor networks are highly dynamic distributed systems, where a consistent view of the global system state is typically not available to the developer. Secondly, the behavior of the sensor network is highly dependent on the physical environment, such that problems might not be easily reproducible. Thirdly, for reasons of energy efficiency, sensor network applications do often perform in-network data processing and aggregation, where raw sensor data is already processed and evaluated inside the network in order to reduce the volume of data which has to be transmitted. To verify and debug such systems, the developer often needs access to both, the raw sensor data and the aggregated output of the sensor network [8].

Currently, research groups working on the development of algorithms and applications for sensor networks typically use arrays of sensor nodes, where each node is connected to a central computer by a serial cable and a serial port multiplexer. While this is possible in small-scale laboratory settings, large-scale field experiments typically cannot be instrumented this way. On the other hand, mobile terminals can give untethered access to groups of sensor nodes in the field for in situ debugging, provided they both use the same radio technology. Here Bluetooth is an ideal candidate.

Sensor Node Heterogeneity. As mentioned earlier, sensor networks should employ in-network data processing and aggregation in order to reduce the amount of data that has to be transmitted within the network. This is desirable in order to achieve energy efficiency and to match the typically limited capacity of the communication channels. However, many research projects came to the conclusion that the computing and memory resources of sensor nodes are often too limited to perform typical signal processing tasks (e.g., FFT, signal correlation). Hence, clustered architectures were suggested [13,27], where the cluster-heads are equipped with more computing and memory resources. Cluster-heads then process and aggregate sensor data collected from the nodes in their cluster. In [13,

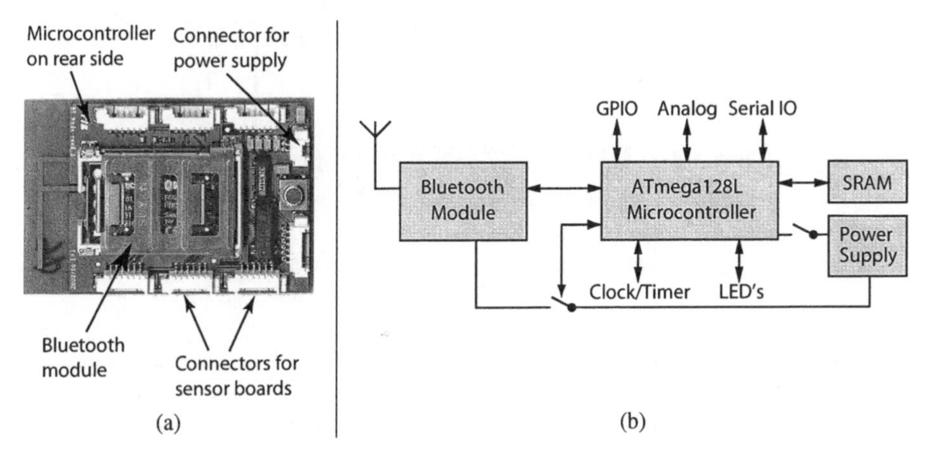

Fig. 1. (a) BTnode hardware (b) BTnode system overview.

27], for example, PDAs with proprietary hardware extensions for wirelessly interfacing the sensor nodes are used for this purpose. With Bluetooth-enabled sensor nodes, off-the-shelf PDAs and laptops can be easily integrated as cluster heads without hardware modification.

3 BTnode Architecture

3.1 Hardware

The BTnode is an autonomous wireless communication and computing platform based on a Bluetooth radio module and a microcontroller. The device has no integrated sensors, since individual sensor configurations are required depending on the application. Instead, with its many general-purpose interfaces, the BTnode can be used with various peripherals, such as sensors, but also actuators, DSPs, serial devices (like GPS receivers, RFID readers, etc.) and user interface components. An interesting property of this platform is its small form factor of 6x4 cm while still maintaining a standard wireless interface.

The BTnode hardware (see Fig. 1) is built around an Atmel ATmega128L microcontroller with on-chip memory and peripherals. The microcontroller features an 8-bit RISC core delivering up to 8 MIPS at a maximum of 8 MHz. The on-chip memory consists of 128 kbytes of in-system programmable Flash memory, 4 kbytes of SRAM, and 4 kbytes of EEPROM. There are several integrated peripherals: JTAG for debugging, timers, counters, pulse-width modulation, 10-bit analog-digital converter, I2C bus, and two hardware UARTs. An external low-power SRAM adds an additional 240 kbytes of data memory to the BTnode system. A real-time clock is driven by an external quartz oscillator to support timing updates while the device is in low-power sleep mode. The system clock is generated from an external 7.3728 MHz crystal oscillator.

An Ericsson Bluetooth module is connected to one of the serial ports of the microcontroller using a detachable module carrier, and to a planar inverted F antenna (PIFA) that is integrated into the circuit board.

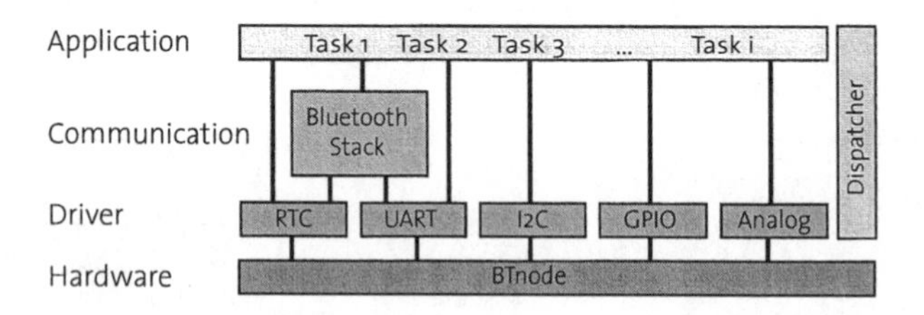

Fig. 2. A lightweight OS framework for WSN applications.

Four LEDs are integrated, mostly for the convenience of debugging and monitoring. One analog line is connected to the battery input and allows to monitor the battery status. Connectors that carry both power and signal lines are provided and can be used to add external peripherals, such as sensors and actuators.

3.2 Lightweight Operating System Support

The BTnode system software (see Fig. 2) is a lightweight OS written in C and assembly language. The drivers, which are available for many hardware subsystems, provide convenient APIs for application developers. The system provides coarse-grained cooperative multitasking and supports an event-based programming model.

Drivers. The drivers are designed with fixed buffer lengths that can be adjusted at compile time to meet the stringent memory requirements. Available drivers include memory, real-time clock, UART, I2C, LEDs, power modes, and AD converter. The driver for the Bluetooth radio provides a subset of the networking functionality according to the Bluetooth specification. Bluetooth link management is performed on Bluetooth's L2CAP layer. RFCOMM, a serial port emulation, provides connectivity to computer terminals and consumer devices, such as cameras and mobile phones. Dial-up connections to a modem server through a mobile GSM phone are easily established with a special-purpose function. All other GSM services (such as file sharing, phone book and calendar) can be utilized through lower-level interfaces.

Process Model. Like most embedded systems, the BTnode does not support a dynamic process model. Only a single application is present on the system at a time. At compile time, applications are linked to the system software, which comes as a library. The resulting executable is then uploaded to the BTnode's Flash memory, effectively overwriting any previous application code. After uploading, the new application starts immediately. However, the BTnode system can also be reprogrammed through the network. To do so, the application currently running on the system needs to receive the new executable, save it to SRAM, and then reboot the system in bootloader mode. The bootloader finally transfers the received executable to Flash memory and starts it.

Programming Model. The system is geared towards the processing of (typically externally triggered) events, such as sensor readings, or the reception of data packets on the

```
1: #include <btnode.h>
   #define THRESHOLD_EV (MAX_SYSTEM_EVENT+1)
3: static u16 connection_id = 0;

5: int main( int argc, char* argv[] ) {
     btn_system_init( argc, argv, /* ... */ );
     btn_bt_psm_add( 101 ); /* accept connections */
8:   btn_disp_ev_reg( BT_CONN_EV, conn_cb, 0 );
9:   btn_disp_run();
     return 0; /* not reached */
   }
12: void conn_cb( call_data_t call_data, cb_data_t cb_data ) {
     connection_id = (u16)(call_data & 0xFFFF);
     bt_sensor_start( BTN_SENSOR_TEMP );
15:  bt_sensor_set_threshold( BTN_SENSOR_TEMP, 30, THRESHOLD_EV );
16:  btn_disp_ev_reg( THRESHOLD_EV, sensor_cb, 0 );
   }
18: void sensor_cb( call_data_t call_data, cb_data_t cb_data ) {
     u8 error = 0;
     u8 databuf = (call_data & 0xFFFF);
     btn_bt_data_send( connection_id, databuf, sizeof(data_buffer) );
   }
```

Fig. 3. A simple BTnode program.

radio. To this end, BTnode applications follow an event-based programming model, common to embedded systems and GUI programming toolkits, where the system schedules application tasks on the occurrence of events. Internally, those tasks (or event handlers) are implemented as C functions.

In the BTnode system, events model state changes. A set of predefined event types is used by the drivers to indicate critical system conditions, such as expired timeouts or data being ready for reading on an I/O device. Applications can define their own event types, which are represented by 8-bit integer values. Individual events can carry type-specific parameters for evaluation in the application task.

A central component of the system is the dispatcher, where applications register event/event-handler-function pairs. After the dispatcher has been started, it also accepts individual events from drivers and application tasks and stores them in a FIFO queue. While the event queue is not empty, the dispatcher starts the corresponding handler (i.e., the previously registered function) for the next event in the queue, passing the events individual parameters.

Events are processed in the order they are received by the dispatcher. Only one event handler can be active at a time and every event handler is always completely executed before the next is scheduled. So every event handler depends on the previous event handler to terminate in time. However, long tasks can be broken into smaller pieces and can be triggered subsequently by the event mechanism (e.g., by application-specific events).

Fig. 3 shows a typical BTnode program, which is waiting for an incoming Bluetooth connection. Once the connection is established, it repeatedly transmits sensor data exceeding a given threshold. During initialization, the program registers a handler for connection events (line 8). It then passes control to the dispatcher (line 9), which enters sleep mode until events occur that need processing. Once the connection is established, the corresponding handler function `conn_cb` is called by the dispatcher (line 12). The program initializes the sensors with a threshold value (line 15) and registers the event handler `sensor_cb` for handling this event (line 16). On the occurrence of the sensor event, the associated data is sent over the connection (line 18 ff).

Portability. The whole system software is designed for portability and is available for different emulation environments (x86 and iPAQ Linux, Cygwin, and Mac OS X) apart from the embedded platform itself. Emulation simplifies application building and speeds up debugging since developers can rely on the sophisticated debugging tools available on desktop systems. Also, the time for uploading the sensor-node application to the embedded target can be saved. Furthermore, different platforms, such as an iPAQ running Linux, can be used as cluster heads, reusing much of the software written for the embedded nodes and making use of the resources of the larger host platform for interfacing, extended computation or storage.

4 System Services

Besides the core system software, the BTnode platform also provides means to integrate sensor nodes into a surrounding communication infrastructure, to exchange and collaboratively process the data of locally distributed sensors, and to outsource computations to nearby devices with more sophisticated resources. These functions constitute the core of the BTnode system services.

4.1 Ad hoc Infrastructure Access

As noted in Sect. 2, sensor networks often need access to a background infrastructure. However, due to their deployment in remote, inaccessible, or undeveloped regions, fixed access to a background infrastructure via stationary gateways usually cannot be assumed. We therefore provide *mobile ad hoc infrastructure access* by means of mobile gateways. The core idea of this approach is depicted in Fig. 4: when handheld devices – such as Bluetooth-enabled mobile phones or PDAs – are placed in the vicinity of the sensor network, BTnodes can utilize these mobile devices to access background services in an ad hoc fashion. Such mobile gateways provide access to an infrastructure via GSM or WLAN, for example.

The BTnode system provides services for utilizing nearby mobile phones as mobile infrastructure gateways. An RFCOMM connection is set up between a BTnode and the nearby phone. AT commands manage and control GSM data connections to a background infrastructure server, which in turn must provide a regular modem or GSM gateway itself. Alternatively, a BTnode can embed sensory data into an SMS message and transfer it over the public phone network. Depending on the application, both approaches have their advantages and disadvantages. While building up a GSM data connection is more suitable for sending larger amounts of data, sending data in an SMS message is asynchronous and does not require connection establishment with the communication partner. The infrastructure provides software for running a GSM gateway and for processing of incoming messages (see Fig. 4).

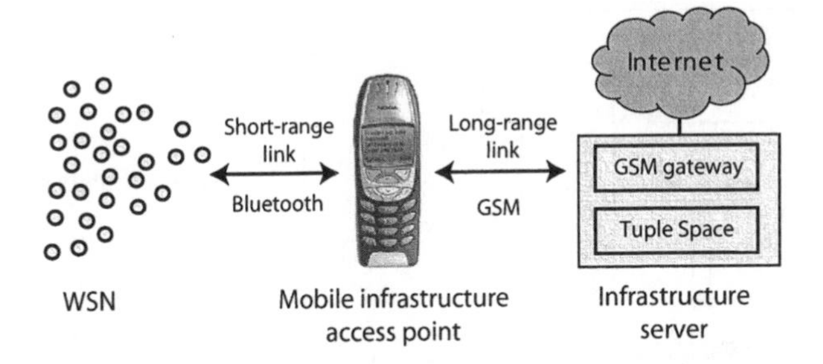

Fig. 4. Ad hoc infrastructure access with handheld devices as mobile gateways.

Sensor networks or parts thereof can be temporarily disconnected from the background infrastructure. Our service infrastructure provides a centralized tuple space [5,7] in order to enable cooperation across such temporarily isolated network patches. When disconnected, nodes buffer sensory data locally and synchronize with the tuple space when they regain connectivity. As infrastructure communication takes place via mobile gateways the delay for relaying the information over the centralized server can become relatively large in the order of tens of seconds.

4.2 Collaborative Processing of Sensory Data

A typical sensor node has only limited processing power, can perceive only a small local subset of its physical environment, and might not have all necessary sensors to solve a complex sensing task on its own. For these reasons, sensor nodes usually need to cooperate with others. The basic system concept that enables such kind of inter-node interaction in the BTnode system services is a distributed tuple space.

The distributed tuple space serves as a shared data structure for a set of sensor nodes which enables them to exchange and process data collaboratively. The main advantage of the tuple space approach is that nodes become able to share their resources, such as sensors and memory, with other nodes. For example, it is possible for nodes to operate on remote sensors as if they were locally attached, and to store data tuples at other nodes with free memory resources.

This distributed tuple space implementation does not require access to the centralized background infrastructure mentioned in the previous section. Instead, it is implemented as a data structure covering multiple sensor nodes. BTnodes read local sensors, embed sensory data into tuples, and write it into the distributed tuple space. Other nodes can then access these tuples and fuse the corresponding sensory data with local measurements. The corresponding result is usually embedded into a tuple that is again written into the space. Our implementation also offers the possibility to register callbacks on certain sensor types, which is very handy for sensor fusion algorithms. For example, a callback could be registered at a remote sensor for a temperature tuple, which would notify the local node each time a tuple is generated. The temperature data from the callback could then be fused with the local node's own temperature readings.

The distributed tuple space also makes it easier for an application programmer to design sensor fusion algorithms because they can operate on a broader common data

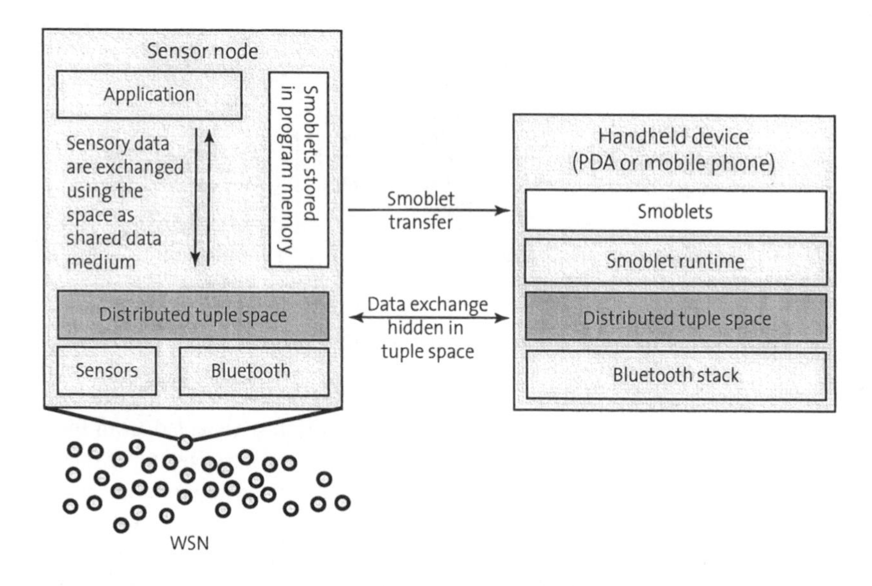

Fig. 5. Ad hoc outsourcing of computations to handheld devices with Smoblets.

basis. From the programmer's viewpoint, the tuple space is a grouping concept based on a node's current situation or on application-specific information. Therefore, sensor nodes that share their data in a distributed tuple space appear like a single node with many sensors and resources. This simplifies the system view for the application programmer.

4.3 Outsourcing Computations

In Sect. 2 we mentioned that some sensor network applications require the use of devices with more computing and memory resources than sensor nodes in order to perform, for example, complex signal processing tasks. Both the use of Bluetooth as the BTnodes' communication technology and the portability of the BTnode software makes it easy to integrate commodity devices such as Bluetooth-enabled PDAs into the sensor network, for example as cluster heads.

Additionally, for outsourcing computations spontaneously to nearby mobile Bluetooth-enabled devices, the BTnode system provides the so-called *Smoblet* service. Smoblets are precompiled Java classes, similar to applets, that are stored in the program memory of BTnodes and transferred to nearby mobile devices, where they are executed to process input from the sensor network. To enable this kind of cooperation, the mobile device joins the distributed tuple space of the sensor node that originally provided the Smoblet code. As the actual origin of sensory data and the actual location of resources becomes to some degree transparent through the tuple space, it becomes irrelevant on which device (handheld or sensor node) the code is actually executed. By transferring code to a nearby handheld device with more resources and by providing access to the same data basis through the distributed tuple space, sensor nodes can save considerable amounts of energy [27,11].

A Smoblet can also contain Java code for a user interface, which – when transmitted to the mobile device – allows the user to adapt certain settings of a sensor node or to control an application. The transmission of code can be initiated by a sensor node itself based on its current context, or by a user by means of a simple user interface on the handheld.

Fig. 5 gives an overview of the Smoblet concept. The BTnode system services provide a set of libraries that allow Java programs to operate on data in a distributed tuple space and to transfer Java code over a Bluetooth connection. An application programmer generates the Smoblet code using these libraries on a host machine and transforms the Java classes (or JAR archive files) into C source files. The programmer can then compile and link these files into the application for the sensor node. When the resulting program file is transferred to the microcontroller of the sensor node, the Smoblets are automatically stored into the program memory of that node. BTnodes themselves cannot execute Java code, the program memory serves merely as data storage for the Smoblets. When a handheld device now enters the range of the sensor node, the Smoblet code can be transferred over a local Bluetooth connection. Thereby, the handheld joins the distributed tuple space of the sensor node and executes the Smoblet code, which can now operate on the shared data structure provided by the distributed tuple space.

5 Applications

BTnodes have been used as a prototyping platform for a number of applications [2,11, 21,25,26]. Below we present two additional applications and discuss how they make use of the various features of our platform. The first application is a sensor network for tracking mobile objects using a remote-controlled toy car as a sample object. The second application is about monitoring the state of products during transportation on a truck.

5.1 Object Tracking

The purpose of this project is to explore the use of Smart Dust for tracking the movements of mobile entities in the open field (e.g., vehicles, animals). Smart Dust [10] are next generation sensor nodes, which provide sensing, computing, and wireless communication capabilities on a single device as tiny as a few cubic millimeters. The focus here is on support mechanisms such as node localization and time synchronization that are suitable for large networks of Smart Dust.

Since Smart Dust is not yet available, we built a prototype system based on BTnodes in order to evaluate the algorithms we developed. We used a remote-controlled toy car as a sample target. A number of BTnodes are randomly deployed in the area of interest and can change their location after deployment. The nodes use attached sensors to detect the presence of the car and send respective notifications to a base station. The base station fuses these notifications in order to estimate the current location of the car. A graphical user interface displays the track and allows to control various aspects of the system. The sensor data fusion requires that all nodes share a common reference system both in time and space, which necessitates mechanisms for node localization and time synchronization. The base station consists of a Linux laptop computer equipped with a Bluetooth radio. A detailed description of the system can be found in [23].

The hardware abstraction provided by the BTnode operating system software allowed us to develop and debug the BTnode software on a Linux platform. Moreover, the base station software executing on the Linux laptop uses the same operating system software as the BTnodes, which makes it particularly easy to port the base station software to another BTnode, which could then send off tracking results to a remote user via a mobile phone.

5.2 Product Monitoring

The goal of this application is to monitor the state of goods during transport such that the owner of a product can query the product's state and is notified whenever it is damaged. This notification should be instantaneous, containing information about when, where, and why a product has been damaged. To simplify prototyping, the application does not rely on extra stationary hardware installed inside vehicles. In our prototype, BTnodes are attached to ordinary products. The nodes themselves are equipped with acceleration and temperature sensors to determine the product's current state and to derive the reason why a product has been damaged. When a product is damaged, the attached BTnode tries to notify the owner of the product.

Background infrastructure access takes place using the Bluetooth-enabled mobile phone of the vehicle's driver. Whenever it is in range of the products, they use it as a mobile access point for communicating with a background infrastructure server (see Sect. 4). Between consecutive infrastructure accesses, BTnodes buffer data and synchronize their state with a background infrastructure service whenever access becomes possible. Additionally, the driver's mobile phone is used as location sensor. When a product is damaged, the corresponding BTnode obtains the current cell id and location area code from the mobile phone. The information about where an event occurred is then embedded into the message sent to the background infrastructure service.

The user interaction through the mobile phone consists of a WAP (Wireless Application Protocol) interface for querying the status of a product and an SMS message to notify the owner by a short text message whenever the product has been damaged. It is implemented as a Servlet that generates WAP pages providing information about the product's current state. User inputs can also be passed on directly to the product when a mobile access point enters the range of the attached BTnode.

6 Energy Consumption

One major criticism of Bluetooth and its use in sensor networks is its high energy consumption. However, as measurements in [18] confirm, the energy consumption per transmitted bit is competitive with dedicated low-power radios. The authors conclude, that Bluetooth is a viable solution energy-wise if the radio is only switched on for short burst transfers and stays turned off otherwise. Note that this fits many sensor network applications quite well, where network activity is rather bursty and short, triggered by rare activities in the environment (e.g., a product being damaged in the product monitoring application). There are two important strategies to switch off the radio frequently: duty-cycling and wake-up radio.

With *duty cycling*, the radio is switched on in a periodical pattern, trading off network latency for energy efficiency. Consider for example the product monitoring application described above. We can estimate the average battery lifetime of a BTnode as follows: A 10 % duty cycle for Bluetooth yields a quite acceptable average power consumption of 6.5 mW and a battery lifetime in the order of weeks on a standard 840 mAh Li-ion battery. Here, 12 mW, 160 mW, and 0.5 mW power consumption were assumed for sensing, communication, and idle modes, respectively.

For *wake-up radio* [24], a second low-power radio is attached to a BTnode. While Bluetooth is normally switched off, the low-power radio handles low-bandwidth data and network management traffic. Bursts of high bandwidth traffic are negotiated on the low-power channel. Both the receiver and sender then switch on their Bluetooth radios, before sending the actual data via Bluetooth. This strategy can save a significant amount of power if the high-power radio is only occasionally needed. We envision a future dual-radio platform, which would allow the implementation of such a strategy.

Additionally, newer Bluetooth hardware has considerably improved power characteristics compared to the modules used for the current generation of BTnodes, and is expected to reduce power consumption in communication mode by a factor of 2-4. However, even with the current hardware, energy consumption is low enough to actually build prototype applications that run for hours or days, enabling us to gain practical experience through experimentation.

7 Related Work

The classical approach to Wireless Sensor Network devices aims at low power and highly integrated node hardware. One of its most prominent representatives is the UC Berkeley Mote [16] that has been commercialized and is widely used by researchers all over the world. Other research groups have developed similar platforms, all based on a low power microcontroller and a custom radio front-end chip [3,16,22]. Common to these architectures is that all protocol processing (baseband and MAC) is done on the host microcontroller, which implies a meticulous design process, knowledge about RT systems, and locks up many resources on the host CPU. Another drawback of these platforms is that they can only communicate with devices equipped with the same radio and protocol processing capabilities. In contrast to that, the BTnodes can interact with any Bluetooth-enabled device without the need to integrate further hardware and software.

Other prototyping systems based on FPGAs, StrongARM processors, and similar hardware components [22] are geared towards protocol and communication development only. Other researches have tried to set up a development platform that contains as many subsystems as possible (e.g., the I-Badge system [6]), inducing considerable system complexity and also interfacing constraints.

Most of the available hardware platforms only provide a thin hardware abstraction layer, on top of which applications are executing. One particular exception is TinyOS [16], a component-oriented operating system designed for the Berkeley Motes. Similar to our approach it provides asynchronous events as a basic programming abstraction.

System services (such as multi-hop routing) as well as applications are implemented as a set of components with well-defined interfaces. A simple declarative language is used to connect these components in order to form a complete application. TinyOS applications are programmed in a custom-made C dialect, requiring special development tools. This is in contrast to our platform, which uses standard C and an unmodified GNU tool chain.

Various projects aim to provide a service infrastructure or middleware which supports the development of complex sensor network applications. TinyDB [15] interprets the sensor network as a distributed database being constantly filled with sensory data such that a simple SQL-like query language can be used to query the sensor network. SensorWare [4] uses an agent-like approach to program sensor networks for complex tasks. A scripting language is used to define small programs which are then distributed to the nodes of the network. Later on, executing scripts can migrate from node to node along with their state information. DSWare [19] uses a real-time variant of an event notification system, which allows to subscribe to specific events generated by sensor nodes. In contrast, we decided to follow an approach based on tuple spaces. We believe that this is a useful abstraction for many sensor applications that depend on fusing information from many nearby sensor nodes. Firstly, a tuple space provides a natural place for collecting and fusing this information. Secondly, since physical phenomena are often local, cooperation in sensor networks is also often local. Hence, local tuple spaces can be efficiently implemented by using broadcast communication over very few hops.

8 Conclusion and Future Work

We have presented the BTnode platform for prototyping sensor network applications. It consists of a Bluetooth-based sensor node, a lightweight operating system with support for several hardware platforms, and a set of system services to support the development of complex sensor network applications. As a proof-of-concept, we presented two complex applications which we developed using this platform.

We motivated our platform by the need for an environment to quickly prototype sensor network applications using commercial off the shelf devices in conjunction with sensor nodes. Additionally, we observed that sensor networks are often part of larger systems consisting of heterogeneous devices. Collaboration of the sensor nodes with these devices requires a common and widely supported networking technology, such as Bluetooth.

Up to now, about 200 BTnodes have been manufactured, which are used by a number of research projects across Europe. As part of the Swiss National Competence Center for Mobile Information and Communication Systems (NCCR-MICS) [29], we intend to develop the BTnode platform into a versatile and widely available prototyping platform for sensor network applications. A BTnode system-software kit consisting of a build environment (avr-gcc cross compiler and standard libraries), source code, debugging support, demo examples, and documentation has been assembled and is available to developers along with a support mailing list and software repository [28].

We are currently considering augmenting the next generation of BTnodes with an additional ultra low-power radio, which is developed by one of our partner institutions within NCCR-MICS. Among other things, this will allow us to build more energy-efficient sensor networks based on Bluetooth by using a wake-up strategy similar to the one described in [24] – thus combining the energy efficiency of ultra low power radios with the interoperability of Bluetooth.

Acknowledgments. The work presented in this paper was conducted as part of the Smart-Its project and the NCCR-MICS. The Smart-Its project is funded by the European Union under contract IST-2000-25428, and by the Swiss Federal Office for Education and Science (BBW 00.0281). NCCR-MICS, the National Competence Center in Research on Mobile Information and Communication Systems, is supported by the Swiss National Science Foundation under grant number 5005-67322.

References

1. I.F. Akyildiz et al. Wireless sensor networks: A survey. *Computer Networks*, 38(4):393–422, Mar. 2002.
2. S. Antifakos, F. Michahelles, and B. Schiele. Proactive instructions for furniture assembly. In *Proc. 4th Int'l Conf. Ubiquitous Computing (UbiComp 2002)*, pages 351–356. Springer, Berlin, Sept. 2002.
3. M. Beigl and H. Gellersen. Smart-Its: An Embedded Platform for Smart Objects. In *Proc. Smart Objects Conference (SOC 2003)*, Grenoble, France, May 2003.
4. A. Boulis, C.C. Han, and M.B. Srivastava. Design and implementation of a framework for programmable and efficient sensor networks. In *Proc. 1st ACM/USENIX Conf. Mobile Systems, Applications, and Services (MobiSys 2003)*, pages 187–200. ACM Press, New York, May 2003.
5. N. Carriero and D. Gelernter. Linda in Context. *Communications of the ACM*, 32(4):444–458, April 1989.
6. A. Chen et al. A support infrastructure for the smart kindergarten. *IEEE Pervasive Computing*, 1(2):49–57, 2002.
7. N. Davies et al. Limbo: A tuple space based platform for adaptive mobile applications. In *Proc. 6th Int'l Conf. Open Distributed Processing/Distributed Platforms (ICODP/ICDP '97)*, pages 291–302, May 1997.
8. J. Elson et al. EmStar: An Environment for Developing Embedded Systems Software. Technical Report 0009, Center for Embedded Networked Sensing (CENS), 2003.
9. D. Estrin et al. Connecting the physical world with pervasive networks. *IEEE Pervasive Computing*, 1(1):59–69, 2002.
10. B. Warneke et al. Smart Dust: Communicating with a Cubic-Millimeter Computer. *IEEE Computer*, 34(1):44–51, Jan. 2001.
11. C. Plessl et al. Reconfigurable hardware in wearable computing nodes. In *Proc. Int. Symp. on Wearable Computers (ISWC'02)*, pages 215–222. IEEE, Oct. 2002.
12. D. Ganesan et al. Complex Behavior at Scale: An Experimental Study of Low-Power Wireless Sensor Networks. Technical Report CSD-TR 02-0013, UCLA, Feb. 2002.
13. H. Wang et al. Target classification and localization in habit monitoring. In *Proc. 2003 Int'l Conf. Acoustics, Speech, and Signal Processing (ICASSP 2003)*, volume 4, pages 844–847. IEEE, Piscataway, NJ, Apr. 2003.
14. J. Heidemann et al. Effects of Detail in Wireless Network Simulation. In *Proc. SCS Multiconference on Distributed Simulation 2001*, Jan. 2001.
15. S. R. Madden et al. TAG: A Tiny Aggregation Service for Ad-Hoc Sensor Networks. In *Proc. 5th Symp. Operating Systems Design and Implementation (OSDI 2002)*, Boston, USA, Dec. 2002.

16. J. Hill et al. System architecture directions for networked sensors. In *Proc. 9th Int'l Conf. Architectural Support Programming Languages and Operating Systems (ASPLOS-IX)*, pages 93–104. ACM Press, New York, Nov. 2000.
17. O. Kasten and M. Langheinrich. First Experiences with Bluetooth in the Smart-Its Distributed Sensor Network. In *Workshop on Ubiquitous Computing and Communications, PACT 01*, Barcelona, Spain, October 2001.
18. M. Leopold, M.B. Dydensborg, and P. Bonnet. Bluetooth and Sensor Networks: A Reality Check. In *Proc. 1st ACM Conf. Embedded Networked Sensor Systems (SenSys 2003)*. ACM Press, New York, Nov. 2003.
19. S. Li, S. H. Son, and J. A. Stankovic. Event Detection Services Using Data Service Middleware in Distributed Sensor Networks. In *IPSN 2003*, Palo Alto, USA, April 2003.
20. A. Mainwaring et al. Wireless sensor networks for habitat monitoring. In *Proc. 1st ACM Int'l Workshop Wireless Sensor Networks and Applications (WSNA 2002)*, pages 88–97. ACM Press, New York, Sept. 2002.
21. F. Michahelles and B. Schiele. Better rescue through sensors. In *Proc. 1st Int'l Workshop Ubiquitous Computing for Cognitive Aids (at UbiComp 2002)*, Sept. 2002.
22. J.M. Rabaey et al. PicoRadio Supports Ad Hoc Ultra-Low Power Wireless Networking. *IEEE Computer*, 33(7):42–48, July 2000.
23. K. Römer. Tracking Real-World Phenomena with Smart Dust. In *EWSN 2004*, Berlin, Germany, January 2004.
24. E. Shih, P. Bahl, and M. Sinclair. Wake on Wireless: An Event Driven Energy Saving Strategy for Battery Operated Devices. In *Proc. 6th ACM/IEEE Ann. Int'l Conf. Mobile Computing and Networking (MobiCom 2001)*, pages 160–171. ACM Press, New York, Sept. 2002.
25. F. Siegemund. Spontaneous interaction in ubiquitous computing settings using mobile phones and short text messages. In *Proc. Workshop Supporting Spontaneous Interaction in Ubiquitous Computing Settings (at Ubicomp 2002)*, Sept. 2002.
26. F. Siegemund and C. Flörkemeier. Interaction in Pervasive Computing Settings using Bluetooth-enabled Active Tags and Passive RFID Technology together with Mobile Phones. In *Proc. 1st IEEE Int'l Conf. Pervasive Computing and Communications (PerCom 2003)*, pages 378–387. IEEE CS Press, Los Alamitos, CA, Mar. 2003.
27. H. Wang, D. Estrin, and L. Girod. Preprocessing in a Tiered Sensor Network for Habitat Monitoring. Submitted for publication, September 2002.
28. BTnodes - A Distributed Environment for Prototyping Ad Hoc Networks. http://www.btnode.ethz.ch.
29. NCCR-MICS: Swiss National Competence Center on Mobile Information and Communication Systems. http://www.mics.org.

A Real-World, Simple Wireless Sensor Network for Monitoring Electrical Energy Consumption

Cornelia Kappler[1] and Georg Riegel[2]

[1] Information and Communication Mobile, Siemens AG, 13623 Berlin, Germany
cornelia.kappler@siemens.com
http://www.siemens.com
[2] dezem GmbH, Lohmeyerstr. 9, 10587 Berlin, Germany
georg.riegel@dezem.de
http://www.dezem.de

Abstract. We present a simple, commercial WSN for monitoring electrical energy consumption. We discuss WSN characteristics, practical problems, constraints and design decisions which mainly are motivated by our concrete application and the need to make our WSN low cost to implement, plug&play to install and easy to maintain. This results in a system and protocols different from many WSNs discussed in the literature. Particularly, we present a multi-hop routing protocol suited to a simple system such as ours.

1 Introduction

WSNs are an active research area (for a review see [1]). However feedback loops to real-world applications are often missing. This paper is a contribution to bridging this gap. At the example of a simple WSN in actual commercial use we discuss WSN characteristics, practical problems, constraints, motivations for design decisions as well as the relation to today's research.

A commercial WSN obviously is primarily driven by the need to make it a good business. It must therefore be low-cost to implement, plug&play to install and easy to maintain. It turns out these drivers make our system different from many of the WSNs discussed in the literature.

WSNs can be very diverse, ranging from networks featuring a very high number of randomly distributed sensors (e.g. dropped from planes) to carefully placed sensors such as ours. Sensor readings may only make sense taken together e.g. for determining areas with specific properties (e.g. the area where temperature > 50^0C), or – in our case – they function stand-alone and only network for transferring their readings to a central processor. Thus, this paper also aims at completing the picture of diversity in possible WSNs, which someday may lead to a classification scheme.

This paper is organized as follows: In Sec. 2, we describe the application area of the WSN discussed in this paper. In Sec. 3, we describe details of the system and its components, explaining design decisions. In Sec. 4, we briefly discuss why we prefer wireless transmission technology to Power Line Communication (PLC), which would

H. Karl, A. Willig, A. Wolisz (Eds.): EWSN 2004, LNCS 2920, pp. 339–352, 2004.

appear to be the technology of choice in our case. In Sec. 5, we present an extension to the dezem system as it exists today to cover multi-hop topologies. As we will explain, multi-hop routing protocols of the form typically discussed in the literature are not applicable to our case as our sensors support only unidirectional communication, and because the functionality we require is comparatively restricted. We therefore present a new routing protocol that is optimized specifically for our requirements. In Sec. 6, finally, we summarize our findings and draw conclusions.

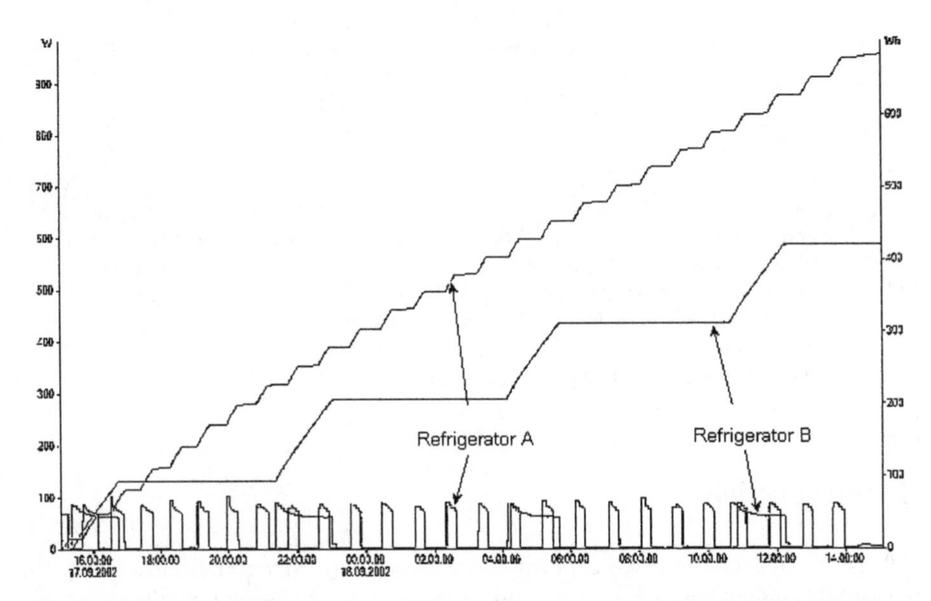

Fig. 1. Consumption profiles for two different refrigerators in an office building. The lower curves are current power [W], the monotonously increasing curves are total energy consumed [Wh], both as a function of time.

2 Motivation

The idea underlying our system is that today large energy saving potentials should be attainable in local, highly distributed energy consumption settings such as in office contexts. For example, it is estimated that 20% or more of electrical energy in office buildings could be saved without much effort. Such potentials however rarely are realized because it is unclear what is actually consuming how much electrical energy. For example, it is known that stand-by consumption can be significant, but just how much one can save by simply switching off an appliance is not known. Instead, the utility company delivers one aggregate number by the end of the year.

Dezem therefore developed a WSN technology for monitoring local electrical energy consumption. It mainly aims for deployment in industrial production and office contexts but can also be used in private households. Data is measured by sensors located either adaptor-like in electrical outlets or at sub-distribution nodes such as fuse-boxes. Measurements are collected by a central server and can be accessed and interpreted in real-time.

Fig. 1 illustrates typical consumption profiles of two different refrigerators in an office building. One can see the energy consumed by fridge B is much lower than that of fridge A, due to more widely spaced – albeit longer - cooling cycles.

The dezem technology, while continually being enhanced, is in commercial use since mid-2002.

3 Dezem System Components

Fig. 2 shows the conceptual system architecture, which is in fact rather straightforward. Sensors, or groups of sensors (at sub-distribution nodes) send their pre-processed data wirelessly to a Central Unit, the CU. They do this either directly, or, if coverage is not sufficient, multi-hop via transceivers (not yet implemented). The CU processes the measurements and sends them, via TCP/IP, to a database on a web server, where they can be accessed by the customer. For a discussion of an alternative transmission technology, see Sec. 4.

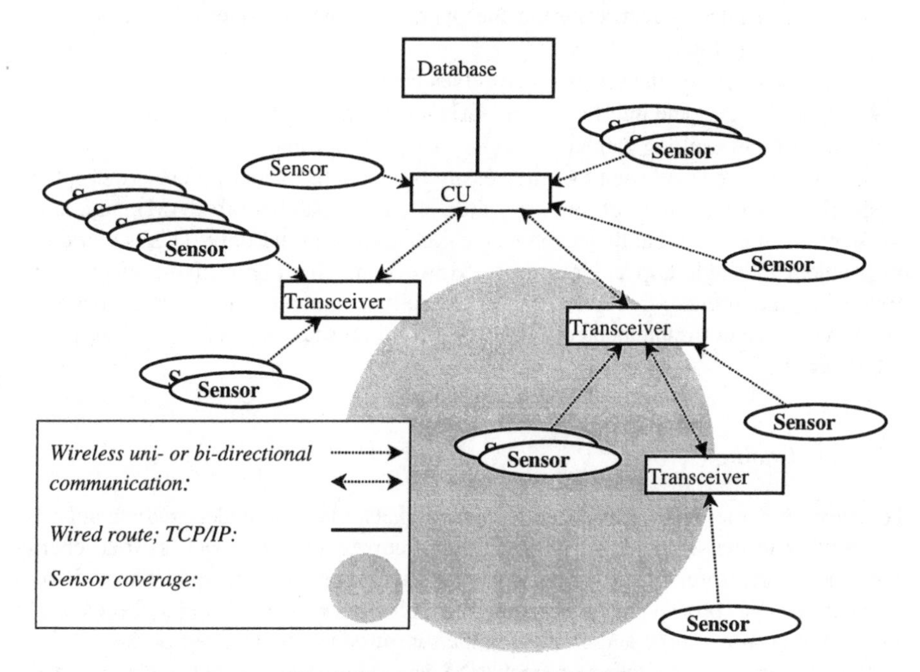

Fig. 2. Conceptual system architecture

A common constraint in WSNs is that sensor energy consumption be minimal, because they live on batteries. Our case is somewhat different, as by definition sensors are attached to the power grid. However, since the goal of the dezem project is enabling energy savings, a sensor whose energy consumption is not minimized would be a contradiction in terms. Our sensors consume on average 60mW while sending.

As already mentioned in the introduction, our system must be low-cost to implement, plug&play to install, and easy to maintain in order to make it a good business. Therefore all functionality is stripped down to a minimum. Communication protocols are very simple and, most importantly, sensors can only send, not receive, as shown in Fig. 2

3.1 Sensors

Sensors measure the electrical energy consumed by the consumer or consumer-aggregate attached to the outlet or sub-distribution circuit they are sensing.

As commonly suggested, sensors wake the sending unit to send their current measurement when certain trigger conditions are met, namely either when energy consumption shows interesting changes, or when a certain time has passed since the last packet was sent (alive-signal). I.e. sensors preprocess data and don't send continuously. In our case, this is important for several reasons:

- there is a regulatory restriction on the overall airtime and power for each sensor (see Sec. 3.3 below);
- energy consumed by the sensors themselves is reduced;
- we avoid clogging the interface to the CU (see estimate in Sec. 3.3 below);
- the size of the resulting database is more easily manageable.

Sensors must be small such that they easily fit inside fuse–boxes. Furthermore, they need to be low-cost. Thus sensors have little storage capacity (1kb RAM), and, since we found it is not vital that they can receive data, they are send-only. This is obviously no problem in single-hop environments. However in multi-hop environments, self-organizing the routing tree with unidirectionally communicating sensors becomes tricky. We have not implemented multi-hop yet, but discuss a possible routing protocol in Sec. 5.

3.2 Data Coding

To provide some protection against lost packets, data is coded redundantly by transmitting in the same packet information on current power as well as total energy consumed. Therefore, when one packet is lost, its content can be approximated from the next one. We protect for recognition of garbled packets by adding a checksum. In addition to power, energy and checksum, the sensor codes its ID in the packet.

Note sensors do not timestamp packets. Since each packet is sent out immediately after the measurement and (single-hop) takes – within the accuracy required for our purpose – the same negligible time to travel to the CU, the CU can add time stamps.

The advantages of this procedure are less logic on the sensors and no need to synchronize them.

3.3 Wireless Interface

We send on the 868 Mhz frequency band assigned to Non-specific Short Range Devices in Europe [2] by the European Radiocommunications Committee (ERC). Their regulations restrict sending power to 25mW, and furthermore allow each sending device to occupy the channel only <0.1 % of time. On the other hand, this frequency band is available for use by everybody without licensing. The restriction in sending power ensures interference between different networks is minimized.

Typical sensor coverage, limited by regulations, ranges up to 35m inside buildings. It is clear that this coverage is not sufficient for large office buildings, hence multi-hop becomes necessary – or a workaround, see Sec. 3.5, or another transmission technology needs to be applied, as described in Sec. 4.

Data is sent with 1,2 kb/s. Whenever a sensor has to send something it does so. However, since sensors cannot detect collisions, colliding packets are not re-sent.

We can accept up to about 10% lost or malformed packets since some information can be recovered, see Sec. 3.2, and consumption profiles still contain enough detail for our purpose. Packet loss and packet corruption may occur because of packet collisions, because of problems with antenna tuning, too large distance between sensor and CU, etc. When the percentage of lost and garbled packets is detected as too high, the human administrator has some options for fixing the problem by e.g. better tuning of antennas etc.

Estimation of Collision Probability. In order to see whether we operate the system in a safe regime, and when it will collapse due to congestion on the air interface we do an estimate of the collision probability for packets. We calculate the probability P that one sensor starts emitting a packet when already another one is doing so. We assume Poisson-distributed sending events issued independently by n sensors. This yields

$$P([t > T]) = (n-1)\,\lambda \int_0^T \exp(-(n-1)\,\lambda t)\,dt = 1 - \exp(-(n-1)\,\lambda T) \approx 1 - \exp(-n\,\lambda T) \qquad (1)$$

Thereby, $T \approx 80$ms is the time needed to emit a packet, $\lambda \approx 1/$ min is the average sending rate, and n, the number of sensors is up to 50. The Poisson distribution actually has a cut-off value at the time the "alive signal" is sent, which is however so much larger than the average sending rate (about once in 15 minutes or less) that the cut-off can be neglected.

Equation (1) does not account for collision of three or more packets. However since we are interested in probabilities for two-packet collisions < 10% (since we can accept a total of up to 10% lost packets), the probability for multi-packet collisions can be neglected.

We want the exponential < 0.9 in Equ. (1), in other words, the system needs to be operated such that $n\lambda < 1.3$ /s. Currently, $n\lambda \leq 0.8$ /s, so according to this estimate we are operating the system up to its limit, with collision probability $P \leq 6\%$. This insight has impact on the multihop-protocol we describe in Sec. 5, as obviously we must minimize the additional load on the air interface due to signaling.

3.4 Transceivers

Will be discussed with multi-hop operation in Sec. 5

3.5 CU and Database

The CU receives packets from sensors, timestamps them, sorts out garbled packets, and forwards them to a database on a web server via LAN or email. Data is immediately accessible via a web interface as shown in Fig. 1 and 2.

Large buildings cannot be covered by a single CU. The simplest approach is employing two or more CUs, which might result in some sensor data being received more than once. This problem is solved by either coding into CUs what sensors they are responsible for, or by having the database sort out duplicate data. The former is not "plug and play", and the latter wastes bandwidth. We therefore developed a concept for multihop operation, see Sec. 5.

The database also correlates sensor ID and the sensor location information, which is provided by the human administrator.

4 Alternative Transmission Technique

The reader may wonder why we are using wireless transmission in an environment where Power Line Communication (PLC) seems the technique of choice: each sensor is attached to the power grid anyways. Indeed this was the first option we explored. We almost altogether abandoned it for the following reasons:

- The commercially available PLC modems we tested at the time (one year back) proved rather unstable in operation.
- PLC modems are about 8 times as expensive than wireless modems.
- Electrical current comes with three phases, only one of which is attached to a given electrical outlet. PLC information rides on this phase and hence can be read only from electrical outlets attached to the same phase. Usually however it is unknown which outlet is connected to what phase, making setting up a system involving electrical outlets at customer premises tedious or sometimes even impossible.
- Energy consumed by PLC is two to four times higher than with wireless communication.

5 Routing Protocol for Multi-hop Operation

In this section we describe a possible multi-hop extension to our system, which would enable operation over distances larger than single-sensor coverage. Currently, when sensor coverage is not sufficient for all sensors to reach the CU, we employ more than one CU, as described above. As a caveat we would like to add that a multi-hop solution must be compared to this multi-CU solution regarding costs and complexity of implementation, maintenance and operation. Such a comparison however is beyond the scope of the conceptual study, which is the focus of this section.

As already noted by others [1], finding multi-hop routes in a WSN may be a different problem than routing in Ad-hoc networks, because not all nodes necessarily need to communicate with each other, and because nodes often are nomadic rather than mobile.

5.1 Related Work

A variety of routing protocols have been proposed for WSNs [1]. The most simple, albeit not scalable, among these is flooding, where each signaling message is re-broadcasted by all transceivers. In more sophisticated variants of flooding, re-broadcasting is e.g. dependent on time, location or on a random number generator (gossiping, cf. [3]). A promising approach to routing is forming clusters of sensors with one node acting as "cluster head" that is responsible for transmitting all sensor data from the cluster towards the receiver. For example, LEACH [4] forms clusters such that global energy usage is evenly distributed and minimized. In [5], TopDisc, a complete topology discovery algorithm is proposed that likewise organizes the WSN into clusters. A different approach is taken by the Directed Diffusion paradigm [6]. It is suitable in systems where the possibly multiple CUs do not need to know all the data – as is the case for us – but rather registers for data of interest to them.

5.2 Design Goals and Design Constraints

Interestingly, none of the WSN routing protocols known to us is applicable to our case. The main difference is that all of the protocol proposals assume sensors capable of bi-directional communication. The IEEE 802.15. Task Group 4 is chartered to investigate solutions for low data rate, low energy consumption, low complexity wireless personal area networks, including sensor networks [7]. However, "low complexity" includes bi-directionally communicating sensors. We believe, however, the economic reasons leading to uni-directionally communicating sensor design are not very particular to our case, and such simple sensors must be accounted for in protocol proposals.

Additionally, our main design constraints are costs and complexity, plus restricting the load on the air interface, whereas usually minimizing energy consumption is the overriding factor. Furthermore, the functionality we need is very restricted. Sensors only need to communicate tree-topology like with the CU. For the CU, in contrast, it is not necessary to know the topology except that all sensor nodes are connected. Consequently, while the basic idea of our protocol is related to the TopDisc algorithm [5], it has less functionality and hence reduced complexity and overhead.

In a multi-hop extension, data is routed ad-hoc like, via zero or more intermediate nodes (transceivers) until reaching the CU, cf. Fig. 1. Transceivers can both send and receive, so building the routing tree up to transceivers is similar to the typical problem studied in the literature. Since our sensors can only send, they don't know whether data they send reaches the CU. Therefore, CU and transceivers must control the routing process.

The routing protocol we describe below is specific to our problem. It builds a tree topology for the routes from sensors to CU, by propagating routing information hop-by-hop from the center (CU) to the outside branches (transceivers far away) The routing protocol minimizes the number of hops on the path between sensor and CU, balances load between transceivers, keeps low the number of messages exchanged and the protocol complexity. It adapts to transceiver failures, nomadic changes in topology, as well as addition and removal of transceivers and sensors. On the other hand, we do not aim at complete topology information, nor do we need to optimize for very fast convergence as in our case the actual time sensing starts is not critical.

5.3 State on CU and Transceivers

The CU is configured to expect data from n sensors. When sensors are added or removed, n is adapted accordingly. Furthermore it is told a time interval in which to check for missing sensors. Also, the CU maintains a routing table with entries for all sensors it receives packets from, containing

- sensor ID,
- number of hops experienced by packets from each sensor,
- transceiver ID of first hop (i.e. first transceiver relative to sensor),
- Time of last packet seen for this sensor.

Each transceiver is given

- a threshold value for acceptable signaling strength,
- a number C for how many sub-threshold packets are acceptable,
- a time interval $[0,\tau]$ from which to pick random sending times – this is important when a group of transceivers is triggered to respond to an event,
- a larger “listening” time interval $[\sigma_i,\sigma_f]$, $\sigma_i > 0$,
- another time constant for scaling an “exponential back off” procedure,
- an integer m. Every m packets more complete information about a sensor is enclosed in a packet.

The above constants presumably are identical for each transceiver and could e.g. be distributed along with routing information.

Also, each transceiver maintains a “previous hop” routing table with entries on all sensors and transceivers it receives packets from, containing

- sensor / transceiver ID,
- total hop count for this entity,
- a flag, telling whether itself or another entity is responsible for relaying packets from this entity.
- a counter c for counting sub-threshold packets from this entity. This counter is initialized to zero.

Additionally, each transceiver maintains a “next hop” routing table with entries on transceivers, which it may use as next hop towards the CU, containing

- transceiver ID,
- number of hops to CU,
- A flag on whether it is currently using this transceiver as next hop to the CU.

From the content of these tables it is clear that neither the CU nor any other node is "omniscient", in the sense that no entity has a complete view of which sensor routes along what transceivers. Rather, this information is distributed: each transceiver just knows from which previous-hop entity it is to forward packets to which next hop. While for routing towards the transceiver omniscient nodes are unnecessary, such nodes would simplify error tracking. Furthermore, "reverse routing state" allowing targeted routing towards a particular sensor does not exist. As it turns out, however, we almost never need to "reverse route". Hence we found these drawbacks don't outweigh the additional load on the air interface incurred by moving about complete routing information.

5.4 Configuration Phase

In the configuration phase, the tables on transceivers and CU are populated. Transceivers and CU send using CSMA.

Connecting Transceivers. It is recommended to first switch on transceivers and only afterwards the CU.

First, each transceiver establishes a least-hop route to the CU. This process is initiated by the CU, which in this simple topology ensures each transceiver finds an optimal route if it is available. To this end, the CU sends a Beacon message advertising zero hops to the CU. Transceivers receiving the Beacon message with an above-threshold signaling strength have found the route to the CU and enter this route in the "next hop" table. They subsequently emit (randomly timed to avoid collisions) a Beacon message themselves, advertising a one-hop route to the CU and giving their transceiver ID.

A transceiver Q receiving such a message above-threshold enters a corresponding entry in its "next hop" table. If this is the first Beacon message transceiver Q receives, it emits a randomly timed Beacon message, advertising a two-hop route. Then it waits a random "listening" time whether another Beacon message will arrive. If another one arrives, it again waits a random time etc. When no further Beacon message arrives, it randomly selects a transceiver P among those offering the least-hop route to the CU. Transceiver Q sends a Connect message containing its transceiver ID to the transceiver P, which means P is to relay Q's packets. Transceiver P adds transceiver Q into its "previous hop" routing table. Since by construction, messages advertising routes with fewer hops usually arrive earlier, it is safe for Q to advertise a 2-hop route when it receives the first Beacon message. We only wait for more Beacon messages in order to achieve load balancing. If Q detects the Connect message collided it resends it. The Connect message must be acknowledged and will be resent (with exponential back off) until an Acknowledgement message arrives. If, unexpectedly, a Beacon message advertising an even better route arrives at transceiver Q after it sent the Connect message, Q will send a Disconnect message to the corresponding transceiver (which must be acknowledged), a Connect to the better next hop, and emit a new Beacon message.

The process continues, finally connecting all transceivers. If a transceiver didn't manage to find an acknowledged "next hop" it might e.g. give a visual alarm signal to simplify error management for the human administrator.

Connecting Sensors. As soon as sensors are connected, they start normal operation sending their data, including sensor ID. Transceivers are ready to route sensor data as soon as they established a next hop transceiver and hence a route to the CU.

When a transceiver Q receives a packet directly from a sensor, it checks whether this sensor already appears in its "previous hop" routing table. If not, Q checks whether the strength of the received signal is above the threshold T. If it isn't, it does nothing. If it is, it enters the sensor as its own responsibility in the "previous hop" routing table, and sends a randomly timed "Claim" message to the other transceivers, claiming this sensor ID, and adding its own transceiver ID, as well as the hop-count for this sensor.

At the same time, transceiver Q expects the same kind of Claim message from other transceivers. When it receives such a message, it enters the corresponding sensor ID and hop-count as sent by the other transceiver into the previous-hop routing table.

When transceiver Q receives a packet from a sensor, which according to the previous-hop routing table was already claimed by another transceiver P, it checks (possibly only with some low probability to reduce overhead) whether the hop count reported by P is higher than that offered by Q. If yes, transceiver Q claims this sensor for itself by updating its previous-hop routing table and sending the Claim message as above to other transceivers. Transceiver P receives this message and overwrites the entry in its previous-hop routing table accordingly.

When Claim messages collide, transceivers notice because of CSMA, and resend their claims, randomly timed. When two transceivers hear the same sensor but cannot hear each other, the CU resolves the conflict. It knows first-hop transceiver ID and hop-count, because they are piggybacked on the first data packets (and again in regular intervals). It thus sends a "Stop this sensor" message towards the first-hop transceiver with the higher hop count (or draws one of them at random if the hop count is equal), meaning it should stop forwarding packets from this sensor. The CU can address this message explicitly to the (from the perspective of the sensor) last hop transceiver. From then on, the message must be flooded outwards to all other transceivers, since no reverse routing state exists, and will thereby also reach the first-hop transceiver, which acts accordingly. We acknowledge this "flooding" is very wasteful, however assume it is a rare occurrence – which needs to be verified.

To summarize, the previous-hop routing table tells a transceiver whether a particular sensor is already taken care of, and if yes, whether itself or another transceiver is responsible for relaying messages by this sensor.

It should be clear from this section that there is always exactly one transceiver responsible for each sensor: If a sensor was not able to connect to a transceiver, the CU notices because it knows the number of sensors that should send data, and will reiterate the route establishment process and alert the administrator if need be. The converse case is that a sensor is connected to more than one transceiver. This can only happen when transceivers didn't hear their respective Claim messages. In this case the CU resolves the conflict as described above.

Load balancing. The routing protocol described above automatically achieves load balancing between transceivers offering the same hop-count on the basis of random number generation: Transceivers claim sensors only after waiting a random time interval, and they randomly pick their next hop transceiver.

5.5 Normal Operation

In normal operation, transceivers receive packets from sensors. If they are responsible for this sensor according to their "previous hop" routing table, they attach their ID as sender information, and broadcast the packet. If the received signaling strength is below threshold, they increase counter c by two (the threshold must be configured conservatively such that, typically, the sensor ID is still legible), if the received signaling strength is above threshold, and if c greater than zero, c is decreased by one. Every m packets, commencing with the first packet, they also attach their ID as immutable "first hop" ID, and their hop-count towards the receiver. When a transceiver receives a packet whose immediate sender is a transceiver for which it is responsible according to its "previous hop" routing table, it replaces the sender information with its own ID, and broadcast the packet. As before, counter c is changed as appropriate. The CU upon receiving a packet, unpacks it, time stamps it, updates the fields in its sensor table (time last packet seen, as well as first hop and hop count if applicable), and prepares the packet to be sent to the data base.

Time stamping. In the single-hop case, delegating time stamping away from the sensors is no problem, since the time between generation of a data packet and its receipt by the CU is negligible. However, in a multi-hop scenario, this is no longer the case. In the literature, powerful synchronization schemes are discussed [8]. It turns out that for us they are not necessary since the accuracy we need – order of seconds – can be achieved by simple means:

In multi-hop, each transceiver adds a presumably constant processing time T_p. There is rarely queuing of packets as in our broadcast medium at all times only one packet can be on its way. The CU hence may continue time stamping packets when they are received, subtracting T_p times number of hops. However, within the accuracy necessary for us, it may very well be that even this is unnecessary. (Note that the "back-off and resend" procedures described above only are used in the route-establishment phase)

5.6 Configuration Changes and Failures

Adding and (Re)moving Transceivers. When a transceiver boots up, it waits a random "listening" time whether it receives a Beacon message. If it doesn't, it broadcasts a Query message. This Query is (randomly timed) replied to by Beacon messages from all transceivers / the CU receiving it if they themselves have already established a route. This allows the newly arrived transceiver to send a Beacon message itself, populate its "next hop" table and send a Connect to the best next-hop transceiver. Beacon messages are processed by transceivers no matter when they arrive in order to always have the best possible route to the CU.

When a transceiver R goes down in a controlled fashion (e.g. it is (re)moved), it broadcasts a "Good Bye" message, prompting all transceivers in scope to remove the corresponding entry in the "next hop" table. All transceivers using R as next hop thereupon send a Connect message to a randomly picked transceiver from the set of least hop transceivers in the "next hop" table. If this implies their hop-count to the CU increases, they send a new Beacon message to allow their previous hops to adapt to the new situation.

Adding and (Re)moving Sensors. Adding a sensor after routing has been established is no problem, as transceivers will instantly notice this sensor is not yet in their database, and issue a Claim message. Likewise removing a sensor is no problem, because transceivers will not notice a particular sensor stopped sending. However when the number of sensors is changed, the sensor count n in the CU needs to be adapted.

Failure of Transceivers or Sensors. In order to cope with transceivers just failing without first sending a Good Bye, and failed sensors, the CU in regular intervals (e.g. 5 times the "alive signal" interval, or even just once per hour or per day) scans whether it still receives data from all sensors. If a sensor is missing, the route establishment process is reiterated by the CU sending a Beacon message in the hope of finding alternative routes for all sensors that were routed via a failed transceiver. If even after this rerouting some sensors cannot be connected (possibly the problem is a failed sensor), an Alarm message is sent to the human administrator.

Variability in Signaling Strength. The signaling strength of packets from a particular transceiver R or sensor S received by a transceiver T may vary e.g. due to changes in environmental conditions, such as electronic device usage, people moving, or whether a particular door is open or closed, which in an extreme case may cause packets to be received incorrectly or not at all. Since for us, each individual sensor is important, we need to protect against this problem. As explained above, counter c counts sub-threshold packets (weighing them against the number of over-threshold packets). When counter c becomes greater than threshold C, transceiver T decides no longer be responsible for S resp. R. It sends a "Disconnect" message to R, prompting it to attach to another transceiver. For S, it broadcasts an "Unclaim" message, prompting another transceiver to step in – the CU will notice if no other transceiver connection for S is established and act accordingly. Then T sets counter $c > C$, to remind itself to never again act as next hop for S resp. R. This rather conservative attitude protects against route flapping due to e.g. doors that are sometimes opened, and sometimes closed.

6 Conclusions

In this paper, we presented a commercial WSN for monitoring electrical energy consumption. We discussed practical problems, constraints, and design decisions. A commercial WSN is primarily driven by the need to make it a good business. It must therefore be low-cost to implement, plug&play to install and easy to maintain. We observe many of the paradigms and solutions discussed in the literature, e.g. typical sensor design including a receiver, the need for complete routing table and sophisticated time stamping, are not necessary and hence too expensive and complex for our application. We presented our own, boiled-down, custom-made solutions. While obviously such boiled-down solutions do not fit all – or may be not even many – other WSNs, we believe this commonly to be the case, for the simple reason that "the WSN" does not exist. WSNs are very heterogeneous. We agree with [9] that work on classification of WSNs (their purpose, topology, parameters, architecture, protocols etc.) would be very helpful. As a step towards this goal we summarize typical properties of the dezem WSN which differentiate it from others:

- each sensor is important and needs to be connected,
- individual sensor readings are meaningful by themselves (i.e. do not need data from other sensors for interpretation),
- up to 10% lost or garbled packets are acceptable,
- time resolution in the order of seconds is sufficient,
- number of sensors is 10s to 100s,
- no data-centric or location-centric routing,
- sensors send data only to the CU, they do not need to communicate among themselves (tree-topology),
- sensors only communicate unidirectionally,
- for multi-hop dedicated transceivers are necessary,
- sensors just send when they have something to send, CSMA is used by transceivers,
- System is optimized for
  - low costs
  - low complexity
  - low load on the air interface
- minimal energy consumption is important however not vital.

Acknowledgements. The authors would like to thank Reiner Raddatz, Logim GmbH, and Peter Pregler, Siemens AG for providing helpful information.

References

[1] I. F. Akyildiz, W. Su, Y. Sankarasubramaniam and E. Cayirci, "A Survey on Sensor Networks", IEEE Comm. Mag., Vol 40.8, Aug 2002.

[2] ERC/REC 70-03, "ERC Recommendation relating to the use of short range devices (SRD), Feb. 2002.

[3] S. Hedetniemi, S. Hedetniemi, and A. Liestman "A Survey of Gossiping and Broadcasting in Communication Networks", Networks, vol. 18, 1988.
[4] W. R. Heinzelman, A. Chandrakasan and H. Balakrishnan, "Energy-efficient Communication Protocol for Wireless Microsensor Networks", Proc. HICSS, Maui, Hawaii, 2000.
[5] B. Deb, S. Bhatnagar and B. Nath, "A Topology Discovery Algorithm for Sensor Networks with Applications to Network Management, Proc. IEEE CAS Workshop on Wireless Communication and Networking, Pasadena, USA, Sept. 2002.
[6] C. Intanagonwiwat, R. Govindan and D. Estrin, "Directed Diffusion: A Scalable and Robust Communication Paradigm for Sensor Networks", Proc. ACM MobiCom'00, Boston, 2000.
[7] E. Callaway, P. Gorday, L. Hester, J.A. Gutierrez, M. Naeve, B. Heile and V. Bahl, "Home Networking with IEEE 802.15.4: A Developing Standard for Low-Rate Wireless Personal Area Networks", IEEE Comm. Mag., Vol 40.8, Aug 2002.
[8] J. Elson and K. Römer, "Wireless Sensor Networks: A New Regime for Time Synchronization", ACM SIGCOMM Computer Communication Review (CCR), 33.1, Jan. 2003.
[9] V. Handziski, A. Köpke, H. Karl and A. Wolisz, "A common wireless sensor network architecture?", Proc. 1st GI/ITG KuVS Fachgespäch "Sensornetze", TKN Technical Reports Series, Berlin, July 2003.

Reliable Set-Up of Medical Body-Sensor Networks

H. Baldus, K. Klabunde, and G. Müsch

Philips Research Laboratories Aachen
Weisshausstr. 2, 52066 Aachen, Germany
{Heribert.Baldus,Karin.Klabunde,Guido.Muesch}@philips.com

Abstract. Wireless medical body sensor networks enable a new way of continuous patient monitoring, extending the reach of current healthcare solutions. They improve care giving by a flexible acquisition of relevant vital sign data, and by providing more convenience for patients.
Applying the general concept of sensor networks to the medical domains requires development of reliable, distributed microsystems with secure and fail-safe system set-up and efficient operation. The system has to work automatically, but has also to be always under explicit control of any clinician.
To address this, we realized a medical body sensor network, covering a set-up procedure that involves dedicated sensors for patient identification, as well as a special configuration pen for reliable, easy system set-up. The system architecture has been validated by an implementation based on the UC Berkeley Motes System. Main advantages that we achieve are interference-free operation of different body sensor networks in close vicinity, as well as intuitive usage by non-technical personnel.

1 Introduction and Motivation

Continuous monitoring of critical vital signs of patients is a key process in hospitals. Today, this is usually performed via different cabled sensors, being attached to the patient and connected to bedside monitors. These solutions have clear limitations – the amount of measurements is limited by available device plugs, patient treatment is disturbed by cabling, and finally patient mobility is severely restricted, as the patient is tight to devices at the bedside.

This opens up a new application area for sensor networks. Small wireless sensors, attached to the patient body, measure vital sign data, and transmit them via the established sensor network to an external observation unit.

However, current sensor network solutions have to evolve to be suited for the medical domain, which requires flexible, but safe set-up of the system as well as reliable, fail-safe operation. Today's sensor networks, mainly used for environmental monitoring, are either statically pre-configured for a certain task, or build spontaneous ad-hoc networks. For medical usage, this system behavior has to be transformed to a reliable and defined system set-up, working automatically but nevertheless being under explicit control of a clinician. For this, two issues have to be resolved: a) a suitable communication protocol is needed for the wide variety of medical sensors, and b) a convenient and fail safe set-up mechanism has to be developed.

H. Karl, A. Willig, A. Wolisz (Eds.): EWSN 2004, LNCS 2920, pp. 353–363, 2004.

This paper introduces the architecture for a distributed medical body sensor network, and then mainly focuses on the system set-up process. The system configuration includes automatic patient identification, and thus allows a safe device association. Clinicians, who will be the users of the sensor network, require a flexible and easy-to-use procedure to set up patient sensors. But they also need to be always in control of the system, which excludes completely automatic discovery and configuration procedures. Thus, our new concept deploys a special pen, personalized for each authorized clinician, for set-up and configuration of the medical body-worn sensor network. This gives the clinician an intuitive means to control association and disassociation of wireless sensors to the network. We present the functional system architecture and the detailed protocol of the set-up procedure.

For our system prototype, that validates the architectural concepts, we deployed as basic sensor elements the UC Berkeley motes, along with the TinyOS software, capable of maintaining a wireless communication and flexible enough to incorporate medical sensors [1][6].

The remainder of the paper is organized as follows: Section 2 describes today's approaches on configuration of sensor networks. Section 3 addresses the specific requirements for medical sensor networks, and section 4 then introduces our system architecture of wireless sensor network and set-up procedure. In section 5, we explain the implementation of our prototype. Finally, we end with conclusions and future work.

2 Related Work

Technology improvements in the field of microelectronics with respect to size and power consumption created a new category of devices: autonomous micro devices. Adding sensor capabilities and radio interfaces leads to wireless sensor networks.

Two kinds of wireless sensor configurations can be seen today: point-to-point and autonomous auto-configuration. In both cases the membership of sensor nodes to a certain network is either statically predefined or subject to a manual association procedure.

2.1 Point-to-Point

In a point-to-point communication a single wireless sensor talks to a single receiving device, typically some type of display device (e.g. wireless weather stations). The association of sensor to receiver is often only maintained by using the same radio frequency or the same radio protocol. Most such systems are operated in the free ISM frequency bands. As long as only few systems are deployed the limited radio range together with the low duty cycle of typical sensor applications guarantees an interference free operation.

With more and more wireless solutions deployed in the open ISM bands, additional mechanisms are needed. The communication itself has to be made reliable in the presence of interference sources; e.g. by using spread spectrum technologies and/or error correction. Using low duty cycles as a simple scheme of interference prevention is currently enforced in the 868Mhz ISM bands in Europe [2].

Beside the communication itself, also the association of sensors to receivers has to be solved reliably. A typical solution is a learning sequence initiated by a time synchronous button press on both devices (wireless sensor and receiver). During the activation of the learning sequence a piece of information is exchanged between wireless sensor and receiver. Often the unique identifier of the wireless sensor is sent to the receiver. After that the receiver will only listen to messages containing the newly learned identifier, preventing false sensor data reception by other nearby sensors.

2.2 Autonomous Auto-configuration

Another type of sensor networks are large-scale wireless sensor networks containing multiple sensor and receiving nodes. Once such a network is deployed, communication from any sensor node to any other node shall be possible. Because the size of such sensor networks is larger than the radio range of a single node, multi-hop routing algorithms are used to cover the whole sensor network. Efficient multi-hop algorithms are subject to many research activities around TinyOS [3][4].

Once a new sensor node enters such a large-scale wireless sensor network it should automatically become part of the network. Typically a predefined network identifier is used to decide on the association of a new node to an existing sensor network. Only nodes with the same network identifier will be member of the same sensor network. Currently those network identifiers are statically defined at the building process of the nodes. With this static definition a wireless sensor node cannot be easily reconfigured for use in a different sensor network.

2.3 Other ad-hoc Communications

Further popular examples for ad-hoc communication systems are wireless LANs and Bluetooth. For wireless LANs again some kind of membership to a certain ad-hoc network has to be defined. This will be configured manually by defining a shared network identifier for all participating devices.

Bluetooth uses another kind of manual network set-up. One device, later to be master of the Bluetooth Piconet, will initiate a device discovery that will be answered by devices in range. The user of the master device will explicitly initiate a communication to one of the found devices. Bluetooth also offers the concept of device pairing. Here two devices will exchange keys (which are typically derived from a PIN entered manually by the operating user) to build a secure relationship [5]. Whenever a service is needed that is related to a paired device, Bluetooth automatically establishes the communication to that device (e.g. using the dial up networking profile of Bluetooth from a notebook which will automatically connect to the paired Bluetooth enabled mobile phone). Nevertheless the pairing has to be done manually and with user intervention.

Another method of device pairing, stemming from the area of ubiquitous computing, is discussed by Holmquist et.al. – 'Smart-Its Friends' technique [8]. They describe a method based on context-proximity where two devices connect if they have the same context, e.g. the same movement pattern. The idea is to take the network nodes (Smart-Its) that shall be connected, impose the same movement pattern on them e.g. by shaking them, and then these network nodes will connect automatically.

The 'Smart-Its Friends' technique uses context-awareness to realize automatic connection set-up between potential network nodes. This does not really fit to the requirements of our medical application scope where a network set-up, which is predictable and explicitly controlled by clinicians, is required.

2.4 Wireless Medical Sensor Networks

First monitoring products with single wireless sensors exist (e.g. wireless 1-lead ECG from Nihon Kohden) but work on systems with several wireless medical sensors is still in its infancy.

One activity in this area has been started by the Fraunhofer Institute Germany [7], which works on a proprietary hard- and software platform for medical Body Area Networks and the development of wireless medical sensors. The process of setting up a sensor network for medical monitoring applications as well as integration of patient and clinician identification is not yet covered.

3 Requirements

The area of patient monitoring poses stringent requirements on a convenient and fail-safe set-up procedure for medical sensor networks.

On the one hand patient safety and convenience for clinical staff have to be ensured, on the other hand the technical requirements as e.g. low power consumption, and reliable set-up and operation have to be met.

In the following we describe the requirements on such a set-up procedure.

The set-up procedure for the wireless sensor network has to be as 'safe as cable', i.e. it must be as intuitive and easy to use by clinicians as today's cabled solutions. The system should allow a flexible sequence of actions and avoid unintentional activation/deactivation of medical sensors.

The hospital environment demands fast but controlled set-up of the sensor network, i.e. the set-up procedure should be automated as far as possible but must always be under control of a clinician. Avoidance of unauthorized set-up is required here, i.e. only clinicians may initiate and control the set-up of a sensor network. Furthermore, for legal reasons, the medical context requires the logging of which clinician performed which action.

The developed set-up procedure must also support flexibility in the sense that dynamically adding/removing sensor nodes to/from a network must be easily possible. There should be no limitation in the number of sensor nodes that can build a network.

It is required that adding/removing of a sensor node can be performed by different clinicians, i.e. one clinician added a sensor to sensor a network, which is later extended by a second clinician, removal of one/all of the sensors could be performed by a third clinician.

During the set-up, interference with other nearby sensor networks must be avoided, i.e. a sensor of a new patient may not accidentally be added to sensor network of another patient. The set-up may not interfere with other patient's sensor networks.

Technical requirements that have to be met and have to be considered during the design of the set-up procedure are low power consumption and low storage usage.

4 Architecture of Wireless Sensor Network and Set-Up Procedure

This section describes the overall system architecture as well as the set-up procedure and protocol we developed. The set-up protocol designed enables fast association of multiple sensors to a patient's sensor network.

4.1 System Overview

Our wireless sensor network consists of three kinds of devices: medical sensors, a special sensor for patient identification acting as medical sensor system controller and a set-up pen. We assume that each sensor has a hospital wide unique address.

Typically, the data of the sensor network will be sent to a bedside monitor or central surveillance system for display and further processing.

Different medical sensors are attached to the human body and are responsible for vital sign data acquisition and processing (e.g. electrocardiogram (ECG), oxygen blood saturation (SpO_2), non-invasive blood pressure (NiBP), ...).

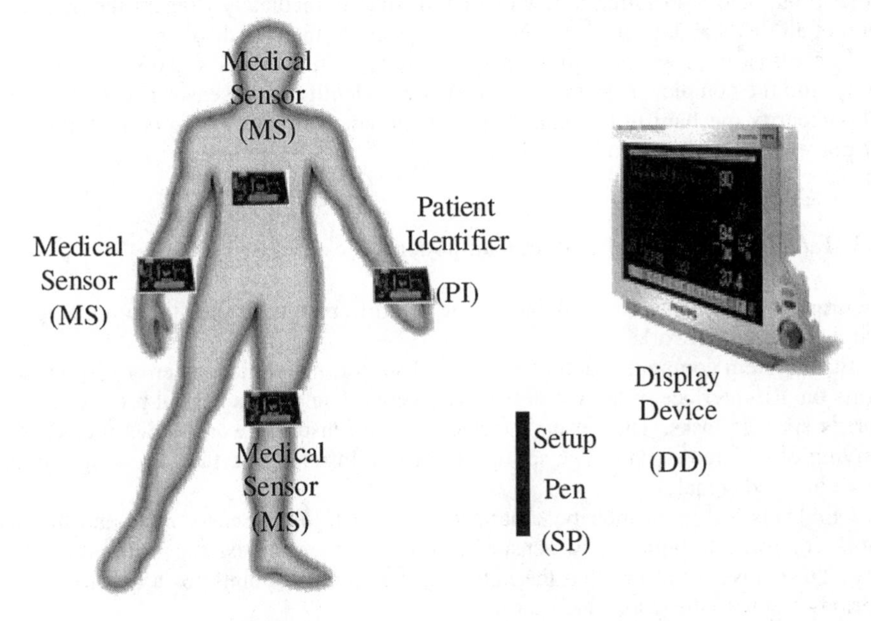

Fig. 1. Sensor network - system components

One of the sensor nodes is the patient identifier, i.e. each patient has a unique hospital wide patient identifier – which is used as unique identifier for the patient's sensor network. All sensors actually connected to a patient form a complete medical monitoring system, which is represented by the patient identifier. I.e. it acts as medical sensor system controller having knowledge about the sensor configuration and who performed which action (adding/removing/configuring sensors).

Upon entering the hospital each patient will be equipped with his patient identifier, which is attached to the patient's body. The patient identifier contains the patient's name and other patient specific data. To add an additional medical sensor or display to the patient, the clinician uses his personal set-up pen and points to the patient identifier and then to the new sensor (or vice versa). A button on the set-up pen activates sending the coded IR signal.

Every clinician has his own personalized portable set-up device, which sends a unique ID using an infrared signal. The ID sent by the set-up device uniquely identifies the clinician. The patient identifier node stores the clinician's identification that started the association.

The medical monitor contains the same communication interface as the other sensor network nodes. It can become part of the wireless sensor network and can, additionally, act as gateway between the wireless sensor network and a central surveillance system connected to the clinical network.

The sensor network as such is completely autonomous and operational even without any connection to an external display device. This means, that a monitored patient can be easily transferred within the hospital. When the first display device is disassociated, the complete sensor network remains active. Transferring the patient to another room, and associating a new display device, immediately triggers the transmission of all medical data of the whole sensor system to the new display.

In contrast to easy association/disassociation of the medical sensors, we immediately stop the complete system when the patient identification sensor is disassociated. This security mechanism prevents that medical data from the wrong patient might be displayed at another patient's monitor.

4.2 Functional Blocks of System Components

Figure 2 shows the functional blocks of the different components comprising the wireless sensor network.

Each system component acting as medical sensor, patient identifier or display supports the RF interface of the wireless sensor network and has a central processing unit for its specific tasks. The set-up interface of the sensor network nodes includes receiving of infrared signals. The set-up pen has an infrared interface allowing sending of an infrared signal.

Clinicians just point their personalized set-up pen to the sensor node and the network controller to build a new sensor network or to add a new sensor to an existing one. To remove a sensor from the network, the clinician points for a defined longer period of time to the respective sensor.

Pointing to the network controller and then to a display device (with the same communication interface) initiates displaying the vital signs and the patient identification on a medical monitor.

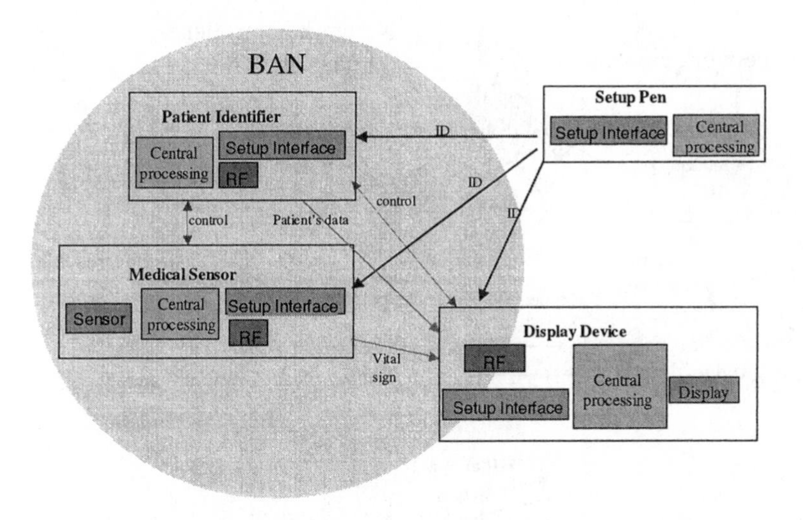

Fig. 2. Functional blocks of system components

In the following we will describe the set-up protocol in more detail (see figure 3).

4.3 Set-Up Protocol

When a sensor node or display device is triggered via its infrared interface and receives the unique code of a set-up pen, it enters a state where it is sending a 'connection request' message for a certain time interval. This message contains the code received via its IR interface and its own unique address.

The patient identifier also enters for a certain amount of time (configurable interval) a state where it is waiting for a 'connection request' from a sensor node or display device. If such a message is received, the patient identifier node checks whether the code received with this message is the same it has received from the set-up pen via its infrared interface. If so, the connection will be established. If a wrong code is received, the message will be ignored.

The unique patient identifier is used as unique id for the sensor network.
The set-up protocol as described here avoids interference between two nearby wireless sensor networks, as only nodes triggered by the same set-up pen will establish a connection. It is therefore not possible that a new sensor will be connected to the wrong sensor network unintentionally.

Using the time interval for receiving/sending the corresponding messages was easy to handle in practice. We used time intervals of 5 to 10 seconds, where we sent between 2 to 4 packets per second.

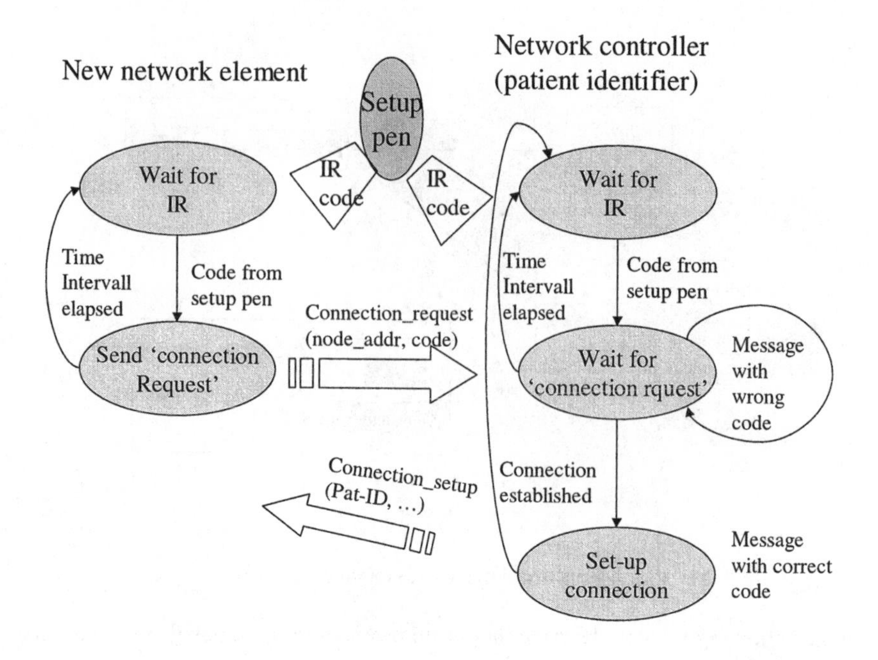

Fig. 3. Details of set-up protocol

4.4 Discussion

Other possible alternatives for the set-up procedure are to build a set-up pen with the same communication interface as used for the different sensor nodes. This would eliminate the need for the IR interface in each node. However, our solution based on infrared is superior to such a solution as the directed infrared signal from the set-up pen ensures sending the signal to exactly the node it is pointing to. This cannot be guaranteed if the set-up pen uses the same radio interface as the other sensor nodes. The signal would be spread over a larger area and thus reach several nodes at the same time. Limiting the radio to an exactly specified shorter range cannot be done reliably.

The current set-up procedure relies on proper sensor communication and networking. We assume that every sensor node has been equipped with a unique address, and that each type of sensor is at most once present in one patient network[1]. To enhance reliability, the set-up procedure can be extended by automatic address assignment, and automatic checking of configuration consistency.

System security is achieved by the individual association of a set-up pen to a certain clinician -via the specific unique ID. This could be further extended, so that a set-

[1] Having e.g. two ECG sensor nodes could from the application point of view lead to ambiguous medical results, and from the communication perspective to overload and collisions in the individual sensor network.

up pen additionally provides a hospital-specific ID, that is being checked by the sensors during the setup procedure. This ensures authentication and authorization for set-up, and prohibits that sensors that are not configured for the hospital might be deployed. This way, we provide the security being required in the envisaged mission-critical, medical environment. Extended security like dedicated communication protection, as being proposed for general ad-hoc sensor networks [9], are not required, as our application only runs in a rather restricted hospital environment.

The following section describes our prototypical implementation and the experiences we collected during our experiments.

5 Prototype Description

The University of Berkeley sensor nodes, also known as Berkeley Motes, have been chosen for validating our concepts. The hardware platform contains an 8 bit micro controller and a radio interface and is optimized for low power operation. The corresponding software platform TinyOS, an event driven and modular runtime for sensor network application, provides a complete wireless networking stack. Example applications are available for demonstrating the use of TinyOS components. This combination of suitable hardware and software was the ideal starting point to prove our concepts.

From the different available hardware platforms, the MICA platform has been used in our prototype. MICA consists of a processor board containing the micro controller, an Atmega169 with 128kB flash ROM and 4kB RAM. The micro controller is operated at 4Mhz and has a peak power consumption of 5mA (2mA idle, 5µA sleep). The processor board we are using also contains a radio transceiver in the 868Mhz ISM band, the RFM TR1001[2]. The MICA platform provides a bit rate of 40kbps. Power consumption ranges between 4.5mA for receiving and 7mA for transmitting. The typical radio range is between 10 to 30 meters (indoors).

The processor board is equipped with a connector used for (re)programming the micro controller. The same connector is used to combine different sensor boards with the processor board. A standard MICA sensor board is available, already equipped with temperature, light, accelerometer and magnetometer.

For our application area, we have implemented additional own sensors, like medical sensors (ECG) and infrared receivers. We used the basic sensor board, which is a prototyping board with connector for the MICA processor board.

5.1 Hardware Components

The following hardware components have been implemented:

Set-Up Pen

A small device was needed to send a unique identifier for starting the set-up procedure between a Patient Identifier and a Medical Sensor. Unlike all other components of our prototype system, the Set-up pen was not implemented using a Berkeley Mote.

[2] The original MICA platform uses a TR1000 suitable for the 915Mhz band in the US.

An 8 bit micro controller with small footprint (ATtiny15L) was chosen to generate a modulated infrared signal representing the unique identifier of the pen. A similar modulation scheme as used by remote controls was chosen in order to make sure that the transmission is reliable and not subject to ambient light changes.

Medical Sensor
The medical sensor has been implemented using a MICA mote and a basic sensor board with an ECG amplifier and an infrared receiver. The ECG amplifier is a standard design based on an instrumentation amplifier and filtering, providing a one lead ECG. We use a sample rate of 100Hz. 10 samples are transmitted per packet.

Display Device
The display device has been implemented using a MICA mote with a basic sensor board containing an infrared receiver. MICA mote and sensor board are then plugged into the programming board that provides a serial connection to a PC. Once the MICA mote has been activated as part of the medical sensor network it acts as a gateway for the communication of the application on the PC to the sensor nodes in the medical body area network.

Patient Identifier
This device is the central controlling entity for the set-up procedure of the medical body sensor network. It consists of a MICA mote plus a basic sensor board with infrared receiver. Currently the Patient Identifier is implemented as a single device, but can also be seen as just a functionality of the medical sensor network, which can be implemented as part of a medical sensor node.

5.2 Experiences and Discussion

The following experiences have been made during the development and implementation of a medical sensor network using the Berkeley Motes.

- It works – The overall concept of using a Set-up Pen to configure a medical sensor network can be shown using our current prototype implementation. The Berkeley Motes with TinyOS are a suitable environment for research work for any kind of wireless sensor networks.
- Realisation of the medical sensor network as a distributed system is eased by TinyOS. Flexible allocation of functions for node management, synchronization and time-slot allocation for real-time vital-sign data streams has been successfully realised.
- Coexistence – Multiple medical sensor networks can be set-up and configured without interfering each other. The use of a unique identifier per Set-up Pen has proven to be a suitable concept.

Beside the overall suitability of the Berkeley Motes and TinyOS, a real medical sensor network will pose additional requirements on such a network not currently met by the Berkeley Mote hardware.

- Radio Interface – The current radio interface is optimized for large-scale sensor networks covering ranges of up to 30 to 100 meters (outdoors). The antenna design is not optimized for operation near the human body. As the range of a medical sensor network is limited to only a few meters the output power of the radio could be significantly decreased.

- The radio interface and the low layer protocols are not resistant concerning interference of further radio sources, e.g. a GSM phone can easily drop down the whole system.
- Power Consumption – The power consumption of the used micro controller is not really low. Ultra-low power consumption is one of the key requirements for medical sensor networks to be practical in a hospital environment. Frequent recharging or exchanging of batteries does not fit into the clinical workflow. Lower power designs are available now. Whether the new low power version of the Berkeley Motes will meet our requirement is still open.

6 Conclusions and Future Work

We realized a patient monitoring system based on wireless body-worn medical sensors. The set-up protocol designed enables fast association of multiple medical sensors to individual patients in an intuitive way. The integrated patient and clinician identification leads to safety and reliability of the overall system.

Currently, each network node is pre-configured with a unique address. Future work includes a dynamic allocation of device addresses, included in the set-up procedure, and a power-efficient communication protocol for operation of the sensor network.

References

1. Jason Hill, Robert Szewczyk, Alec Woo, Seth Hollar, David Culler, Kristofer Pister: System architecture directions for network sensors. ASPLOS 2000, Cambridge, November 2000
2. CEPT/ERC Rec 70-03
3. Chris Karlof, Yaping Li, and Joseph Polastre: ARRIVE: An Architecture for Robust Routing In Volatile Environments. Technical Report UCB//CSD-03-1233, University of California at Berkeley, Berkeley, CA, Mar. 2003
4. Deepak Ganesan, Bhaskar Krishnamurthy, Alec Woo, David Culler, Deborah Estrin, Steven Wicker: An Empirical Study of Epidemic Algorithms in Large Scale Multihop Wireless Networks. (IRP-TR-02-003), Mar 2002
5. Thomas G. Xydis, Simon Blake-Wilson: Security Comparison: Bluetooth™ Communications vs. 802.11. Bluetooth Security Experts Group, 2002
6. David E. Culler, Jason Hill, Philip Buonadonna, Robert Szewczyk, and Alec Woo: A Network-Centric Approach to Embedded Software for Tiny Devices. EMSOFT 2001, Oct 2001.
7. Hans-Joachim Mörsdorf : BAN – Langzeiterfassung und drahtlose Übertragung von Patiententendaten. In: Kroll M, Lipinski H-G, Melzer K (Hrsg.), Mobiles Computing in der Medizin, Köllen Druck und Verlag, Bonn, 2003
8. Holmquist, L.E., Mattern, F., Schiele, B., Alahuhta, P., Beigl, M. and Gellersen, H-W. Smart-Its Friends: A Technique for Users to Easily Establish Connections between Smart Artefacts. In: Proceedings of UbiComp 2001 (Tech note), September 30 - October 2, Atlanta, Georgia, USA
9. Sasha Slijepcevic, Miodrag Potkonjak, Vlasios Tsiatsis, Scott Zimbeck, and Mani B. Srivastava: On communication security in wireless ad-hoc sensor networks. In: Proceedings of Eleventh IEEE International Workshops on Enabling Technologies: Infrastructure for Collaborative Enterprises. WET ICE 20022, Pittsburgh, PA, USA, 10-12 June 2002

Author Index

Lecture Notes in Computer Science

For information about Vols. 1–2843
please contact your bookseller or Springer-Verlag

Vol. 2844: J.A. Jorge, N.J. Nunes, J.F. e Cunha (Eds.), Interactive Systems. Proceedings, 2003. XIII, 429 pages. 2003.

Vol. 2846: J. Zhou, M. Yung, Y. Han (Eds.), Applied Cryptography and Network Security. Proceedings, 2003. XI, 436 pages. 2003.

Vol. 2847: R. de Lemos, T.S. Weber, J.B. Camargo Jr. (Eds.), Dependable Computing. Proceedings, 2003. XIV, 371 pages. 2003.

Vol. 2848: F.E. Fich (Ed.), Distributed Computing. Proceedings, 2003. X, 367 pages. 2003.

Vol. 2849: N. García, J.M. Martínez, L. Salgado (Eds.), Visual Content Processing and Representation. Proceedings, 2003. XII, 352 pages. 2003.

Vol. 2850: M.Y. Vardi, A. Voronkov (Eds.), Logic for Programming, Artificial Intelligence, and Reasoning. Proceedings, 2003. XIII, 437 pages. 2003. (Subseries LNAI)

Vol. 2851: C. Boyd, W. Mao (Eds.), Information Security. Proceedings, 2003. XI, 443 pages. 2003.

Vol. 2852: F.S. de Boer, M.M. Bonsangue, S. Graf, W.-P. de Roever (Eds.), Formal Methods for Components and Objects. Proceedings, 2003. VIII, 509 pages. 2003.

Vol. 2853: M. Jeckle, L.-J. Zhang (Eds.), Web Services – ICWS-Europe 2003. Proceedings, 2003. VIII, 227 pages. 2003.

Vol. 2854: J. Hoffmann, Utilizing Problem Structure in Planning. XIII, 251 pages. 2003. (Subseries LNAI)

Vol. 2855: R. Alur, I. Lee (Eds.), Embedded Software. Proceedings, 2003. X, 373 pages. 2003.

Vol. 2856: M. Smirnov, E. Biersack, C. Blondia, O. Bonaventure, O. Casals, G. Karlsson, George Pavlou, B. Quoitin, J. Roberts, I. Stavrakakis, B. Stiller, P. Trimintzios, P. Van Mieghem (Eds.), Quality of Future Internet Services. IX, 293 pages. 2003.

Vol. 2857: M.A. Nascimento, E.S. de Moura, A.L. Oliveira (Eds.), String Processing and Information Retrieval. Proceedings, 2003. XI, 379 pages. 2003.

Vol. 2858: A. Veidenbaum, K. Joe, H. Amano, H. Aiso (Eds.), High Performance Computing. Proceedings, 2003. XV, 566 pages. 2003.

Vol. 2859: B. Apolloni, M. Marinaro, R. Tagliaferri (Eds.), Neural Nets. Proceedings, 2003. X, 376 pages. 2003.

Vol. 2860: D. Geist, E. Tronci (Eds.), Correct Hardware Design and Verification Methods. Proceedings, 2003. XII, 426 pages. 2003.

Vol. 2861: C. Bliek, C. Jermann, A. Neumaier (Eds.), Global Optimization and Constraint Satisfaction. Proceedings, 2002. XII, 239 pages. 2003.

Vol. 2862: D. Feitelson, L. Rudolph, U. Schwiegelshohn (Eds.), Job Scheduling Strategies for Parallel Processing. Proceedings, 2003. VII, 269 pages. 2003.

Vol. 2863: P. Stevens, J. Whittle, G. Booch (Eds.), «UML» 2003 – The Unified Modeling Language. Proceedings, 2003. XIV, 415 pages. 2003.

Vol. 2864: A.K. Dey, A. Schmidt, J.F. McCarthy (Eds.), UbiComp 2003: Ubiquitous Computing. Proceedings, 2003. XVII, 368 pages. 2003.

Vol. 2865: S. Pierre, M. Barbeau, E. Kranakis (Eds.), Ad-Hoc, Mobile, and Wireless Networks. Proceedings, 2003. X, 293 pages. 2003.

Vol. 2866: J. Akiyama, M. Kano (Eds.), Discrete and Computational Geometry. Proceedings, 2002. VIII, 285 pages. 2003.

Vol. 2867: M. Brunner, A. Keller (Eds.), Self-Managing Distributed Systems. Proceedings, 2003. XIII, 274 pages. 2003.

Vol. 2868: P. Perner, R. Brause, H.-G. Holzhütter (Eds.), Medical Data Analysis. Proceedings, 2003. VIII, 127 pages. 2003.

Vol. 2869: A. Yazici, C. Şener (Eds.), Computer and Information Sciences – ISCIS 2003. Proceedings, 2003. XIX, 1110 pages. 2003.

Vol. 2870: D. Fensel, K. Sycara, J. Mylopoulos (Eds.), The Semantic Web - ISWC 2003. Proceedings, 2003. XV, 931 pages. 2003.

Vol. 2871: N. Zhong, Z.W. Raś, S. Tsumoto, E. Suzuki (Eds.), Foundations of Intelligent Systems. Proceedings, 2003. XV, 697 pages. 2003. (Subseries LNAI)

Vol. 2873: J. Lawry, J. Shanahan, A. Ralescu (Eds.), Modelling with Words. XIII, 229 pages. 2003. (Subseries LNAI)

Vol. 2874: C. Priami (Ed.), Global Computing. Proceedings, 2003. XIX, 255 pages. 2003.

Vol. 2875: E. Aarts, R. Collier, E. van Loenen, B. de Ruyter (Eds.), Ambient Intelligence. Proceedings, 2003. XI, 432 pages. 2003.

Vol. 2876: M. Schroeder, G. Wagner (Eds.), Rules and Rule Markup Languages for the Semantic Web. Proceedings, 2003. VII, 173 pages. 2003.

Vol. 2877: T. Böhme, G. Heyer, H. Unger (Eds.), Innovative Internet Community Systems. Proceedings, 2003. VIII, 263 pages. 2003.

Vol. 2878: R.E. Ellis, T.M. Peters (Eds.), Medical Image Computing and Computer-Assisted Intervention - MICCAI 2003. Part I. Proceedings, 2003. XXXIII, 819 pages. 2003.

Vol. 2879: R.E. Ellis, T.M. Peters (Eds.), Medical Image Computing and Computer-Assisted Intervention - MICCAI 2003. Part II. Proceedings, 2003. XXXIV, 1003 pages. 2003.

Vol. 2880: H.L. Bodlaender (Ed.), Graph-Theoretic Concepts in Computer Science. Proceedings, 2003. XI, 386 pages. 2003.

Vol. 2881: E. Horlait, T. Magedanz, R.H. Glitho (Eds.), Mobile Agents for Telecommunication Applications. Proceedings, 2003. IX, 297 pages. 2003.

Vol. 2882: D. Veit, Matchmaking in Electronic Markets. XV, 180 pages. 2003. (Subseries LNAI)

Vol. 2883: J. Schaeffer, M. Müller, Y. Björnsson (Eds.), Computers and Games. Proceedings, 2002. XI, 431 pages. 2003.

Vol. 2884: E. Najm, U. Nestmann, P. Stevens (Eds.), Formal Methods for Open Object-Based Distributed Systems. Proceedings, 2003. X, 293 pages. 2003.

Vol. 2885: J.S. Dong, J. Woodcock (Eds.), Formal Methods and Software Engineering. Proceedings, 2003. XI, 683 pages. 2003.

Vol. 2886: I. Nyström, G. Sanniti di Baja, S. Svensson (Eds.), Discrete Geometry for Computer Imagery. Proceedings, 2003. XII, 556 pages. 2003.

Vol. 2887: T. Johansson (Ed.), Fast Software Encryption. Proceedings, 2003. IX, 397 pages. 2003.

Vol. 2888: R. Meersman, Zahir Tari, D.C. Schmidt et al. (Eds.), On The Move to Meaningful Internet Systems 2003: CoopIS, DOA, and ODBASE. Proceedings, 2003. XXI, 1546 pages. 2003.

Vol. 2889: Robert Meersman, Zahir Tari et al. (Eds.), On The Move to Meaningful Internet Systems 2003: OTM 2003 Workshops. Proceedings, 2003. XXI, 1096 pages. 2003.

Vol. 2890: M. Broy, A.V. Zamulin (Eds.), Perspectives of System Informatics. Proceedings, 2003. XV, 572 pages. 2003.

Vol. 2891: J. Lee, M. Barley (Eds.), Intelligent Agents and Multi-Agent Systems. Proceedings, 2003. X, 215 pages. 2003. (Subseries LNAI)

Vol. 2892: F. Dau, The Logic System of Concept Graphs with Negation. XI, 213 pages. 2003. (Subseries LNAI)

Vol. 2893: J.-B. Stefani, I. Demeure, D. Hagimont (Eds.), Distributed Applications and Interoperable Systems. Proceedings, 2003. XIII, 311 pages. 2003.

Vol. 2894: C.S. Laih (Ed.), Advances in Cryptology - ASIACRYPT 2003. Proceedings, 2003. XIII, 543 pages. 2003.

Vol. 2895: A. Ohori (Ed.), Programming Languages and Systems. Proceedings, 2003. XIII, 427 pages. 2003.

Vol. 2896: V.A. Saraswat (Ed.), Advances in Computing Science – ASIAN 2003. Proceedings, 2003. VIII, 305 pages. 2003.

Vol. 2897: O. Balet, G. Subsol, P. Torguet (Eds.), Virtual Storytelling. Proceedings, 2003. XI, 240 pages. 2003.

Vol. 2898: K.G. Paterson (Ed.), Cryptography and Coding. Proceedings, 2003. IX, 385 pages. 2003.

Vol. 2899: G. Ventre, R. Canonico (Eds.), Interactive Multimedia on Next Generation Networks. Proceedings, 2003. XIV, 420 pages. 2003.

Vol. 2901: F. Bry, N. Henze, J. Maluszyński (Eds.), Principles and Practice of Semantic Web Reasoning. Proceedings, 2003. X, 209 pages. 2003.

Vol. 2902: F. Moura Pires, S. Abreu (Eds.), Progress in Artificial Intelligence. Proceedings, 2003. XV, 504 pages. 2003. (Subseries LNAI).

Vol. 2903: T.D. Gedeon, L.C.C. Fung (Eds.), AI 2003: Advances in Artificial Intelligence. Proceedings, 2003. XVI, 1075 pages. 2003. (Subseries LNAI).

Vol. 2904: T. Johansson, S. Maitra (Eds.), Progress in Cryptology – INDOCRYPT 2003. Proceedings, 2003. XI, 431 pages. 2003.

Vol. 2905: A. Sanfeliu, J. Ruiz-Shulcloper (Eds.), Progress in Pattern Recognition, Speech and Image Analysis. Proceedings, 2003. XVII, 693 pages. 2003.

Vol. 2906: T. Ibaraki, N. Katoh, H. Ono (Eds.), Algorithms and Computation. Proceedings, 2003. XVII, 748 pages. 2003.

Vol. 2910: M.E. Orlowska, S. Weerawarana, M.P. Papazoglou, J. Yang (Eds.), Service-Oriented Computing – ICSOC 2003. Proceedings, 2003. XIV, 576 pages. 2003.

Vol. 2911: T.M.T. Sembok, H.B. Zaman, H. Chen, S.R. Urs, S.H.Myaeng (Eds.), Digital Libraries: Technology and Management of Indigenous Knowledge for Global Access. Proceedings, 2003. XX, 703 pages. 2003.

Vol. 2913: T.M. Pinkston, V.K. Prasanna (Eds.), High Performance Computing – HiPC 2003. Proceedings, 2003. XX, 512 pages. 2003.

Vol. 2914: P.K. Pandya, J. Radhakrishnan (Eds.), FST TCS 2003: Foundations of Software Technology and Theoretical Computer Science. Proceedings, 2003. XIII, 446 pages. 2003.

Vol. 2916: C. Palamidessi (Ed.), Logic Programming. Proceedings, 2003. XII, 520 pages. 2003.

Vol. 2918: S.R. Das, S.K. Das (Eds.), Distributed Computing – IWDC 2003. Proceedings, 2003. XIV, 394 pages. 2003.

Vol. 2920: H. Karl, A. Willig, A. Wolisz (Eds.), Wireless Sensor Networks. Proceedings, 2004. XIV, 365 pages. 2004.

Vol. 2922: F. Dignum (Ed.), Advances in Agent Communication. Proceedings, 2003. X, 403 pages. 2004. (Subseries LNAI).

Vol. 2923: V. Lifschitz, I. Niemelä (Eds.), Logic Programming and Nonmonotonic Reasoning. Proceedings, 2004. IX, 365 pages. 2004. (Subseries LNAI).

Vol. 2927: D. Hales, B. Edmonds, E. Norling, J. Rouchier (Eds.), Multi-Agent-Based Simulation III. Proceedings, 2003. X, 209 pages. 2003. (Subseries LNAI).

Vol. 2928: R. Battiti, M. Conti, R. Lo Cigno (Eds.), Wireless On-Demand Network Systems. Proceedings, 2004. XIV, 402 pages. 2004.

Vol. 2929: H. de Swart, E. Orlowska, G. Schmidt, M. Roubens (Eds.), Theory and Applications of Relational Structures as Knowledge Instruments. Proceedings. VII, 273 pages. 2003.

Vol. 2932: P. Van Emde Boas, J. Pokorný, M. Bieliková, J. Štuller (Eds.), SOFSEM 2004: Theory and Practice of Computer Science. Proceedings, 2004. XIII, 385 pages. 2004.

Vol. 2937: B. Steffen, G. Levi (Eds.), Verification, Model Checking, and Abstract Interpretation. Proceedings, 2004. XI, 325 pages. 2004.

Vol. 2950: N. Jonoska, G. Păun, G. Rozenberg (Eds.), Aspects of Molecular Computing. XI, 391 pages. 2004.